Recent Advances in Composite Materials

Recent Advances in Composite Materials

In Honor of S.A. Paipetis

Edited by

EMMANUEL E. GDOUTOS
Democritus University of Thrace,
Xanthi, Greece

and

ZAIRA P. MARIOLI-RIGA
Hellenic Aerospace Industry,
Schimatari, Greece

SPRINGER-SCIENCE+BUSINESS MEDIA, B.V.

A C.I.P. Catalogue record for this book is available from the Library of Congress.

ISBN 978-90-481-6294-9 ISBN 978-94-017-2852-2 (eBook)
DOI 10.1007/978-94-017-2852-2

Printed on acid-free paper

Table of Contents

4. Metal Matrix Composites

5. Structural Analysis

6. Applications

Editors' Preface

This book contains 31 papers presented at the symposium on "Recent Advances in Composite Materials" which was organized in honor of Professor Stephanos A. Paipetis. The symposium took place at Democritus University of Thrace, in Xanthi, Greece on June 12-14, 2003. The book is a tribute to Stephanos A. Paipetis, a pioneer of composite materials, in recognition of his continuous, original diversified and outstanding contributions for half a century.

The book consists of invited papers written by leading experts in the field. It contains original contributions concerning the latest developments in composite materials. It covers a wide range of subjects including experimental characterization, analytical modeling and applications of composite materials.

The papers are arranged in the following six sections: General concepts, stress and failure analysis, mechanical properties, metal matrix composites, structural analysis and applications of composite materials.

The first section on *general concepts* contains seven papers dealing with composites through the pursuit of the consilience among them, computation and mechatronic automation of multiphysics research, a theory of anisotropic scattering, wave propagation, multi-material composite wedges, a three-dimensional finite element analysis around broken fibers and an in situ assessment of the micromechanics of large scale bridging in ceramic composites.

The second section on *stress and failure analysis* contains five papers dealing with a review of durability analysis of polymer matrix composites over the last twenty-five years, an experimental and numerical analysis of matrix cracking in brittle composites, delamination detection in laminated composites by Lamb waves, a perforation study of Zylon sheets and an analysis of the strain redistribution in composite laminates resulting from off-axis ply cracking.

The third section on *mechanical properties* contains six papers dealing with mechanical properties of nanocomposite multilayers, measurement of matrix stiffness of oxide/oxide composites by ultrasonic, mechanical characterization of fiber reinforced photopolymers, correlation between micromechanical and macromechanical properties of composites and microfibrillar composites and the long term behavior of continuous fiber oxide/oxide composites under thermal exposure

The fourth section on *metal matrix composites* contains three papers dealing with on a relationship between crack opening displacement and strain energy density, an experimental study of the dependence of fracture characteristics on the specimen thickness and a theoretical evaluation of strength degradation during monotonic and cyclic loading of metal matrix composites.

The fifth section on *structural analysis* contains three papers dealing with the design of composite joints, core failure mechanisms of composite sandwich beams and a theoretical analysis of buckling of cylindrical shells under external pressure.

Finally, the sixth section on *applications* contains seven papers dealing with design and manufacturing of structural components by resin transfer molding, high friction, low wear composites for tribological applications, the design and analysis of a pedestrian bridge composed of a concrete deck and truss girders made of IPC sandwich panels, strengthening of historic structures and concrete beams with fiber reinforced polymers, the optimum design of fiber reinforced concrete-steel composite columns and the contact delamination of composite patches used to repair metal structures.

We consider it an honor and priviledge we have had the opportunity to edit this volume. We wish to thank very sincerely the authors who have contributed to this volume and all those who participated in the symposium on 'Recent Advances in Composite Materials" to honor Stephanos A. Paipetis, our mentor and thesis advisor to one of use, a loyal friend and colleague, an international figure in the mechanics community, a pioneer who is credited with the organization of the first prestigious center of excellence in composites materials in the university of Patras in Greece and played a key role in the research and education of many Greek scientists and engineers in the discipline of composite materials, whose interests cover a wide area of disciplines including mechanics, poetry, philosophy, cookery, just to mention a few of them.

We would like to thank Mrs L. Adamidou of Democritus University of Thrace for editing this volume. Finally, a special word of thanks goes to Dr. Arno Schouwenburg and Ms Nathalie Jacobs of Kluwer Academic Publishers for their interest in publishing this work and their kind collaboration and support.

January, 2003

Xanthi & Schimatari
Greece

Emmanuel E. Gdoutos
Zaira P. Marioli-Riga
Editors

OPTICKS
PRINCIPIA

Biography of Stephanos A. Paipetis

Stephanos A. Paipetis was born on February 1, 1938 in the ancient city of Corfu (Kerkyra), Greece, the only child of Alcibiades and Alexandra Paipetis. Old documents dating back to 1315 and earlier show that, in that era of feudalism, his family possessed lands and had full rights of civil transactions. In the times to follow, until World War II, the family included important politicians and wealthy merchants. The family probably originates from Venice, a great power, which occupied Corfu for several centuries.

Corfu, an island of great natural beauty and first class tourist attraction, lies in the Ionian Pelagos, northwest of the Greek mainland. Its history goes back to the very remote antiquity. Mythology has it that out of the union of Korkyra, a beautiful nymph, daughter of Assopos river, with Poseidon (Neptune), the God of sea, Pheax, the first to inhabit the island was born. After him, Corfu is still referred to as the island of Pheakes (pl. of Pheax). By this name the island is mentioned by Homer in his *Odyssey*, as the first hospitable place, where Ulysses landed on his long journey home and his last station before Ithaca, his final destination. Corfu was colonized by the Eretreans of Euboia in the period 775-750 BC, followed by the Corinthians in the 6th century BC, while in ca. 330 BC, was put under the protection of Alexander the Great. In the sequence it was conquered by the Syracusians and the Romans, became part of the Byzantine Empire (337-1267 AD), of the Venetian Republic (1386-1797), of the French Republic (1797-1799), under Turkish-Russian protection (1799-1807), under Imperial French (1807-1814) and British dominion (1814-1864), until the final union with Greece in 1864. All these adventures and tribulations the island fortunately enabled Corfu to escape from the fate of the Greek mainland, suffering for four centuries under Turkish occupation, and contributed substantially to the development of an advanced culture: Great writers, poets, philosophers and composers were born in Corfu and created some of the oldest cultural institutions and societies of modern Greece.

The town of Corfu was founded in the 8th century BC and its architectural style is affected by the long Venetian presence, with tall buildings, narrow, stone-paved streets and little squares in front of churches within the walls, built by the Venetians on top of the Byzantine walls, forming, along with its two fortresses and the huge squares in the Eastern part, an incredible defense system. A number of historical buildings create a special character of the town.

Stephanos' boyhood was associated with the difficult conditions of the Second World War and the German-Italian occupation of Greece. This period was spent in a small town 8 km south of the city of Corfu. His father was a Bank employee and his mother, an exceptionally strong and fearless person, was a school teacher. Till the end of the war the family had to reside in a small room of the school building, since their home was taken over by the occupation forces. For the ensuing six years he attended his elementary school there, and the exquisite nature of Corfu had a profound effect on his character and way of thinking.

In 1950 Stephanos entered, after due examinations, the Gymnasium (High School) in the city of Corfu, which he attended for the next six years. The school followed a classical curriculum, including among others, six years of ancient Greek, and four years of Latin, English and French, however, for the last two years he chose to attend the Physics-Mathematics curriculum. This resulted from his decision to pursue studies in Engineering. The National Technical University in Athens – the only technical school in the country at the time – was a real challenge, as its entrance exams were the toughest of all. Stephanos took the exams in the Fall 1956, and it was one of the greatest days of his life when he heard that he passed, ranked fourth among several hundreds of candidates.

He was registered at the School of Mechanical and Electrical Engineering of the National Technical University (Metsovion Polytechnion). He was already attending lectures for a month when more good news came: In view of the difficulty of entrance to the Metsovion, candidates used to go for entrance exams at the School of Science of the Kapodistrian University of Athens as well, as an alternative in case of failure at Metsovion. The results of those exams were published and, at the Department of Physics, Stephanos was ranked first among nearly 1,000 candidates. Although this was a great success, he preferred to continue with his engineering studies.

Studies at Metsovion were indeed tough. Classes ran from 6 to 10 hours a day, six days a week. Summers were devoted to gaining practical experience, and Stephanos chose to work at various industries in Germany and Austria, which he believes was important for his getting in the heart of Mechanical Engineering. His most preferred area was Machine Design, and he remembers with gratitude his late Professor Nicholas Theophanopoulos, who was always like a father to all of his students, current and past. In fact he appointed Stephanos as an unsalaried assistant practically immediately after his graduation and for the period of his (compulsory) military service. He received his Degree in Mechanical and Electrical Engineering in June 1961 and next October he joined the then Royal Hellenic Navy to serve as Reserve Ensign till January 1964. This was long but fruitful, since he gained further experience, both technical, at the Royal Shipyard of Salamis, and administrative, at the General Navy Staff.

Immediately after his dismissal, he was appointed a Senior Assistant at the then Chair of Machine Design (or of Machine Elements, as the official title was), dealing with teaching and research. It was at that time, middle sixties, when he discovered the great challenge that composite materials presented for future engineering structures. In fact he developed a special section of the existing laboratory, to deal with composites and associated experimental techniques, such as stress analysis, photo elasticity, dynamic phenomena and vibration and wave problems.

It was at that time when his family life started. In 1964 he married Potoula, nee Kapogianni, a philologist, and their first child, Alexandra, now a philologist, was born. In 1967 their second child was born, Alcibiades, now a Mechanical Engineer, PhD holder of the University of London and a Professor at the Hellenic Naval Academy. Their third son, Assimakis, arrived later in 1971.

During the period 1964-69 he worked as consulting engineer for a large hospital in Piraeus, the Public Power Corporation (the electricity Company of Athens), and with a number of public authorities, and had the opportunity to design a large number of engineering installations and plants. He was also appointed as expert by the Greek Ministry of Justice and had substantially contributed to the investigation of, among others, a number of major industrial accidents.

During the period 1967-1974 Greece suffered under the dictatorship with all its consequences. Despite some health problems of the children, Stephanos left Greece and went to England, where he was registered as postgraduate student at the Department of Mechanical Engineering of Imperial College. He started working for an MSc Degree, but soon after he was transferred to MPhil and eventually to PhD, which he managed to complete in practically less than a total of 2 calendar years, the minimum required. This was a record time for the Department, as far as people could remember. The title of his thesis was "Dynamic Properties Viscoelastic Composites," where he developed one of the first "exact" micromechanical theories for determination of complex moduli for fibre and particle reinforced composites under dynamic loading.

Following his return home in September 1970, Stephanos shifted to the Laboratory for Strength of Materials of the Metsovion Polytechnion, directed by the late Professor P. S. Theocaris, where he worked first as senior assistant and then Associate Professor till 1979. That Laboratory was a great place, fully equipped with all kinds of modern testing and experimental mechanics facilities along with a huge annual budget. He was placed in charge of the testing section, where he introduced a whole series of new tests for materials and structures and issued several thousands of test certificates for industrial and construction companies. His research extended in several areas, around the central core of experimental mechanics: It included composite materials with polymeric and ceramic matrices (physical, mechanical, thermal, dielectric etc. characterization and design, inclusion and interface problems etc.), experimental mechanics (optical methods: moiré, photoelasticity, method of caustics, interference methods, electro-optical phenomena), cracking of polymers, dynamics of materials and structures (damping properties, stress, waves, periodic elasticity, methods for eigenfrequency determination, multistage dampers etc), experimental determination of stress fields around inclusions due to circumferential stress waves etc. These years were productive from the point of view of research, as a number of problems, unsolved so far, were fully investigated, although satisfaction was sometimes dimmed by Theocaris' authoritarian character and behaviour.

The Laboratory was constantly receiving visitors and Stephanos had the opportunity to meet and cooperate with distinguished scientists from all over the world. Professor Werner Goldsmith of the University of California at Berkeley spent his sabbatical there during the academic year 1974-75 and they developed a life-long friendship. Professor Goldsmith spent a second sabbatical in Greece, in the Academic year 1984-85, when Stephanos was already a Professor in Patras, and they both enjoyed working together. In June 2002, Professor Goldsmith visited Patras again, to receive his Honorary Doctorate Degree from the University of Patras, awarded to him in recognition of his outstanding scientific contributions in general, and his service to Greek universities, in particular.

In 1978 Stephanos received a post-doctoral award and spent three months at Imperial College, where he developed a new "exact" micromechanical theory for particle composites in the presence of chemical interaction between the constituents.

In 1980 Stephanos was elected Professor of Mechanics at the School of Engineering of the University of Patras, a position which he holds up to now. Immediately after his appointment he instituted the Laboratory of Applied Mechanics (AML), well known for its multiple activities. Currently, AML employs, besides the Director, 4 Associate Professors, over 35 postgraduate students and several members of administrative and technical personnel.

In 1983 Stephanos was appointed by the Greek Ministry of Defense as the National Representative at the Structures and Materials Panel of NATO/AGARD (Advisory Board of Aerospace Research and Development), a highly reputed scientific agency. In the period 1994-96 he was elected Deputy Chairman of the Panel. He still holds the same position at the Applied Vehicle Technology Panel (AVT), the new RTO (Research and Technology Organization), which replaced AGARD since 1997. This appointment gave him the opportunity to create an extensive network of cooperating research institutes, industries and universities, as for example DLR (German Aerospace Institute), British Aerospace, Aerospatiale, Dassault Industries, Phillips, Physical Acoustics, US, Daimler-Benz, University of Kaiserslautern, Germany, University of Rome "La Sapienza", Free University of Brussels, etc, through a large number of R&D projects. Since 1985 the latter were linked to relevant European Union projects, which provided ample financial support and modern research equipment for AML. The network included a number of distinguished professors and scientists of several Greek universities and companies, as for example, Professor Emmanuel Gdoutos of Democritus University of Thrace and Dr. Zaira Marioli-Riga of Hellenic Aerospace Industry. In addition, many distinguished scientists from all over the world visited AML and delivered lectures on various subjects. The effort resulted to a massive technology transfer into the country and its further development in the respective areas.

The activities of the Laboratory are generally centered in composite materials (with polymeric and ceramic matrices) and the development of new Nondestructive testing and nondestructive evaluation methodologies. The latter, along with other dynamic methods, especially acoustic emission, and the development of suitable computer codes, were successfully applied to failure prediction of aerospace material and structures. Acoustic emission was also applied to investigate the stability of geological structures and earthquake prediction, which led to substantial results and made a couple of successful predictions. However, a problem of such immense complexity can possibly be solved by using simultaneously several methods based on precursory phenomena.

During the period 1999-2003, Stephanos served as Chairman of the Department of Mechanical Engineering, and succeeded to have it enlarged and renamed it to "Department of Mechanical Engineering and Aeronautics". The effort was greatly assisted by the above network and, despite some problems at the beginning, it seems now to proceed most satisfactorily.

He is or was a Member of the Editorial Board of many professional journals, such as "Fibre Science and Technology" (UK), "Advanced Composite Materials – International Transactions of The Japan Society of Composite Materials (Japan), "Science and Engineering of Composite Materials" (UK), Mechanics of Composite Materials and Structures (US). He is member of several international and Greek professional societies to the foundation of which he contributed, such as the Hellenic Society of Theoretical and Applied Mechanics (Founding Member), the Hellenic Acoustical Society (Founding Member), the Hellenic Society of Composite Materials (Founding President) and the Hellenic Society for Non Destructive Testing (Past President).

Stephanos organized and chaired a considerable number of successful conferences and symposia. The COMP series of International Symposia on Composites has developed almost to an institution: COMP'86, COMP'88, COMP'90, COMP'95 and COMP'03 were organized in the respective years with very good attendance and great success (the last two of them were held in Corfu). Several national conferences on composites and NDT methods, as well as AGARD/RTO Meetings were organized with equal success.

In August 2001 Stephanos organized an International Symposium on *"Extraordinary Machines and Structures in Antiquity"*, which took place in the premises of the International Olympic Academy, adjacent to the archaeological site of Ancient Olympia. His contributions dealt with engineering analysis of the shields of Achilles and Ajax – high-toughness laminated structures, which, as proved by the analysis incorporated advanced scientific knowledge, as the one used in modern applications today. Also with an acoustical analysis of the subterranean Chamber of the Dead at the *Nekyomanteion* (an oracle practicing necromancy in about 2000-600 BC) of the Acheron River, which provided substantial insight into the fully unknown and unexplored ritual processes performed therein. It is interesting that such a construction, e.g. a long tunnel dug in solid rock with identical arches at regular intervals along its axis, exhibits a remarkable sound directivity indicating deep knowledge of room acoustics, comparable to that of ancient theaters, although at much earlier times. Indeed, the investigation of admirable structures in ancient Greece as well as descriptions of structures and other relevant themes tend to establish a new approach to the Engineering capabilities of the ancient Greeks, contrary to the prevailing idea, i.e. that they were dealing with spiritual matters only and detested any activities involving labour. Stephanos is currently organizing another international conference, titled "Science and Technology in Homeric Epics", which is expected to shed light in the potentially advanced knowledge of the Mycenaean era (1500-1200 BC).

Along with his professional activities, Stephanos is working in a wide range of themes of general interest. He has translated to Greek many works of foreign writers and poets, mainly of the English language, such as John Keats, William Blake, Walt Whitman etc. and has published a number of his translations already, along with his own poetry and prose. One of his recent works is the translation of *"Brave New World Revisited"* by Aldous Huxley, accompanied by a thorough analysis of historical, sociological, anthropological etc. aspects, proving that the conclusions of this prophetic essay concerning the grim future of humanity are being verified much earlier than the great writer and philosopher assumed. In philosophy, he is working on the relation of modern physics with ancient philosophy. He is currently preparing a book, titled *"The Chessboard of the World"*, dealing with the philosophy of Dualism, which had a strong influence on human thinking, viewed in parallel with the energy principles of contemporary physics. Moreover, language as a phenomenon of the human nature, is another of his favorite subjects, and has developed a special elective course, which he is teaching for two semesters in the line of cultural subjects provided by the curriculum of the Department of Mechanical Engineering and Aeronautics of the University of Patras.

Stephanos is active with the current social, national and international problems, not hesitating to criticize politics and government, whenever he thinks he has a reason for it, and has published scores of relevant articles in the local and national press (sample titles are provided in the list of his publications that follows). Currently, he is preparing to issue, as editor-in-chief, a periodical to deal with the great problems of modern society and the world, in general.

Besides all of the above, Stephanos enjoys his family life. His children are not children anymore, but Myrto, his grand daughter from Alexandra, occupies considerable part of his free time. He is also a gourmet cook and has published a book under the humorous title *"Beautiful Pasta With Seafood,"* containing, among others, personal recipes given to him by several fellow-scientists from all over the world. The book became a great success and media were dealing for long with it. Stephanos has a great sense of humor and can tell jokes ad infinitum, something that all of his family share and participate to, along with the rest of his interests. Potoula, his wife, when asked to describe him, says: "All I can tell is that you will never get bored with him", a view that his friends and colleagues share.

Stephanos is now approaching retirement (compulsory in Greece at the age of 67). By thinking back on his career and contribution, he thinks that most important was that he has managed to guide and support young people to develop into important figures in science. The number of his former students that are now professors in reputed Greek and foreign universities, is indeed impressive and makes him very proud. In his Laboratory, which was started with a part-time assistant 25 years ago, he will be able to leave in his place four full Professors, all distinguished scientists. Moreover, he certainly hopes to continue with his work, probably more efficiently, having got rid of the burden of administrative, etc. duties that an active position entails.

List of Publications by Stephanos A. Paipetis

A. BOOKS

In English:

1. A.D. Dimarogonas and S.A. Paipetis, "Analytical Methods in Rotor Dynamics," *Applied Science Publishers,* London 1983.
2. S.A. Paipetis and G.S. Holister, Editors, "Photoelasticity in Engineering Practice," *Elsevier Applied Science Publishers,* London 1985.
3. S.A. Paipetis and G.C. Papanicolaou, Editors, "Engineering Applications of New Composites," *Omega Scientific,* Oxford, UK, 1988.
4. S.A. Paipetis and G.C. Papanicolaou, Editors, "Phase Interaction in Composite Materials," *Omega Scientific*, Oxford, UK, 1992.
5. S.A. Paipetis and T.P. Philippidis, Editors, "Advanced Composites in Emerging Technologies", *AMATEC Publications,* Patras, Greece, 1992.
6. S.A. Paipetis and A.A. Youtsos, Editors, "High Technology Composites in Modern Applications," *University of Patras Publications,* Patras, Greece, 1996.
7. S. A. Paipetis and E.E. Gdoutos, Editors, "Composite Materials and Structures", Proceedings, 1st National Conference, Democritus University of Thrace, *Kyriakides Bros Publishers,* Thessaloniki, Greece, 1997.
8. S.A. Paipetis, Editor, "Extraordinary Machines and Structures in Antiquity", *Peri Technon Publishers,* Patras, Greece, 2003

In Greek:

1. N. Theophanopoulos and S.A. Paipetis, "Machine Elements for Motion Transformation," *National Technical University of Athens,* 1964.
2. S.A. Paipetis, "An Introduction to Dynamic Measurements", *National Technical University of Athens,* 1968.
3. S.A. Paipetis, "The Technique of Stroboscopic Balancing", *Tech. Annals,* 1966.
4. S.A. Paipetis, "Engineering Mechanics: Statics", *5th Edition, Ion Publishers,* Athens 2001.

5. S.A. Paipetis: "Engineering Mechanics: Dynamics", *5th Edition, Ion Publishers,* Athens, 2002.
6. S.A. Paipetis and D. Polyzos: "Vibrations and Waves", 3nd Edition, *Ion Publishers,* Athens, 2003.
7. S.A. Paipetis, Editor: General Introduction and Comments, in the re-edition of the classic "Natural Science in Homeric Epics" by K. Zeggelis (1893), *University of Patras Publications,* 1997.
8. S.A. Paipetis, Dialogues of Aristoboulos and Orthagoras, *Philosophical Discourses, University of Patras Publications,* 1997 (in Greek).
9. S.A. Paipetis, "Otho the Great", a play by John Keats, translated. in Greek, *Ion Publishers,* Athens, 2000.
10. S.A. Paipetis, «Short Everyday Stories», *Peri Technon Publishers,* Patras, Greece (a selection of 15 symbolic stories on ontological, political and social problems)
11. S.A. Paipetis, «The Songs of a Changing World», *Peri Technon Publishers,* Patras, Greece (a collection of poems from all over the world on the essence of things)

TECHNICAL ARTICLES

1. P.S. Theocaris and S.A. Paipetis, "Vibration Measurements by the Moire Method", *Journal of Physics E (Scientific Instruments)* ***5***, 1972, 497-513.
2. S.A. Paipetis, "On the Motion of a Linear Viscolastic Oscillator", *Scientific Yearbook,* Nat. Tech Univ. of Athens, **5**, 1971, 497-513 (in Greek).
3. S.A. Paipetis and P. Grootenhuis, "Dynamic Properties of Viscoelastic Composites", *Meeting of the British Acoustical Society on "Scattering Phenomena in Acoustics",* London, February 1971.
4. P.S. Theocaris and S.A. Paipetis, "Shrinkage Stresses in Three-Dimensional Two-Phase Systems", *Journal of Strain Analysis,* ***8***, 1973, 286-273.
5. P.S. Theocaris and S.A. Paipetis, "Plane Kinematics of Moire Fringes", *Journal of Physics E (Scientific Instruments,* ***6***, 1973, 978-990.
6. P.S. Theocaris and S.A. Paipetis, "Thermal Crazing of Hot-setting Plasticized Epoxy Polymers," *Fibre Science and Technology,* ***7***, 1974, 33-44.
7. P.S. Theocaris, S.A. Paipetis and C. Liakopoulos, "On the Theory of Spatial Moire Fringes", *Scientific Yearbook, Nat. Tech Univ. of Athens,* **3**, 1972, 63-79.
8. P.S. Theocaris and S.A. Paipetis, "Spatial Kinematics of Moire Fringes", *Optik* ***40***, 1974, 296-309.
9. P.S. Theocaris, S.A. Paipetis and J.M. Tsangaris, "Thermal Crazing Phenomena in Epoxy Resins", *Polymer* ***15***, 1974, 441-445.
10. P.S. Theocaris, S. A. Paipetis and G. Vroulis, "Circumferential Waves Around Inhomogeneities in Plane Viscoelastic Solid Systems", *Journal of Sound and Vibration* ***35***, 1974, 521-533.
11. P.S. Theocaris, S.A. Paipetis and C. Liakopoulos, "Some Aspects on the Formation of Spatial Moire Fringes", *Optik* ***42***, 1975, 269-286.

12. P.S. Theocaris and S.A. Paipetis, "Stress Concentration in Elastic Composites with Limiting Shear Properties," *Journal of Composite Materials, **9**,* 1975, 244-250.
13. S. A. Paipetis, G. Vroulis and P.S. Theocaris, "Circumferential Stress Waves Around a Circular Cavity in a Polymeric Plate", *Acustica **34***, 1975, 34-63.
14. P.S. Theocaris, S.A. Paipetis and G.C .Papanicolaou, "Dynamic Properties of Metal-Filled Epoxies", *Fibre Science and Technology, **8***, 1975, 221-242.
15. P.S. Theocaris and S.A. Paipetis, "State of Stress Around Inhomogeneities by the Method of Caustics", *Fibre Science and Technology, **9**,* 1976, 19-40.
16. P. S. Theocaris and S.A. Paipetis: "State of Stress Around Inhomogeneities by the Method of Caustics", *Fibre Science and Technology, **9**,* 1976, 19-40.
17. G.C. Papanicolaou, S.A. Paipetis and P.S. Theocaris, "Crosslinking Studies in Plasticized Epoxies by Means of Dynamic Measurements", *Journal of Applied Polymer Science, **20**,* 1976, 903-910.
18. P.S. Theocaris and S.A. Paipetis: "Constrained Zones at Singular Points of Inclusion Contours", *International Journal of Mechanical Sciences, **18***, 1976, 581-588.
19. G.C. Papanicolaou, S.A. Paipetis and P.S. Theocaris, "Thermal Properties of Metal-Filled Epoxies", *Journal of Applied Polymer Science, **21**,* 1977, 689-701.
20. S.G. Paolinelis, S.A. Paipetis and P.S. Theocaris, "Three-Point Bending at Large Deflections," *Journal of Testing and Evaluation, **5**,* 1977, 427-346.
21. S.A. Paipetis and P.S. Theocaris, "Stress Concentration in Three-dimensional Composites with Limiting Shear Properties" in Developments in Composite Materials, G. S. Holister, Editor, *Applied Science Publishers Ltd.*, 1977, 197-207.
22. G.C. Papanicolaou, S.A. Paipetis and P.S. Theocaris, "Indentation Studies in Plasticized Epoxies" *Journal of Applied Polymer Science, **22**,* 1978, 1417-1430.
23. P.S. Theocaris, S.A. Paipetis and G.C. Papanicolaou, "Indentation Studies in Aluminium-Filled Epoxies" *Journal of Applied Polymer Science, **22**,* 1978, 2245-2252.
24. S.A. Paipetis and P.S. Theocaris, "Propagation of Stress Waves in Plasticized Epoxies," *Acustica, **41**,* 1978, 122-127.
25. P.S. Theocaris, S.A. Paipetis and G.C. Papanicolaou, "The Concept of Boundary Interphase in Composite Mechanics," *Colloid and Polymer Journal, **256**,* 1978.
26. S.A. Paipetis, P.S. Theocaris and C. Stassinakis, "Effect of Geometry and Imperfect Bonding in Composite Systems with Limiting Shear Properties," *Fibre Science and Technology, **11**,* 1978, 335-352.
27. S.A. Paipetis, A. Marchese and P.S. Theocaris, "Dynamic Properties of Plastically Prestrained Aluminium," *Materialpruefung, **26**,* 1978, 378-380.
28. S.A. Paipetis, S.G. Paolinelis and P.S. Theocaris, "Three-Point Bending at Large Deflections of Beams with Different Moduli in Tension and in Compression," *Journal of Testing and Evaluation, **7**,* 1979, 177-181.

29. S.A. Paipetis, C. Stassinakis and P.S. Theocaris, "On the Forced Motion of Viscoelastic Timoshenko Beams," *National Technical University of Athens,* pp. 94, 59 illustrations, Athens, 1978.
30. S.A. Paipetis and P. Grootenhuis," The Dynamic Properties of Viscoelastic Fibre Composites," *Fibre Science and Technology, **12**,* 1979, 353-376.
31. S.A. Paipetis and P. Grootenhuis, "The Dynamic Properties of Viscoelastic Particle Composites," *Fibre Science and Technology, **12**,* 1979, 377-393.
32. S.A. Paipetis, A. Marchese and P.S. Theocaris, "The Dynamic Properties of Plasticized Epoxies over a Wide Frequency Range," *Colloid and Polymer Journal, **257**,* 1979, 478-485.
33. S.A. Paipetis, "The Dynamic Properties of Plastically Pretorsioned Mild Steel", *Materialpruefung, **21**,* 1979, 198-201.
34. S.A. Paipetis, "The Mechanical Behaviour of Particle Composites in the Presence of Shrinkage Stresses", *Colloid and Polymer Journal, **257**,* 1979, 934-941.
35. S.A. Paipetis, C. Stassinakis and P.S. Theocaris, "Complex Moduli Derived from the Vibrations of a Timoshenko Beam," *Acustica,* **1**, 1980, 27-34.
36. S.A. Paipetis, "Thermomechanical Properties of Unidirectional Composites in Their Transition Region," *Colloid and Polymer Journal, **258**,* 1980. 42-50.
37. S.A. Paipetis, "Thermomechanical Properties of Particle Composites in Their Transition Region," *Fibre Science and Technology, **13**,* 1980, 449-463.
38. S.A. Paipetis, "Mathematical Modelling of Composites" in Developments in Composite Materials-2 (Stress Analysis), G.S. Holister, Editor, *Applied Science Publishers Ltd,* London, 1981, 1-37.
39. S.A. Paipetis, "Experimental Modelling of Composites," in Developments in Composite Materials-2 (Stress Analysis), G.S. Holister, Editor, *Applied Science Publishers Ltd,* London, 1981, 39-73.
40. S.A. Paipetis, "A Dunkerley Procedure for Higher Modes," *Acustica, **49***, 1981, 73-76.
41. S.A. Paipetis, G.M. Tsangaris and J.M. Tsangaris "Dielectric Properties of Aluminium-Filled Epoxies," *Polymer Communications, **24**,* 1983, 373-375.
42. S.A. Paipetis and M. Kroustalis, "Approximate Determination of Eigenfrequencies in Damped, Lumped-Mass Elastic Systems," *Acustica, **53**,* 1983, 269-302.
43. S.A. Paipetis, "A Generalized Dunkerley-Graeffe, Procedure for Complex Mechanical Systems," *Journal of Sound and Vibration, **93**,* 1984, 173-180.
44. D. Vlissidis and S.A. Paipetis, "The Mechanical Strength of Complete Maxillary Dentures," *Journal of Biomedical Materials Research, **18**,* 1984, 413-425.
45. S.A. Paipetis, "Interfacial Events and Reinforcing Mechanis of Rubber/Carbon Black Composites," *Fibre Science and Technology, **21**,* 1984, 107-124.
46. S.A. Paipetis, "A Note on the Approximate Determination of Equal Eigenfrequencies in Elastic Systems," *Acustica, **56***, 1984, 53-56.

47. S.A. Paipetis and A. Vacakis, "A Method of Analysis for Unidirectional Viscoelastic Vibration Isolators With Many Degrees of Freedom", *Journal of Sound and Vibration, **98**,* 1, 1985, 13-23.
48. A. Vacakis and S.A. Paipetis, "Transient Response of a Unidirectional Vibration Isolator With Many Degrees of Freedom" *Journal of Sound and Vibration, **99**,* 1985, 557-562.
49. S.A. Paipetis, "Conventional 3-D Photoelasticity: A Review of Principles and Materials" in "Photoelasticity in Engineering Practice," S .A. Paipetis and G. S. Holister, Editors, *Elsevier Applied Science Publishers,* London 1985.
50. A. Vacakis and S.A. Paipetis, "The Effect of a Viscously Damped Dynamic Absorber on a Linear Multi-Degree-of-Freedom System," *Journal of Sound and Vibration, **105**,* 1986, 49-60.
51. S.A. Paipetis, D. Polyzos and C. Gregoriou, "On the Determination of the Mechanical Moduli of Elastic Particle Composites," Proceedings of the First National Conference of Mechanics, *Hellenic Society of Theoretical and Applied Mechanics,* Athens, June 13-18, 1986.
52. S.A. Paipetis, G.M. Tsangaris and A.J. Kontopoulos, "Dielectric Properties of Aluminium-Filled Epoxies Over a Wide Frequency Range", Proceedings, COMP'86 International Symposium on "Engineering Applications of New Composites," S.A.Paipetis and G. C. Papanicolaou, Editors, *Omega Scientific,* Fynamore, UK, 1987, 217-231.
53. S.A. Paipetis and D. Polyzos and V. Kostopoulos, "Analytical Modelling of Interfaces", Proceedings, COMP'86 International Symposium on Engineering Applications of New Composites", S.A.Paipetis and G.C. Papanicolaou, Editors, *Omega Scientific,* Fynamore, UK, p. 37, 1987.
54. S.A. Paipetis and N. Malakis, "Investigation of Seismic Activity by Real-Time Acoustic Emission Signal Processing", Proceedings, International Symposium on "Engineering Aspects of Earthquake Phenomena," A.Koridze, Editor, *Omega Scientific,* Fynamore, UK, 1987.
55. S.A. Paipetis, N. Melanitis and V. Kostopoulos, "Thermomechanical Properties of Rubber/Carbon Black Composites," *Journal of Materials Science,* **23**, 1988, 251-358.
56. K. Vlattas, V. Kostopoulos and S.A. Paipetis, "Further Developments in the Analytical Modelling of Interfaces," Proceedings, COMP'88 2nd International Symposium on "Phase Interaction in Composites", S.A.Paipetis and G.C.Papanicolaou, Editors, *Omega Scientific,* Fynamore, UK, 1989.
57. S.A. Paipetis, S. Pantelakis, G.C.Papanicolaou, G. Pissinou and V. Schulze, "Characterization and Processing of New Advanced Thermoplastic Composites", Proceedings, COMP'88 2nd International Symposium on "Phase Interaction in Composites," S.A.Paipetis and G. C. Papanicolaou, Editors, *Omega Scientific,* Fynamore, UK, 1989.
58. Z. Marioli-Riga and S.A. Paipetis, "Acoustic Emission Monitoring during fatigue tests of Thermosetting Carbon Fibre Reinforced Laminates", Prooceedings, COMP'88 2nd International Symposium on "Phase Interaction in

Composites", S.A.Paipetis and G.C.Papanicolaou, Editors, *Omega Scientific,* Fynamore, UK, 1989.

59. Z. Marioli-Riga and S.A. Paipetis, "Modelling Procedures for the Prediction of Fatigue Failure in Composites by Acoustic Emission," Proceedings COMP '90, Third International Symposium on "Advanced Composites in Emerging Technologies", S. A. Paipetis and Th. P. Philippidis, Editors, *AMATEC Publications,* Patras Greece 1992.
60. S.A. Paipetis and Z. Marioli-Riga, "Fatigue Life Prediction of Structural Composites by Acoustic Emission," *Proceedings Tallinn,* Estonia, 1986.
61. Z.P. Marioli-Riga, A.N. Karanika, T.P. Philippidis and S.A. Paipetis, Damage Detection by Acousto-Ultrasonic Location (AUL) in Debonding/Delamination of Composites, *AGARD Conference Proceedings 530*, pp.24.1-3, 1992.
62. K. Friedrich, S.A. Paipetis and G.C. Papanicolaou, "Effect of Interfacial Bond Quality on Fracture Toughness and Resistance to Fatigue Crack Propagation of Glass Fibre/Thermoplastic Polyester Composites,",Proceedings, COMP'88 2nd International Symposium on "Phase Interaction in Composites", S.A.Paipetis and G.C.Papanicolaou, Editors, *Omega Scientific,* Fynamore, UK, 1989.
63. M. Valavanidis, D. Polyzos and S.A. Paipetis, "Frequency-Dependent Dynamic Properties of Elastic Particle Composites," Proceedings of the Second National Conference of Mechanics, *Hellenic Society of Theoretical and Applied Mechanics,* June 1989.
64. V. Kostopoulos, L. Vellios and S.A. Paipetis, "Dynamic Thermomechanical Properties of Composites Based on Thermoelastic Wave Scattering," *Ingenieur Archiv, **60**,* 1990, pp. 431-443.
65. C. Vordos, D. Vlisidis and S.A. Paipetis, "The Effect of Remelting on The Mechanical Properties on Cr-Co Dental Alloys", *Metall.* 1992.
66. D. Polyzos, M. Valavanidis and S.A. Paipetis "Dynamic Properties of ellipsoidal - Particle Composites", *Science and Engineering of Composite Materials,* **2**, pp. 11-27, 1990.
67. D. Polyzos, M. Valavanidis and S.A. Paipetis "Dynamic Properties of Ellipsoidal Particle Composites," *Science and Engineering of Composite Materials,* **3**, pp. 11-27 (1991).
68. Z.P. Marioli-Riga, A.N. Karanika, T.P. Philippidis and S.A. Paipetis, "Damage Detection by Acousto-Ultrasonic Location (AUL)," in Debonding/Delamination of Composites, *AGARD Conference Proceedings No 530,* pp. 24.1-3, 1992.
69. G.S. Roupakias, T.P. Philippidis and S.A. Paipetis, Enhancing strength in laminated composites by adhesive interleaves, *Proceedings of the 1st National Congress on Computational Mechanics,* Athens Vol. 1, pp. 140-147, 1992.
70. B. Stamos, V. Kostopoulos and S.A. Paipetis, "Identification of Delamination by Eigenfrequency Degradation - An Inverse Problem," *AGARD Conference Proceedings 530,* pp. 24.1-3, 1992.
71. S.A. Paipetis and V. Kostopoulos, "Jet Fuel Absorption and Dynamic Mechanical Analysis of Carbon Fibre Composites," *AGARD Report 785,* pp. 24.1-3, 1992.

72. V. Kostopoulos, L. Vellios, T. Philippidis, S.A. Paipetis and A. Schoberth, "Dynamic Mechanical Analysis of a 2-D Carbon/Carbon Composite", *Journal of Materials Science, **28**,* pp. 5495-5499, 1993.
73. V. Kostopoulos, L.Vellios and S.A. Paipetis, "Thermal Expansivites of Anisotropic Constituents and Varying Interphase Properties", *Journal of Composite Science & Technology,* **2**, 1993, pp. 195-210.
74. S.A. Paipetis, V. Kostopoulos and L. Vellios, "Damage Identification by means of Eigenfequency Degradation and Internal Damping Measurements", Proceedings, *Symposium on Advanced Technology on Design and Fabrication of Composite Materials and Structures,* Torino (Italy), 1993.
75. L. Vellios, V. Kostopoulos and S.A. Paipetis, "Fatigue Damage Growth Monitoring of UD-Composites Using Vibration and Acoustic Emission Techniques", presented at the Ninth International Conference on Composite Materials (ICCM/9), *Convention Center, Madrid (Spain),* July 12- 16, 1993.
76. A.A. Anastassopoulos, D. Van Hemelrijck, W.P. De Wilde, S.A. Paipetis and B. Pirson, Defect Characterisation in Carbon/Epoxy Plates via Signal Processing and Pattern Recognition Analysis of Ultrasonic A-Scan Signals, *Proceedings, Ultrasonics International 93 Conference,* July 1993, pp. 835-838.
77. S.A. Paipetis, D .Polyzos and M.Valavanidis, "Constitutive Relations of Periodic Laminated Composites with Anisotropic Dissipation," *Archive of Applied Mechanics,* 64, pp. 32-43, 1993.
78. G.C. Christopoulos, S.A. Paipetis and K. Friedrich, "Interlaminar Fatigue Crack Propagation in Mode I of Carbon Fiber/PEEK-Composites," *Advanced Composite Letters,* **2**, 1993, pp. 5-9.
79. L.Vellios, V.Kostopoulos and S.A. Paipetis, "Fatigue Effect on the Dynamic Properties of CFRP Composites", *Advanced Composites Letters,* 1995.
80. G.C. Christopolos, R. Walter, S.A. Paipetis and K. Friedrich, "Effect of Testing Temperature on Interlaminar Fatigue Crack Growth in Fabric Reinforced Thermoplastic Composites," *Journal of Thermoplastic Composite Materials, 8* 1995, pp. 193-207.
81. K. Friedrich, A.M. Haeger, Y. Dzenis and S.A. Paipetis, "Erosive Wear Resistance of Advanced Polymer Composites," Proc. Int. Conf. Comp. Materials ICCM-10, Whistler, Canada, August 13-18, 1995, *Woodhead Publ. Ltd.,* Cambridge, U.K., Vol. IV, pp. 155-162.
82. C. Koimtzoglou, V. Kostopoulos, N.E. Melanitis and S.A. Paipetis, "The Effect of Fatigue on the Interfacial Properties of Model Composites," COMP'95, *5th International Symposium on "High Technology Composites in Modern Applications,,* 18-22 September, 1995, Corfu, Greece.
83. L. Vellios, Y. Z. Pappas, V. Kostopoulos and S.A. Paipetis, "Fatigue Damage Characterization of 3-D Sic/sic Composites by Non Destructive Techniques," COMP'95, *5th International Symposium on "High Technology Composites in Modern Applications,"* 18-22 September, 1995, Corfu, Greece.
84. S. E. Kattis, D. Polyzos and S. A. Paipetis, "Dynamic Boundary Element Method for Composite Materials using Homogenization Techniques,"

COMP'95, *5th International Symposium on "High Technology Composites in Modern Applications"*, 18-22 September 1995, Corfu, Greece.

85. Y.P. Markopoulos, G.C. Christopoulos, V. Kostopoulos and S.A. Paipetis, "Deformation and Fracture Behaviour of Carbon Epoxy Woven Fabrics," COMP'95, *5th International Symposium on "High Technology Composites in Modern Applications"*, 18-22 September 1995, Corfu, Greece.
86. S.V. Tsinopoulos, D. Polyzos and S.A. Paipetis, "Elastic Wave Scattering by 3-D Interfacial Cracks", COMP'95, *5th International Symposium on "High Technology Composites in Modern Applications,"* 18-22 September, 1995, Corfu, Greece.
87. A.A. Anastassopoulos, T.P. Philippidis and S.A. Paipetis, "Failure Mechanism Identification in Composite Materials by Means of Acoustic Emission: Is it Possible?," Non Destructive Testing, *A.A. Balkema Publisher,* Rotterdam, 1996, pp.143-149.
88. K. Friedrich, F. Haupert and S.A. Paipetis, "Advanced Processing of Thermoplastic Composites," *1st Hellenic Conf. on Composite Materials & Structures,* Univ. of Thrace, Xanthi, 2-5 July, 1997.
89. S.A. Paipetis, V. Kostopoulos and M. Kyriakopoulos, "On the Impact Response of the Shield of Achilles," *1st Hellenic Conf. on Composite Materials & Structures, University of Thrace,* Xanthi, Greece, 2-5 July, 1997.
90. S.A. Paipetis, "Non-Destructive Characterization of Ceramic Matrix Composites, International Workshop on Advanced Materials for Functional Manifestation of Frontier and Environmental Consciousness," Tokyo *International Forum, Tokyo,* Japan, 17-19 September, 1997 (invited paper).

General articles:

1. S.A. Paipetis, "Poetry and Technology", *Proceedings 1st Symposium of Modern Greek Poetry,* University of Patras, 2-5 July 1981, Gnosis, Athens 1982, 211-221 (in Greek).
2. P.S. Theocaris and S.A. Paipetis, "Higher Education in Greece and its Needs," *Transactions of the Academy of Athens 50,* 1975, 43-72 (in Greek).
3. S.A. Paipetis, "The Way of Quest and the Way of Poetry", *Proceedings, 20th Symposium on Greek Poetry, University of Patras,* July 2000.
4. S.A. Paipetis, "The Metaphysical and Mystical Element in Lorenzo Mavili's Poetry", *21st Symposium on Greek Poetry,* University of Patras, July 2001.

ARTICLES ON GENERAL MATTERS (SAMPLE TITLES)

On national anniversaries:

The Meaning of 28th October 1940 (the day that Italy declared war against Greece)
What Freedom and Sacrifice Means (on the 1821 Revolution that created the new Greek state)
From the Desire for Freedom to the Union (on the Occasion of the 180 Anniversary of the Union of the Ionian Islands with Greece.)

(b) Political criticism:

A Variation on a Well-Known Theme (on absent Representatives at the Parliament)
In the Echo of the Royal Wedding
A Letter to a Politician
A Letter to a Greek Citizen (criticizing political corruption)
The Samuel Syndrome (on the excesses in the action of labour unions)
Democracy in Life, in the Society and in Nature (philosophical approch to Democracy, Nationalism: Expression of Dissension or of Unity?
The Great Teachers of the Nation" (on political corruption)
The Guards of Democracy (on the improper behaviour of Parliament Representatives).

(c) On educational matters:

A Requiem for Higher Education
Total Ecology Engineering Studies: A New Proposal
Academic Shelter (Asylum) and Democratic Awareness
A National Debate on Education Policy: Why and What For?
Connecting Research and Industry in Western Greece
Ultimate Decadence

(d) Of general interest:

On the Threshold of the 21st Century: What to Expect
The Dark Veils of Future
Technology and Culture
We Are Getting Fewer (on the demographic problem and traffic accidents)

(e) Open round table discussions chaired:

Vocations Surviving in Time
Death: a Natural Phenomenon and a Human Problem (2 meetings).

1. General Concepts

FROM COMPOSITES TO CONSILIENCE

Y. F. MISSIRLIS
Biomedical Engineering Laboratory
University of Patras
GR-265 00 Patras, Greece

Abstract. Since the Ionian Times (Thales of Miletus) thinkers strived to reduce all knowledge to certain essential parts. The explosion of knowledge in recent times resulted in overspecialization. A call for interdisciplinarity and even an holistic approach to addressing scientific problems is ever present. Attempts to linking all knowledge, from natural sciences to social sciences and humanities should not ignore the historical fact that the social environment is of paramount importance for human conscious actions.

1. INTRODUCTION

Apart from the meaning of "composite" in the materials science and engineering terminology the word composite refers to one of the five classical orders of Architecture (consisting of elements of the Ionic and Corinthian orders), as well as to a plant family: Compositae (examples being the dandelion and the daisy). Many plant composites are highly successful as weeds [1].

That the term composite is used for technical materials, for technical structures, for biological materials, plants, expression of ideas (composition), for music, architecture etc. led me to the decision to attempt to express some ideas about a composite approach (an holistic, more or less approach) to addressing scientific as well as sociocultural problems. This is the way, I think, that Stefanos Paipetis is contemplating lately as well.

While thinking over these matters I came across an interesting book, written by Edward Wilson [2] with the title: Consilience, and a subtitle: The Unity of Knowledge. It was William Whewell who first spoke of Consilience in his 1840 work *The Philosophy of Inductive Sciences* where he said, "The consilience of inductions takes place when an induction, obtained from one class of facts, coincides with an induction, obtained from another different class" [3].

Therefore I ventured to connect the partial (composites) to the general (consilience) in this essay in a provocative way so as to generate some discussions among the "hard science" scientists.

E.E. Gdoutos and Z.P. Marioli-Riga (eds), Recent Advances in Composite Materials, 3-8

2. ANALYSIS

In fact our biological world is full of manifestations of the efficient use of composite materials and structures. One may suggest that as living organisms evolved on our planet and faced an ever more complex environment, which influenced directly their survival and proliferation endeavors, their design mechanisms combined existing materials to make structures responsive to their evolving needs. The interplay between simplicity and complexity is amazing. Take for example the mammalian proteins: hundreds of thousands exist and all are made from a pool of 20 aminoacids (AA). Of course at a high energetic cost that directs the right combination of the AA for each particular protein. Furthermore all the connections (bonds) of the AA, the so-called peptidic bonds are composed of the 4 basic atoms that exist in life: C-N-O-H. Four basic units (the same number 4 exists in the human DNA: Adenine-Cytosine-Guanine-Thymine), associate 20 other basic units at another level leading to an enormous specificity of proteins.

At the same time the mammalian organism in constructing an artery, for ex., utilises at least three different biomacromolecules, i.e. collagen fibers, elastin fibers and mucopolysaccharide matrix. Such a composite material and structure provides for sufficient strength, elasticity and overall mechanical performance in dealing with all physiological challenges that the particular artery is faced up with.

Plants and animals alike have used this method of combining different materials with divergent properties to construct "new" materials and structures with the required properties necessitated by their environmental constraints. Humans also learned this method and by studying more the marvels of Nature they might become more ingenious (but hardly innovative) in developing composite materials and structures. Examples of natural composite materials and structures such as biomembranes, soft and hard tissues, and biomaterials that are used in artificial organs are given elsewhere [4].

3. SYNTHESIS

It was said before that an artery is composed mainly of three biomolecules. The resulting composite structure is "passive" in the sense that responds mainly to external mechanical stimuli. The artery however consists, in addition, of "active", i.e. living cells of various types: those that regenerate the passive components when the need arises (such as fibroblasts) and those that participate actively in the working of the artery, for example by enlarging or restricting its lumen in response to neuro-mechano- or chemo stimulators (such as the smooth muscle cells). Furthermore the internal wall of an artery is covered by specific cells (endothelial cells) which form the natural nonthrombogenic environment for the circulating blood. The artery itself is part of the arterial system, which is part of the vascular system which along with the heart and the lungs make up the cardiopulmonary system of a mammalian organism. This system provides the fuel (oxygen) and removes the waste product (carbon dioxide) from the organism .This is its major role but not its sole role.

To recapitulate: an organism consists, in descending order, of organ systems, organs, tissues, cells, subcellular organelles, molecular assemblies, molecules, atoms...

The international scientific community has accepted, especially in recent historical times, that the continuous surge of new knowledge necessitated the division of the subject under study into segmental parts while specific scientific disciplines emerged to fulfil this situation. However, in doing so, a heavy price has been paid. This has to do with loss of information which exists in the interconnectivity areas of the whole. In other words, while by focusing in the segmented part of a system, say of a biological organism, this part is studied in greater detail and important knowledge is accumulated, at its boundaries everything becomes blurred, less defined, and maybe indeterminable. Furthermore each scientific discipline developed its own "language" with associated symbols, methodologies and techniques and it became apparent pretty soon that difficulties in understanding and communications hampered meaningful collaboration to address interdisciplinary problems.

In its tortuous and exiting development towards more and more disciplines there were always instances that required crosstalk so that the pure sciences started to blur into each other: physics was not only divided into atomic physics-solid state physics-high energy physics etc. But chemical physics-biophysics-geophysics etc. emerged as well.

The same phenomena manifest themselves in engineering. The technological explosion resulted in the need for new engineering specializations, which in turn fed back and advanced further the technological applications. However the overspecialization had the effect of educating engineers who designed and constructed specific structures but when the need for such structures disappeared, their knowledge was difficult to be diverted to new structures. At the same time, scant knowledge of environmental processes or limited exposure to humanities and sociological education led to uses of material and energy sources in such a way that a multitude of side effects of the human intervention started to negatively affect our planet.

Such concerns have led policy makers, educators and society in general to think anew regarding engineering education, in such a way that the practicing engineer should be able to have a global perspective of the impact of his work and that he is not a mere "technocrat" any more.

4. CONSILIENCE

All I have tried to say so far is that there is a real antithesis: the existing scientific knowledge is enormous while our capacity to grasp, comprehend and contribute to it is limited. If the whole knowledge can be represented in 2D space where on the abscissa we insert the various disciplines and on the ordinate the depth (height) of the specific knowledge what is the optimum approach that anyone should seek: broad and shallow knowledge or narrow and deep one? And to what degree?

If one contemplates on this question several corollaries follow: Has the rate of new knowledge and its overall accumulation been matched by our (brain) capacity to follow it? Are the two processes, i.e. the generation of new knowledge and the reorganization of the nervous connections in our brains so as to "adsorb" this new knowledge "in phase" as to their rates and amounts? Or the chasm between our ability to receive the new information, assess it, comprehend it and transform it to knowledge and the enormous amount of the generated information widens and diverts hopelessly to unknown dimensions?

The overwhelming proliferation of scientific publications, electronic information sites, along with dubious attempts by "popularising" media to bring the semicomprehended knowledge to the lay people manifest the difficult point at which the human society finds itself and the need to "stop for a moment" and reflect and assess its priorities and its goals.

I believe that similar thoughts are being voiced all too often lately expressing the agony of concerned scientists that something must be done.

Coming back to the science and engineering of composite materials and structures it is apparent that several disciplines must join forces so as to address intelligently the design and construction of such materials and structures: materials science, chemistry, textile or metallurgy, the science of interphasial phenomena et. al. In the emerging area of tissue engineering, where the need to create scaffolds from resorbable composite materials while biomolecules and biological cells are to be present and functioning, one should add cell biology, biomaterials science etc.

Such a coworking of natural sciences leads to the more or less accepted notion that the linkages between natural sciences are ever present. Indeed biology blurs into chemistry, which blurs into physics. Organisms can be reduced to molecules whose properties are subjected to the laws of chemistry, and the elements that compose those molecules follow the laws of quantum physics. Such a reductionist approach, termed *consilience by reduction* by Wilson [2] or top-down approach, has indeed been the working method in many if not all the natural sciences and despite some gaps of knowledge at the interfaces of the reductionist entities it may be credited for the advancement of knowledge so far.

5. OVERALL CONSILIENCE ?

How about going the reverse way (bottom-up)? Starting from simple (molecules) to more complex (the brain or the whole ecosystem), from the general to more specific? This is the essence of Wilson's proposal: all knowledge, from arts and humanities through the social sciences to the natural sciences can be unified! In his words [2]: the central idea of the consilience world view is that all tangible phenomena, from the birth of stars to the workings of social institutions, are based on material processes that are ultimately reducible, however long and tortuous the sequences, to the laws of physics.

This fascinating idea, especially I think for those practicing natural and biological sciences and engineering, but rather a horrifying idea for the rest, is, maybe, the natural extension of the authors' earlier approach to explain human behavior and cultural development from first biological principles [5]. Especially being conscious of the hard criticism that his sociobiology based explanation of individual-social-national behaviors have received, now he proposes that behavior is guided by *epigenetic rules*. Epigenesis means that an organism develops under the joint influence of heredity and environment [2]. Nevertheless, a definite bias is evident in all the (numerous and diverse) cases he sites. And this bias is a result of his contempt of the methods of scientific approach of both humanities and social sciences, as well as on excluding from his explanations the fact that many of the important problems of humanity and the biosphere are the result of the appetite of profits and more profits of the ruling classes of most nations. Not only that, but implicitly he assigns a biological-gene determined trait for such a phenomenon!

Definitely there is a need for collaboration between scientists and scholars so as to advance our knowledge of the great epiphenomena of human activity in the hope of becoming better inhabitants of this planet and avoiding its destruction.

As progress in all the physical sciences and technological innovations aid in revealing more details about the workings of the conscious brain, i.e. more details on the nerve circuitry, neurotransmitters and their association during specific actions or thoughts etc. We shall learn more about specifics. However it is an oversimplification to say that eventually the problems of consciousness or the subjective feelings or our ethical values are based on the genetic evolution, even via epigenetic rules.

There is also an opposite view [6]. It states that the borders between traditional intellectual disciplines represent the natural stress lines between our domains of cognitive competence. I.e., our inability to think clearly about some phenomena might underlie our tendency to draw boundaries where we do: the humanities to one side (telling me how you feel) and the sciences to the other (telling me that your brain does the feeling). Not to mention that our cognitive capacity may have limits so that any overall consilience is just a metaphysical exercise.

6. CONCLUSION

It may be that at some state of the evolution environmental reasons distinguished the *Homo* from the rest of creatures. It may be that certain environmental necessities (apart from mere survival) evolved the *Homo sapiens* and since then, and mainly in the historical times, the most active or intelligent persons believe that man can know everything. However a time span of 80 years of one's life is an infinitesimal part of the history of nature. Let us touch down.

The reality is that everything in the world is interconnected and in the process of continuous change. Our history is full of episodes that provide us with magnificent facts of altruism and wars of destruction, of sympathy and cruelty, of harmonious cohabitation and suppressive occupations, of universally acclaimed human rights and their abolition for self-interest etc.

The crux of the matter is that people, nations, scientists, scholars have to talk to each other, to try to understand each other. In that sense consilience, or linking together to attempt to solve the great challenges that humanity and our earth faces is a must. Otherwise, history (and biology) also has shown, that the new comes from the destruction of the old.

7. REFERENCES

1. The Oxford English Reference Dictionary, 2nd edition, 1996
2. Edward O.Wilson," *Consilience, The Unity of Knowledge*", Vintage Books, New York,1998
3. http://www.wikipedia.org
4. Yannis F. Missirlis, "*Composites in Biomedical Engineering*" in *Failure Analysis of Industrial Composite Materials*, ed. by E.E.Gdoutos, K.Pilakoutas and C.A. Rodopoulos, McGraw-Hill, New York,2000
5. David P. Barash, *Sociobiology and Behavior*, Elsevier, New York, 1977
6. H.Allen Orr, The Big Picture, *Boston Review*, Oct./Nov. ,1998

COMPUTATIONAL AND MECHATRONIC AUTOMATION OF MULTIPHYSICS RESEARCH FOR STRUCTURAL AND MATERIAL SYSTEMS

J. G. MICHOPOULOS
U. S. Naval Research Laboratory
Special Projects Group, Code 6303
Computational Multiphysics Systems Laboratory
Materials Science and Technology Division
Washington DC 20375-5000, USA

Abstract. A review of NRL's application of an automation of research paradigm is presented. The rampant computational technology evolution along with automated experimentation are utilized as the means for developing a characterization methodology of continuum systems under the action of multidimensional multiphysics fields. Computationally controlled robotic, multiple degree-of-freedom mechatronic systems are used to greatly accelerate the rate of performing mechanical experiments along loading paths embedded in multidimensional loading spaces. The massive data collections are utilized for the inexpensive and accurate determination of bulk material non-linear constitutive behaviour models as a consequence of multi field generalized loading through parameter identification/estimation inverse approach methodologies. Developments in computational simulation technologies that utilize advances in visualization and automated program synthesis have made possible inexpensive and fast characterization analysis of complex structures for design and qualification needs. The example of how this computational and mechatronic infrastructure is used to create conceptual models for describing and predicting material performance is demonstrated in the case of virtual wind tunnel environment.

1. INTRODUCTION

1.1 Intensions and Context

The present paper is attempting to provide evidence that computational technologies along with computational mathematics -both symbolic and numerical- in conjunction with mechatronic technologies, not only have provided opportunities and realizations for altering the mode of scientific method when applied in physical sciences for modelling of continuous systems such as composites, but also are the drivers behind justifications of why such a transition of the scientific method is both inevitable and preferable.

These intensions will be described in terms of research conducted at NRL for the past forty years within the context of the material constitutive behaviour characterization general problem, when the material is considered under the influence of generalized (multiphysics) and multidimensional loading conditions.

E.E. Gdoutos and Z.P. Marioli-Riga (eds), Recent Advances in Composite Materials, 9-23

This work is both connected to some of professor Paipetis' research work but mainly is directly connected with the underground principles of intellectual independence, freedom, continuous self examination and optimum utilization of physical and cognitive resources for solving problems of significant importance; principles that the author with immense appreciation was able to trace back to his interactions with professor Paipetis, while the later was the author's thesis and dissertation advisor at the National technical University of Athens in the late 70s.

An exposure of the motivational drivers for the scientific discipline of multiphysics system characterization is presented initially. The epistemology of scientific method as a function of the available technology will be subsequently described to justify why research community worldwide appears to have been working within the bounds of an operational paradigm that is not commensurable to our technological opportunities. It will continue with the opportunities for using computational and mechatronic technologies in applied sciences and engineering as followed by individual groups and investigators. Finally, NRL's approach for characterizing multiphysics behaviour of composites will be described and the new and emerging role of the researcher-designer will be sketched and discussed as it applies to the utilization of a virtual wind tunnel environment.

1.2 Motivation for Automation in Research of Multiphysics Systems

The need to be able to predict the behaviour of continuous systems under complex generalized multiphysics conditions is primarily driven from the need to design and utilize such systems in various areas of human technological endeavours. The means of addressing this need through the ages has been generally encapsulated by an approach that seeks the development continuous system modelling within the context of continuum mechanics. Successful models have been traditionally considered to be those that can faithfully and accurately reproduce actual (usually experimentally established) behaviour of existing systems.

The demands -especially in industrial and government research and development environments- for cost and time reduction, multi-mission design requirements, and increased systemic complexity suggested from the need for realistic systems predictive simulation (usually resulting in very computation intensive mathematical models), constitute the technology-push motivators for automation of research.

On the other hand, the innate human tendency for satisfying curiosity and for engineering elegant and efficient solutions that solve problems both on a societal or/and a personal scale, along with the rampant but predictable [1,2] evolution of computational and other associated technologies are the primary science pull motivators for attempting to automate the research process.

It appears that the push aspects are requirements driven, while the pull aspects are driven from an exploitation of the available opportunities perspective.

The main thesis of the present work is that there are two major areas where researchers might look for improving the ability to produce better, faster and cheaper products that except for a very few cases worldwide, have not been recognized widely as areas of opportunity. One is the realization of the role of the researcher relative to the scientific method and the exploitation of the available intellectual, computational and physical resources. The other is the distinct enormous potential of the computational technology and other associated with it technologies. In the following section we present some elements of a computational epistemology of scientific method as it relates to the constitutive characterization of materials in order to prepare the ground for the subsequent section of NRL's approach on the problem.

2. COMPUTATIONAL EPISTEMOLOGY OF THE SCIENTIFIC METHOD FOR PHYSICAL SYSTEM CHARACTERIZATION

2.1 Scientific Method and Scope of a Theory or a Model

The history of science suggests that there are two major schools of thought on the scientific method. Both had as a goal the codification of the rational procedures of science in a way that would free them from arbitrary, unfounded, or superstitious assumptions and ground them in a logically sound manner on the properties of clear and distinct fact. These schools have been traditionally described as [3]:

- The inductivist (empiricist) approach, which states that scientists begin by doing experiments and then derive their theories from the data. This branch was started by the preoccupation of John Locke and Francis Bacon with empirically observed facts.
- The hypothetico-deductive (rationalist) approach, which states that scientists begin with hypothetical theories and then do experiments to test them. This branch was started by Rene Descartes, who believed that all phenomena of physics could be accounted for by a single fully comprehensive mathematical theory, based on Euclidean foundations and conforming to his own deductivist principles. It was Newton, however, who clearly practiced the hypothetico-deductive method for the first time.

The first firm definition of a "theory" within the context of the scientific method appeared in the 1920's and 1930's. This was by the *Vienna Circle* of philosophers, who advocated the doctrine of logical positivism or logical empiricism. It combined empiricist epistemology with the techniques of formal logic that had been developed by Frege, Russel, and Whitehead [3,4]. According to logical positivists, scientific theories are to be understood as sets of axioms in formal deductive systems. Theories are confirmed by deducing their consequences from the axioms, and by checking whether the predictions (hypotheses) hold. In contrast to earlier empiricist views that carried the Baconian theme, such as those of John Stuart Mill [5], this methodology has been called hypothetico-deductive because it emphasizes the use of hypotheses to make predictions, rather than derivation of laws from observations.

The views of Popper evolved around the same time as those of the logical positivists, and were also hypothetico-deductive, but differed primarily in that he saw the main role of prediction to be the attempt to falsify theories and not to confirm them [5]. The general hypothetico-deductive scheme start with a hypothesis

H; suggests the use of logic to deduce predicted observation O and finally if O is observed, then H is confirmed (Hempel [6]), but if not-O is observed then H is falsified (Popper). The outcome of the last two alternatives suggests a non-terminating iterative process.

The dramatic improvement of computational technology in the last 30 years has enabled the NRL group and some other (admittedly very few) researchers to transition from these 18th century approaches which are limited by the performance of the human brain, to that of an industrialized inductivist approach. This approach, which uses the tools of logical positivism within an empirical framework, can be followed because increased computational power allows both automation of experimentation for massive collection of observations and also the assimilation of these observations into a compact and analytically useful form. The general industrialized inductive scheme first asks the researcher to identify observables to be measured; then to collect massive amounts of data spanning the control and observation spaces of the physical system; subsequently to generate a theory representing all these observations and finally to model and simulate the systemic behavior. The outcome of this scientific method is always terminating though it can be (time consuming). The scope of a theory or a systemic model is now to make sure that it is always consistent with the observed facts. A theory in this approach is not verifiable by any experiment, but can be refuted and destroyed by a negative experiment.

2.2 Behavior Characterization for Prediction

Physical system modeling can be abstracted as the activity of constructing a mathematical system defined by a relationship on a set of parameters or observables [4] that span a parameter space. According to systems identification theory [4,7], when there are observables that are measurable, any relationship among them is potentially identifiable. In cases where there is a pre-assignment of the special role of the participating parameters, the mathematical relations between the input and the output state parameters take a functional form. The endeavor of determining this functional form is what traditionally is known as system identification and the resulting mathematical representation is called a model of the natural system [4]. An identified system implies that a model has been established that allows the computation of outputs (or the exact state) of a system under a known set of input and control variables. This is the essence of prediction of future system behavior. Thus, behavior is equated with knowledge of the outputs of a system under the influence of known inputs.

2.3 The computational technology opportunity.

The functional similarity between a human neuron and a transistor as switching devices that can alter between two states (the bit states) at certain rates suggests that bit-state-changes-per-second (BSC/sec) can be thought of as a fundamental metric for characterizing the raw metabolic power for implementing the rational and computational process. A network of such devices (like the human brain or a microprocessor) can be evaluated in terms of this BSC/sec metric.

We have established elsewhere [1,2] that computational technology has surpassed human brain performance in terms of BSC/sec since 1974, the era of the Intel 8080 microprocessor, and is currently approximately 1-2 million times higher. The rate of improvement is a doubling of performance every 20 to 30 months. Some technology limitations suggest that a value of 10^{18} BSC/sec for a single chip will be an impassable barrier [8] provided the integrated circuitry technology is not substitutes by a more capable one. This limit will correspond to an increase in performance 108 times higher than that of the human brain. The presented opportunity is enormous, and it can be extended beyond arithmetic processing.

As computational technology evolved, more and more tasks that had traditionally required the human brain were implemented in computer software. Although it appears that software evolution has always lagged behind and been pulled on by hardware evolution, the history of various quantitative indices for software evolution indicate dramatic progress in software technologies. Considering "time needed to acquire skills to develop a new application using off the shelf software", we have transitioned [9,10] from the period of the 1950's through the 1970's where the average time to acquire skills was about 10 years, to the period after 1985 where it took only about a month to acquire the skills to develop a software application. As a consequence of this change, the number of people involved with software development has increased from a population of 50 thousand people to that of 50 million [10]. New software technologies like declarative programming, object oriented programming, computer aided software engineering, visual programming, and automated program synthesis have been some of the major contributing technologies for these trends. This trend appears to correspond to the change in the role of researchers in their effort to generate new tools and to abstract and capture models of the physical world in computer applications.

Once computers enabled researcher to solve problems more quickly than had been possible by hand they started realizing that their new role was to make use of computed answers.

The high rates of evolution of hardware and software have not only been affected by the rational process and its implementation, but have had a strong effect on them as well. Some goals that motivated the growth of computational technology have been the automation of information handling such as the paperless office and the integrated laboratory. Although the degree to which these goals has been a controversial subject, the fact is that their implementation has put a very high demand on the human ability to absorb and use information. In the pre-computer era of the 1800's the amount of information available for consumption worldwide was doubling every 50 years with annual percentage increase of the order of 5 to 10% [10]. Today we are in a track of worldwide information doubling every 3 years with an annual percentage increase of the order of 35% [10]. It appears that we are faced with the irony of success in using computational technology to generate information without a corresponding increase in human ability to extract value from it.

This experience has motivated the creation of alternative information representation technologies that utilize the visual, aural and tactile avenues in an attempt to enhance the bandwidth of information conveyed to the human brain. The products of these technologies can be found today in the advanced physical reality simulation tools such as NRL's Data Driven Virtual Wind Tunnel (DDVWT) environment the first stages of which will be presented later.

1. NRL'S AUTOMATION OF RESEARCH PARADIGM

3.1 Mechatronic Automation of Experimentation

To follow the industrialized-deductive scientific method, requires acquiring massive amounts of facts in the form of experimental data. This suggests that a capability to amplify the human ability to perform experiments is necessary. In this spirit, mechanical, hydraulic, electric, and computational power have been combined to amplify not only the human ability for deforming material specimens, but also the ability to gather and process sensor data faster than the actual experiment can take place. The degree of automation employed in material testing through the combination of computational and testing machine technologies has evolved extensively in the last 40 years. Robotic testing machines were developed to achieve industrial rates of acquisition of facts about material behaviour.

The first documented events [11] marking the marriage of custom made (analogue) computer technology with a servo-hydraulic testing machine was in 1962 at NRL (see Figure 1-left) and in 1963 when a hybrid computer was used to control a machine with the same intensions (see Figure 1-right).

Figure 1. First feedback controlled testing machines via analogue (left) and hybrid (right) computer.

Both of these machines were used to control crack velocity when measuring the dynamic characteristics of plexiglass in real time. This effort initiated the automation of experimentation allowing a few parameters to be measured and controlled at the same time to subject the specimen under test to a load path that was impractical to achieve by human operation alone. This technology involved a single degree of freedom and the time for preparing the specimen and setting it up in the test fixture was long. A human operator was still required and was the slowest part of the experimental process. The great majority of today's testing machines are descendants of this first computer-machine combination. Human involvement and the high cost (in time and money) of specimen preparation make this kind of technology useful only for non-industrialized low efficiency operations.

The first representative of the next generation of automated testing machines was the In Plane Loader System (IPLS) capable of 3 degrees of freedom loading. This machine, built at NRL in 1974, is shown in Figure 2.

Figure 2. In-plane loader system general view (left) and grip area view (right)

This system was capable of applying two translations and one rotation, all in the plane of the specimen and all displacement controlled. Simultaneous measurement of the boundary tractions and displacements allowed the experimental measurement of energy absorbed during the accumulation of strain-induced damage in real time. The system evolved to accept a stack of 30 specimens at a time and is being used today for the experimental determination of the Dissipated Energy Density (DED) function for organic matrix composite materials for the purpose of material characterization. An evolutionary and more complete description of the IPLS and the associated procedures is given in [1,12-14].

An entire IPLS test takes about 10 seconds while 5 of these seconds are spent positioning the specimen in the grips. As a result, the specimen-testing rate can be 360 specimens/hour. To characterize a material 120 specimens are required. At this rate data can be collected for 12 different materials per hour. The overall daily production rate for an 8-hour day is therefore 960 specimens, or 96 different materials, or 24 materials systems (since 4 layup angle combinations are used for each fiber-resin combination). The total number of experimental points per fiber-resin combination is 6000. Each loading path corresponds to 2.4 Kbytes of data while 288 Kbytes are acquired for each material system. The daily throughput of the acquisition process approaches 20.74 Mbytes/day.

In order to address issues of larger specimens, out of plane loading and whole field strain measurement, a new generation of six degree of freedom testing machines was developed at NRL. The first version used six actuators with analogue controlled valves, mounted on an I-beam frame. This machine was completed and tested in 1983 (see Figure 3-left). A robotic arm was used to insert and remove specimens. Both data acquisition and control processes were increased to six degrees of freedom. The large open frame of this machine allowed actuators to be placed parallel to three orthogonal axes to simplify the algebraic representation of the kinematics of the movable grip.

Starting in 1993, yet another new six degree of freedom loader system (6DLS) was designed and constructed at NRL (see Figure 3-right). It was built in a hexapod architecture configuration developed originally for flight simulator platforms (Stewart platform) and it is classifiable as a 6-6p parallel robotic mechanism. The hexapod architecture made the machine more compact, far stiffer (less prone to energy storage by deformation of the machine itself), and easier to disassemble for modification or relocation than the previous machine.

Figure 3. First (left) and second (right) generation 6DLS s.

In 1996, in fact, the original analogue controlled actuators of the prototype machine were replaced with longer stroke, digitally controlled actuators with minimal changes needed in other structural parts of the machine. This is a system that has been designed to evolve to the third generation of testing machines; a multidimensional generalized loader system (MDGLS) in that additional capability for measuring non-mechanical conjugate pairs of material state variables will be added. This will afford the opportunity of using the loader to identify physical systems under the simultaneous action of thermal, electromagnetic, substance diffusion, and mechanical loading effects.

3.2 Formulation of Coupled Multiphysics of Material Structural Continua

Multiphysics formulations have been developed primarily from two main perspectives: Multi-field coupled or uncoupled constitutive modelling of single continua, and single or multi field models of interacting continua. The first category usually reflects a relation between the conjugate pairs of state variables, while the second reflects coupled or uncoupled field equations (usually PDEs) governing the evolution of the state variable fields within continua (interacting or not). The automated generation and solution of both constitutive and field equations has been pursued.

Fluid dynamics, as a specific discipline of continuum mechanics provides fundamental equations as symbolic conceptual models of dynamic fluid continua. Axiomatic continuum mechanics of deformable solids on the other hand, in conjunction with thermodynamics and electromagnetism provide additional equations for modelling their behaviour. Usually these equations are field evolution PDEs and constitutive response algebraic equations or ordinary differential equations. Conservation laws (of momentum, energy, fluxes etc.) as well as jump, continuity and conservation conditions across boundaries of interacting continua complete the set of equations needed for coupled multiphysics modelling of continua.

The main scientific issues arising out of the plethora of potential methods for generating and combining these equations are the ability to deal with the horrendous symbolic complexity of these representations and the ability to answer the questions of theory similarity, equivalence, and simplification between any two given such theories. A third issue is the choice of the optimization scheme architecture and parameters that are needed for determining any free coefficients participating in the analytical representation of the functions that capture the dissipative constitutive characterization of the continuum at hand.

There are several mathematical theories for describing the constitutive behaviour of mostly non-interacting systems under multi-field conditions such as magnetothermoelasticity, thermoelectroelasticity, hygrothermoelasticity [15], and electromagnetic hygrothermoelasticity. Generalized approaches for deriving such theories have also been given both from an electrodynamics [16-18] and thermodynamic perspective [15, 16, 19, 20]. Equational systems of this sort form the traditional basis of symbolic models for the behaviour of continua under multiphysics stimulus. In the context of solving continuum mechanics boundary value problems, it is traditional to define as field equations the set of PDEs that can be produced to describe field spatial and time evolution when all constitutive equations have been eliminated through term rewriting. There are many methodologies available for constitutive [21-24] and field equation derivation [25-30] applied to single physics applications. All of these approaches attempt to address the first issue above (symbolic complexity) but never the second (theory equivalence). We have recently attempted to address the second issue via the foundation of an alternative theory representation methodology based on algebraic directed graphs for the case of all theories for anisotropic elastic bodies in contact [31]. The third issue (optimization scheme) has scarcely been addressed for single physics applications through associative memory techniques like neural network technology [32,33]. A rigorous methodology for addressing this issue has been developed by NRL in representing an energy dissipation potential as a sum of basis functions, the coefficients of which can be determined via constrained least squares optimization [34-37].

The fundamental difference between all of these approaches and our approach is based on an effort to use the data-driven experience, as it is contained in the experimentally acquired datasets in the multiphysics context. This is achievable trough a meta-theoretic symbolic algebra substrate that permits continuous field theories generation and comparison by considering two specific approaches.

The first is a thermodynamic procedure for deriving constitutive equations using a thermodynamic potential (i.e. internal energy, or Gibbs potential) defined in terms of free coefficients of the state variable monomials [15,16]. The second is a conservative mechanics procedure based on forming a Lagrangean weakened by compensating fields with free coefficients. This can account for nonlinear irreversibility in the state space of the modelled system such as the approaches used in Gauge Theory modelling of damage [38]. The free coefficients can be adaptively determined through minimization of the error between predicted and actual (through experimentation) datasets, by employing standard adaptive optimization methodologies. Another difference between traditional and our techniques is the use of a symbolic algebra substrate in a theorem proving fashion to derive automatically the constitutive and field equations required for numeric solution of specific boundary value problems and feeds them directly to existing finite element infrastructures. In particular, a very suitable infrastructure for automating the field equation generation process is provided by the term pattern matching power of Mathematica's [39] rewriting engine, along with the tensor calculus implementation of Mathtensor [40] and the predicate logic, simplifier, natural induction and Grobner basis theorem prover capabilities of Theorema [41].

3.3 Automation of PDE solving and Visualization and Simulation

The encapsulated coupled PDEs need to be solved eventually for the particular problem and geometry of choice. This implies that one or a suite of numerical solution methodologies can be used for this purpose. The main issues associated with this effort are (a) optimal semi-discretization schemes (i.e., finite elements, finite volumes, boundary elements, finite differences, etc.) for the governing PDEs as well as the coupling transmission conditions across non-matching multidisciplinary interfaces, (2) efficient staggered time-integrators for the fast and yet practical solution of the resulting coupled ODEs, (3) robust mesh partitioning techniques and scalable domain decomposition based iterative algorithms for the parallel solution of the underlying algebraic systems of equations.

The historical efforts for addressing the first three issues are classified into two main categories: (1) those that utilize pre-existing code libraries or programs that implement techniques for the numerical discretization of the differential operators in the participating PDEs (i.e. finite elements, finite differences etc.) and (2) those that exploit symbolic algebra to perform automated program synthesis in terms of automatically generating source code in target languages that when compiled and run, yields solutions to the modelled problems. We have been utilizing three-field formulation (fluid structure-moving grid) of coupled fluid-structure interaction problems that is quite general from the University of Colorado [35-37].

Our current efforts are focusing in encapsulating legacy code (programs and/or libraries) functionality in the form of Java component technology. In particular we are extending the field multiphysics capability from its current form to that of general coupled multi-field aero-structural problems with material nonlinearity. Initial samples of this approach have been encapsulated in our Virtual Wind Tunnel Environment effort.

3.4 Application: Material Softening Issues in a Virtual Wind Tunnel Environment

The Data Driven Virtual Wind Tunnel (DDVWT) environment [35-37] is a an object oriented distributed computational environment under development at NRL, that allows the user to succesfully address these requirements through the automated characterization of material constitutive response in conjunction with the appropriate finite element infrastructure for solving the coupled PDEs associated with the involved multiphysics.

The coupled transient aero-structural problem has been formulated as a three-field problem describing the spatio-temporal displacement evolution for the fluid, the structure and the dynamic mesh in terms of the following coupled semisiscrete PDEs :

$$\begin{aligned}
&\frac{\partial}{\partial t}(A(x,t)W(x,t)) + \tilde{F}_c^c(W((x,t),x,\dot{x})) = \tilde{F}^d(W(x,t)) \\
&\tilde{M}\frac{\partial^2}{\partial t^2}x + \tilde{D}\frac{\partial}{\partial t}x + \tilde{K}_c x = K_c u(W(x,t)) \\
&M\frac{\partial^2}{\partial t^2}u + f^{\text{int}}(u) = f^{ext}(W(x,t))
\end{aligned} \tag{1}$$

where: W is the fluid state vector, A results from the finite element/volume discretization of the fluid equations, $\tilde{F}_c^c$ is the convected vector of numerical convective fluxes, $\tilde{F}^d$ is the vector of numerical diffusive fluxes, u is the structural displacement vector, f^{int} is the vector of internal forces on the structure, f^{ext} is the vector of external forces on the structure, M is the finite element mass matrix of the structure, $\tilde{M}$ and $\tilde{D}$, and $\tilde{K}$ are fictitious mass, damping and stiffness matrices associated with the fluid moving grid and constructed to avoid parasitic interaction between the fluid and its grid, or the structure and the moving fluid.

In an attempt to account for material nonlinear behavior due to strain induced damage the DDVWT has been used to solve the coupled fluid/structure problems of the AGARD 445.6 wing. The wing was modeled as a composite structure and analysis was performed to obtain strain fields that were subsequently used to compute dissipated energy density for each loading condition as a specific combination of angle of attack and far field mach number. These basis cases were evaluated on 5 processors of NRL's *SGI Origin 2000* high performance computing parallel multiprocessor system via Finite Element Tear and Interconnect multi-domain methods. Figure 4 depicts typical distributions of the steady state solutions for the

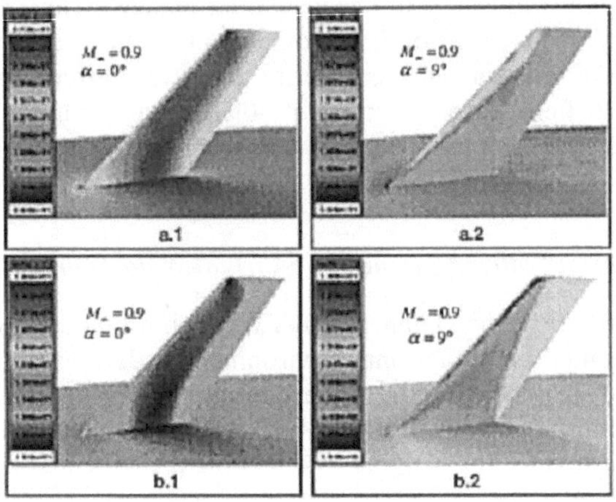

Figure 4. Mach number (top) and pressure (bottom) distributions on AGARD wing

fluid speed (Figure 4a.x) and pressure (Figure 4b.x) on the boundary of the fluid formed by the bottom of the wing, for the case of far field mach $M_{\infty} = 0.9$ and for angle of attack α, at 0° (Figures. 4a.1, 4b.1) and 9° (Figures. 4a.2, 4b.2).

Figure 5 shows the corresponding distributions of material softening in terms of

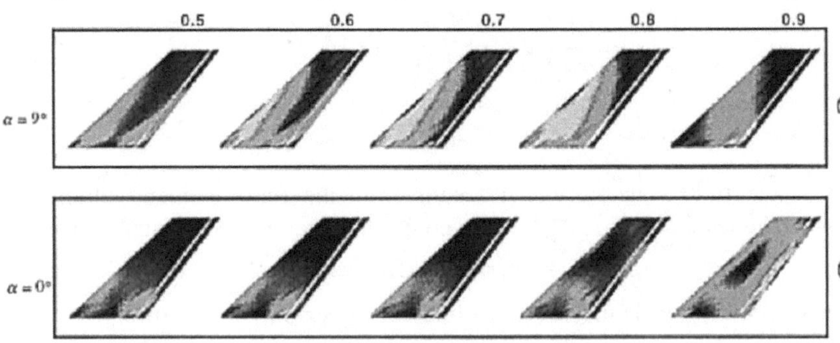

Figure 5. Flight induced material softening distributions

dissipated energy density in relative color scale, as a function of the mach number M_∞ for angles of attack $\alpha = 0°$ and $\alpha = 9°$. Here it is evident that the DED exhibits a hot spot shifting effect as a function of the far field fluid speed. In the case of angle of attack $\alpha = 9°$ the shifting occures from the root of the wing near the trailing edge to the leading edge and then back to the root.

4. CONCLUSIONS

Clearly systems like DDVWT allow the dynamic reconstruction of stimulus conditions based on flight condition parameterization instead of traditional finite element nodal boundary condition descriptions. The computational efficiency of the process is based on the dynamic superposition of basis cases solutions of the coupled fluid-structure interaction problem and provides the user/designer with the ability to dynamically reconstruct a plethora of material and structural response scenarios based on mission parameter choices. This helps harnessing the benefits of a computationally driven industrial-inductive scientific method for predicting material and structural behaviour under multiphysics environments in a natural way.

Dedication: To professor Stephen Paipetis with immense appreciation and respect because his deeds always provided inspiration and ammunition to those Don-Quixotian knights in the battle for disproving the windmill of the universal principle of conservation of aggravation: "Cheap, fast, good. Choose any two".

5. REFERENCES

1. P. R. Factory, *Effects of Computational Technology on Composite Materials Research: The Case of the Dissipated Energy Density*, Keynote Lecture, in Proceedings of the First Hellenic Conference on Composite Materials and Structures, Democritus Univ. of Thrace, Greece, July 2-5,1997, pp 465-494, Kyriakidis Brothers S.A. Publishing House, Thessaloniki, 1997.
2. J. Michopoulos, *Epistemology of Computational Technology for Engineering Applications*, NRL-Report, to appear, 2002.
3. P. Thagard, *Computational Philosophy of Science*, MIT Press, Cambridge, MA, 1988.
4. J. L. Casti, Reality Rules: I & II, John Wiley & Sons, Inc., New York, 1992.
5. K. Popper, *Unended Quest*, Fontana, London, 1976.
6. C. G. Hempel, *Aspects of Scientific Explanation*, Free Press, New York, 1965.
7. D. Graupe, *Identification of Systems*, Robert E. Krieger Publishing Co., Malabar, FL, 1972.
8. P. Gelsinger, P. Gargini, G. Parker and A. Yu, *A Microprocessor Odyssey*, Technology 2001 The Future of Computing and Communications, D. Leebaert, ed. MIT Press, 1991.
9. P. Strassman, *The Politics of Transition of Software from Craft to Product*, NeXTSTEP East Coast Developer Conference 1994, Keynote Address, Washington, DC, 1994.
10. J. Martin, *Hyperdocuments & How to Create Them*, Prentice Hall, Englewood Cliffs, NJ, 1990.
11. M. L. Sebek and A. A. Braun, *The Evolution of Testing Equipment Used in Fracture Mecahnics Research*, MTS monograph, Minneapolis, MN, 1994.
12. P.W. Mast, J.G. Michopoulos, R.W. Thomas, R. Badaliance, I. Wolock, Characterization of strain-induced damage in composites based on the dissipated energy density: Part I - Basic scheme and formulation, *Int. Jnl. of Theor. and Applied Fract. Mech.*, 22, 71-96,(1995).
13. P.W. Mast, J.G. Michopoulos, R.W. Thomas, R. Badaliance, I. Wolock, Characterization of strain-induced damage in composites based on the dissipated energy density: Part II - Composite specimens and naval structures, *Int. Jnl. of Theor. and Applied Fract. Mech.*, 22, 97-114, (1995).

14. P.W. Mast, J.G. Michopoulos, R.W. Thomas, R. Badaliance, I. Wolock, Characterization of strain-induced damage in composites based on the dissipated energy density: Part III -General material constitutive relation, *Int. Jnl. of Theor. and Applied Fract. Mech.*, 22, 115-125, (1995).
15. G.C. Sih, J.G. Michopoulos, S.C. Chou, *Hygrothermoelasticity*, Martinus Nijhoff Publishers (now Kluwer Academic), 1986
16. J.G. Michopoulos, G. C. Sih, *Coupled Theory of Temperature Moisture Deformation and Electromagnetic Fields*, Institute of Fracture and Solid Mechanics report IFSM-84-123, Lehigh University, February 1984.
17. A.C. Eringen and G.A. Maugin. *Electrodynamics of Continua I. Foundations and Solid Media.* Springer-Verlag, New York, (1990).
18. L.D. Landau, E.M. Lifshitz, *Electrodynamics of Continuous Media*, Pergamon Press, Oxford, 1960.
19. C. Truesdell, R. Toupin, *The Classical Field Theories*, in Handbuch der Physik (Herausgegeben con S. Flugge) Bd. III/1, Springer-Verlag, Berlin 1960.
20. A. E. Green; P. M. Naghdi: A unified procedure for construction of theories of deformable media. I,II,III. Classical continuum physics, *Proc. Roy. Soc. London Ser.* A 448 (1995), no. 1934, 335--388.
21. H.Q. Tan, X. Dong, *Symbolic Derivation of Visco-Plastic Material Equations*, in Proc. Beijing Int. Conf. on System Simulation and Scientific Computing, 1989, Beijing, China, pp. 969-973.
22. R. H. Lance, The potential of symbolic computation in engineering education: the opportunity and the challenge. In *Symbolic computations and their impact on mechanics*, ed. A. K. Noor, I. Elishakoff, and G. Hulbert, pp. 339-60. New York: American Society of Mechanical Engineers, 1990.
23. S. M. Arnold, A.F. Saleeb, H.Q. Tan, Y. Zhang, Wxplicit Robust Schemes for Implementation of General Principal Value-Based Constitutive Models, *Comp. & Struct.*, Vol. 53, No. 6, pp. 1437-1451, 1994
24. V.A. Palmov, Large Strains in Viscoplacity, *Acta Mechanica*, Vol 125, pp. 129-139, 1997.
25. A. Niemunis, M. Cudny, On Hyperelasticity for Clays, *Computers and Geotechnics*, Vol. 23, pp. 221-236, 1998.
26. H. Fujio, S. Doi, *FEEL: A Problem-Solving Environment for Finite Element Analysis*, NEC Research and Development, Vol. 39, No. 4, pp. 491-496, 1998
27. E. Kant, F. Daube, W. MacGregor, J. Wald, *Scientific Programming by Automated Synthesis*, Chapter 8 in Automatic Software Design, M. Lowry and R. McCartney, eds, AAAI Press/MIT Press, Menlo Park, CA, 169- 205, 1991.
28. R. van Engelen, L. Wolters, and G. Cats, *Ctadel: A Generator of Efficient Code for PDE-based Scientific Applications*, Technical Report 95-26, Department of Computer Science, Leiden University, 1995.
29. G.O. Cook, Jr. and J.F. Painter, *ALPAL: A Tool to Generate Simulation Codes from Natural Descriptions*, in Expert Systems for Scientific Computing, E.N. Houstis, J.R. Rice, and R. Vichnevetsky (eds), Elsevier, 1992.
30. D. Eyheramendy, T. Zimmerman, Object-Oriented Symbolic Derivation and Automatic Programming of Finite Element Mechanics, *Enginnering with Computers*, Vol. 15, pp. 12-36, 1999
31. J.G. Michopoulos, *Pathology of Symbolic Dimensionality on Continuous Multiphysics Theory Representation and Utilization*, invited talk at the 8th Int. Conference on Applications of Computer Algebra on June 25-28 University of Thessaly Volos Greece.
32. R.M. Pidaparti, M.J. Palakal, Material model for composites using neural networks. *AIAA Journal*, 31 (8), pp. 1533-1535,1993
33. H.S. Shin, G. N. Pande. On self-learning finite element codes based on monitored response of structures, *Computers and Geotechnics* 27, pp. 161-178, 2000.
34. J. Michopoulos, R. Badaliance, T. Chwastyk, L. Gause, P. Mast, R.A. Foerch, *Data Driven Non-linear Optimization for Automated User Material Subroutine Computation and Utilization*, invited paper in the Proceedings of the ANSYS User Group by Malett Technologies Inc., October 2 2001.
35. J. Michopoulos, R. Badaliance, T. Chwastyk, L. Gause, P. Mast, C. Farhat, M. Lesoinne, *Strain Induced Material Nonlinearity for AGARD Wing 445.6 in a Virtual Wind Tunnel Environment: Results from Coupled Multiphysics Simulation,* DoD HPC User Group Conference 2000, Albuquerque New Mexico.
36. J. Michopoulos, R. Badaliance, T. Chwastyk, L. Gause, P. Mast, C. Farhat and M. Lessoine, *Coupled Multiphysics Simulation of Composite Material Softening in a Virtual Wind Tunnel Environment,*

invited paper in Proceedings of Sixth U.S. National Congress on Computational Mechanics, U.S. Association for Computational Mechanics, Dearborn MI, pp 521, 1-3 August 2001.

37. J. Michopoulos, C. Farhat, M. Lesoinne, P. Mast, R. Badaliance, T. Chwastyk, L. Gause, Material *Softening Issues in a Multiphysics Virtual Wind Tunnel Environment*, AIAA Paper 2002-1095, 40th Aerospace Sciences Meeting and Exhibit, Reno, Nevada, January 14-17 (2002)
38. D.C. Lagoudas, A Gauge Theory of Damage, *Int. J. of Engng. Science*, 29 No. 5, pp. 597-606, 1991
39. S. Wolfram, *The Mathematica Book*, 4th Edition Wolfram Media/Cambridge University Press, 1999
40. L.Parker, S.M. Christensen, *MathTensor: A System for Doing Tensor Analysis by Computer*, by Addison-Wesley, 1994
41. B. Buchberger, *Theorema: Computer-Supported Mathematical Proving*. Annual Meeting of the Mathematics Department of the University of Wales, Gregynod, 1998.

A FIRST ATTEMPT TOWARDS A THEORY OF ANISOTROPIC SCATTERING

G. DASSIOS and K. KARADIMA
Division of Applied Mathematics
Department of Chemical Engineering
University of Patras, and
ICEHT/FORTH, GR-265 04 Patras, Greece

Abstract. Integral representations for the scattered field, as well as for the scattering amplitude, for scalar scattering in a completely anisotropic space are derived. These results are obtained via Green's integral formula, the derivation of the fundamental solution for the anisotropic Helmholtz equation and standard asymptotic analysis.

1. INTRODUCTION

Within the enormous literature on scattering theory [1], that was written in the 130 years since Rayleigh's first attempt to postulate the problem of wave-obstacle interaction, we can hardly identify papers that deal with anisotropic media [11]. There is of course an extended literature on wave propagation in anisotropic media [3,4,7-10,12], but almost always no obstacle is present.

In the work at hand an attempt is made towards a general theory of scalar anisotropic scattering based on the introduction of a redirected gradient operator, which carries all directional characteristics of the anisotropic medium. In other words, this medium gradient encodes the behaviour of the directional variations of the space and allows to obtain the fundamental solution for the related anisotropic Helmholtz equation. This is explained in Section 2. Once the fundamental solution is known, Green's second integral theorem is used, in Section 3, to produce the integral representation for the scattered as well as for the total field. Finally, in Section 4, the integral representation of the scattering amplitude is obtained, from the leading asymptotic approximation of the scattered field as the observation point recedes to infinity.

2. THE FUNDAMENTAL SOLUTION

Let $\tilde{\mathbf{A}}$ be the symmetric and positive definite constant dyadic that characterizes the anisotropic medium, which fills the three-dimensional Euclidean space. Then the spectral form of the scalar wave equation [11] for such a medium is written as

E.E. Gdoutos and Z.P. Marioli-Riga (eds), Recent Advances in Composite Materials, 25-34

$$\nabla \cdot \tilde{\mathbf{A}} \cdot \nabla \, U(\mathbf{r}) + k^2 U(\mathbf{r}) = 0 \tag{1}$$

where k is the wave number. Our goal is to derive the fundamental solution for this anisotropic version of Helmholtz's equation, following techniques developed in [2,5,6].

Since $\tilde{\mathbf{A}}$ is positive definite it has a positive definite square root, i.e. there exists a positive definite symmetric dyadic $\tilde{\mathbf{B}}$ such that

$$\tilde{\mathbf{B}}^2 = \tilde{\mathbf{B}} \cdot \tilde{\mathbf{B}} = \tilde{\mathbf{A}} \,. \tag{2}$$

The dyadic $\tilde{\mathbf{B}}$ incorporates the anisotropic character of the medium via the transformation

$$\mathbf{r}_\alpha = \tilde{\mathbf{B}}^{-1} \cdot \mathbf{r} \tag{3}$$

where the existence of $\tilde{\mathbf{B}}^{-1}$ is secured by the positive definiteness of $\tilde{\mathbf{B}}$. Indeed, since $\tilde{\mathbf{B}}$ is positive definite, all its eigenvalues are positive. Hence, its third invariant is positive, a fact that implies its invertibility.

Actually (3) describes how the anisotropy of the medium deforms (through scaling and rotation) the actual position vector $\mathbf{r}$ to the anisotropic point $\mathbf{r}_\alpha$. Positive definiteness restricts the angle of rotation of $\mathbf{r}$ to be less than $\pi/2$. Straightforward chain rule calculations show that

$$\nabla_\alpha = \tilde{\mathbf{B}} \cdot \nabla \tag{4}$$

where ∇_α stands for the anisotropic gradient. As it was expected, the effect of the directional variation of the medium is to change the direction, as well as the magnitude, of the maximum increase of the physical quantities.

From the mathematical point of view, we observe that the isotropic gradient

$$\nabla = \sum_{i=1}^{3} \hat{\mathbf{x}}_i \frac{\partial}{\partial x_i} \tag{5}$$

introduces three orthogonal directions $\hat{\mathbf{x}}_i$, $i = 1,2,3$ along which the spatial variation of any function enters with the same weight. On the other hand, the symmetry of $\tilde{\mathbf{B}}$ implies that there are three orthogonal directions $\hat{\mathbf{s}}_i$, $i = 1,2,3$ (the eigendirections of $\tilde{\mathbf{B}}$) along which

$$\nabla_\alpha = \sum_{i=1}^{3} s_i \hat{\mathbf{s}}_i \frac{\partial}{\partial s_i} \tag{6}$$

where s_i, $i = 1,2,3$ are the eigenvalues of $\tilde{\mathbf{B}}$. Consequently, the role of $\tilde{\mathbf{B}}$ is to impose different weights to the three partial derivatives, i.e. the weights that the medium establishes.

In view of the transformation (3), (4) the anisotropic spectral wave equation (1) is written as

$$(\Delta_\alpha + k^2)u(\mathbf{r}_\alpha) = 0 \tag{7}$$

where the anisotropic Laplacian is defined as

$$\Delta_\alpha = (\tilde{\mathbf{B}} \cdot \nabla) \cdot (\tilde{\mathbf{B}} \cdot \nabla) = \nabla \cdot \tilde{\mathbf{B}} \cdot \tilde{\mathbf{B}} \cdot \nabla = \nabla \cdot \tilde{\mathbf{A}} \cdot \nabla = \tilde{\mathbf{A}} : \nabla \otimes \nabla \tag{8}$$

with the double contraction [5] defined by

$$\mathbf{a} \otimes \mathbf{b} : \mathbf{c} \otimes \mathbf{d} = (\mathbf{a} \cdot \mathbf{c})(\mathbf{b} \cdot \mathbf{d})\,. \tag{9}$$

The fundamental solution for (7) has to satisfy the source equation

$$(\Delta_\alpha + k^2)G^A(\mathbf{r}_\alpha) = -\delta(\mathbf{r}) \tag{10}$$

and since the transformation

$$\mathbf{r} = \mathbf{f}(\mathbf{r}') \tag{11}$$

transforms the Dirac measure as

$$\delta(\mathbf{r}) = \frac{1}{J}\delta(\mathbf{r}') \tag{12}$$

where J is the Jacobian of $\mathbf{f}$, it follows that in the $\mathbf{r}_\alpha$-system (the system imposed by the medium) (10) reads as

$$(\Delta_\alpha + k^2)G^A(\mathbf{r}_\alpha) = -\frac{\delta(\mathbf{r}_\alpha)}{\det \tilde{\mathbf{B}}}\,. \tag{13}$$

The determinant of $\tilde{\mathbf{B}}$ is of course the Jacobian of the linear transformation (3), that is

$$\det \tilde{\mathbf{B}} = \det\left(\frac{\partial \mathbf{r}}{\partial \mathbf{r}_\alpha}\right). \tag{14}$$

In the medium system now, equation (13) accepts the solution

$$G^A(\mathbf{r}\ ) = \frac{1}{4\pi(\det\tilde{\mathbf{B}})}\frac{\exp\{ik\,|\,\mathbf{r}\ \ |\}}{|\,\mathbf{r}_\alpha\,|} \tag{15}$$

and since

$$|\,\mathbf{r}_\alpha\,| = \sqrt{\mathbf{r}_\alpha\cdot\mathbf{r}_\alpha} = \sqrt{(\tilde{\mathbf{B}}^{-1}\cdot\mathbf{r})\cdot(\tilde{\mathbf{B}}^{-1}\cdot\mathbf{r})}$$

$$= \sqrt{(\tilde{\mathbf{B}}^{-1})^2:\mathbf{r}\otimes\mathbf{r})} = \sqrt{(\tilde{\mathbf{B}}^2)^{-1}:\mathbf{r}\otimes\mathbf{r})} = \sqrt{\tilde{\mathbf{A}}^{-1}:\mathbf{r}\otimes\mathbf{r})} \tag{16}$$

we obtain the following expression in the **r**-system (the Euclidean system)

$$G^A(\mathbf{r}) = \frac{1}{4\pi\sqrt{\det\tilde{\mathbf{A}}}}\frac{\exp\left\{ik\sqrt{\tilde{\mathbf{A}}^{-1}:\mathbf{r}\otimes\mathbf{r}}\right\}}{\sqrt{\tilde{\mathbf{A}}^{-1}:\mathbf{r}\otimes\mathbf{r}}} \tag{17}$$

where again the invertibility of $\tilde{\mathbf{A}}$ comes from its positive definiteness.

Obviously, for a point source at the position $\mathbf{r}'$ the fundamental solution is written as

$$G^A(\mathbf{r},\mathbf{r}') = \frac{1}{4\pi\sqrt{\det\tilde{\mathbf{A}}}}\frac{\exp\left\{ik\sqrt{\tilde{\mathbf{A}}^{-1}:(\mathbf{r}-\mathbf{r}')\otimes(\mathbf{r}-\mathbf{r}')}\right\}}{\sqrt{\tilde{\mathbf{A}}^{-1}:(\mathbf{r}-\mathbf{r}')\otimes(\mathbf{r}-\mathbf{r}')}} \tag{18}$$

and solves the equation

$$(\tilde{\mathbf{A}}:\nabla\otimes\nabla + k^2)G^A(\mathbf{r},\mathbf{r}') = -\delta(\mathbf{r}-\mathbf{r}')\,. \tag{19}$$

3. THE INTEGRAL REPRESENTATION

Let V^- be a closed bounded, connected set in $I\!R^3$ with a smooth boundary S and let V denotes its open exterior. The region V^- specifies the scatterer, while V denotes the exterior to the scatterer anisotropic medium of propagation characterized by the symmetric and positive definite dyadic A.

The scattering problem we want to postulate consists of finding a solution of the anisotropic Helmholtz equation (7), which satisfies one of the standard boundary conditions on S [1] and is consistent with the Sommerfeld radiation condition at infinity.

If u denotes the scattered field in V, then a well known procedure [1] based on Green's second identity, applied to the scattered field and the corresponding fundamental solution, as well as an appropriate limiting process produces the integral representation.

$$u(\mathbf{r}) = \int_S \left[u(\mathbf{r}')\frac{\partial}{\partial n'} G^A(\mathbf{r},\mathbf{r}') - G^A(\mathbf{r},\mathbf{r}')\frac{\partial}{\partial n'} u(\mathbf{r}') \right] ds(\mathbf{r}') \tag{20}$$

which holds for $\mathbf{r} \in V$.

Furthermore, if $u^{(i)}$ denotes an incident field that has no finite singularities and

$$u^+(\mathbf{r}) = u^{(i)}(\mathbf{r}) + u(\mathbf{r}), \qquad \mathbf{r} \in V \tag{21}$$

denotes the total field, then similar arguments that lead to (20), imply that

$$u(\mathbf{r}) = \int_S \left[u^+(\mathbf{r}')\frac{\partial}{\partial n'} G^A(\mathbf{r},\mathbf{r}') - G^A(\mathbf{r},\mathbf{r}')\frac{\partial}{\partial n'} + u^+(\mathbf{r}') \right] ds(\mathbf{r}') . \tag{22}$$

In both representations (20) and (22) G^A is given by (18).

Some delicate but straightforward calculations lead to

$$\nabla_{\mathbf{R}} \sqrt{\tilde{\mathbf{A}}^{-1} : \mathbf{R} \otimes \mathbf{R}} = \frac{\mathbf{R} \cdot \tilde{\mathbf{A}}^{-1}}{\sqrt{\tilde{\mathbf{A}}^{-1} : \mathbf{R} \otimes \mathbf{R}}} \tag{23}$$

where

$$\mathbf{R} = \mathbf{r} - \mathbf{r}' \tag{24}$$

and

$$\nabla_{\mathbf{R}} G^A(\mathbf{R}) = \left(ik - \frac{1}{\sqrt{\tilde{\mathbf{A}}^{-1} : \mathbf{R} \otimes \mathbf{R}}} \right) \frac{\mathbf{R} \cdot \tilde{\mathbf{A}}^{-1}}{\sqrt{\tilde{\mathbf{A}}^{-1} : \mathbf{R} \otimes \mathbf{R}}} G^A(\mathbf{R}) . \tag{25}$$

Consequently,

$$\frac{\partial}{\partial n'} G^A(\mathbf{r},\mathbf{r}')$$

$$= \hat{\mathbf{n}}' \cdot \nabla_{\mathbf{r}'} G^A(\mathbf{r},\mathbf{r}')$$

$$= -\left[\frac{ik}{\sqrt{\tilde{\mathbf{A}}^{-1} : (\mathbf{r}-\mathbf{r}') \otimes (\mathbf{r}-\mathbf{r}')}} - \frac{1}{\tilde{\mathbf{A}}^{-1} : (\mathbf{r}-\mathbf{r}') \otimes (\mathbf{r}-\mathbf{r}')}\right] (\hat{\mathbf{n}}' \cdot \tilde{\mathbf{A}}^{-1} \cdot (\mathbf{r}-\mathbf{r}')) G^A(\mathbf{r},\mathbf{r}') \quad (26)$$

where $\hat{\mathbf{n}}'$ stands for the outward unit normal on S.

If we introduce the anisotropic distance function

$$d_\alpha(\mathbf{r}-\mathbf{r}') = \sqrt{\tilde{\mathbf{A}}^{-1} : (\mathbf{r}-\mathbf{r}') \otimes (\mathbf{r}-\mathbf{r}')} \quad (27)$$

with corresponding isotropic version

$$d_\alpha(\mathbf{r}-\mathbf{r}') = |\mathbf{r}-\mathbf{r}'| \quad (28)$$

then

$$G^A(\mathbf{r},\mathbf{r}') = \frac{\exp\{ikd_\alpha(\mathbf{r}-\mathbf{r}')\}}{4\pi d_\alpha(\mathbf{r}-\mathbf{r}')\sqrt{\det \tilde{\mathbf{A}}}} \quad (29)$$

and

$$\frac{\partial}{\partial n'} G^A(\mathbf{r},\mathbf{r}') = -\left[\frac{ik}{d_\alpha(\mathbf{r}-\mathbf{r}')} - \frac{1}{d_\alpha^2(\mathbf{r}-\mathbf{r}')}\right] (\hat{\mathbf{n}}' \cdot \tilde{\mathbf{A}}^{-1} \cdot (\mathbf{r}-\mathbf{r}')) G^A(\mathbf{r},\mathbf{r}') \quad (30)$$

and the representation (22) reads

$$u(\mathbf{r}) = -\int_S \left[u^+(\mathbf{r}') \frac{ikd_\alpha(\mathbf{r}-\mathbf{r}')-1}{d_\alpha^2(\mathbf{r}-\mathbf{r}')} \hat{\mathbf{n}}' \cdot \tilde{\mathbf{A}}^{-1} \cdot (\mathbf{r}-\mathbf{r}') + \frac{\partial}{\partial n'} u^+(\mathbf{r}')\right] G^A(\mathbf{r},\mathbf{r}') ds(\mathbf{r}') \quad (31)$$

for each $\mathbf{r} \in V$. Equation (31) provides the scattered field as a superposition of the fields due to a monopole and a dipole distribution of sources over the boundary S of the scatterer.

Note that for $\tilde{\mathbf{A}} = \tilde{\mathbf{I}}$ we recover the corresponding representation [1] for isotropic scattering

$$u(\mathbf{r}) = -\int_S \left[u^+(\mathbf{r}') \frac{ik|\mathbf{r}-\mathbf{r}'|-1}{|\mathbf{r}-\mathbf{r}'|^2} \hat{\mathbf{n}}' \cdot (\mathbf{r}-\mathbf{r}') + \frac{\partial}{\partial n'} u^+(\mathbf{r}')\right] G(\mathbf{r},\mathbf{r}') ds(\mathbf{r}') \quad (32)$$

where

$$G(\mathbf{r},\mathbf{r}') = \frac{\exp\{ik|\mathbf{r}-\mathbf{r}'|\}}{4\pi |\mathbf{r}-\mathbf{r}'|}. \quad (33)$$

4. THE FAR-FIELD

In order to analyze the behaviour of the scattered field in the radiation zone we have to perform an asymptotic expansion of (31) as $r \to \infty$.

To this end we first derive

$$(\mathbf{r}-\mathbf{r}')\otimes(\mathbf{r}-\mathbf{r}') = r^2\left(\hat{\mathbf{r}}-\frac{\mathbf{r}'}{r}\right)\otimes\left(\hat{\mathbf{r}}-\frac{\mathbf{r}'}{r}\right) = r^2\left[\hat{\mathbf{r}}\otimes\hat{\mathbf{r}}-\frac{\hat{\mathbf{r}}\otimes\mathbf{r}'+\mathbf{r}'\otimes\hat{\mathbf{r}}}{r}+\frac{\mathbf{r}'\otimes\mathbf{r}'}{r^2}\right] \quad (34)$$

and

$$\tilde{\mathbf{A}}^{-1}:(\mathbf{r}-\mathbf{r}')\otimes(\mathbf{r}-\mathbf{r}')$$

$$= r^2(\tilde{\mathbf{A}}^{-1}:\hat{\mathbf{r}}\otimes\hat{\mathbf{r}})\left[1-\frac{1}{r}\frac{\tilde{\mathbf{A}}^{-1}:(\hat{\mathbf{r}}\otimes\mathbf{r}'+\mathbf{r}'\otimes\hat{\mathbf{r}})}{\tilde{\mathbf{A}}^{-1}:\hat{\mathbf{r}}\otimes\hat{\mathbf{r}}}+\frac{1}{r^2}\frac{\tilde{\mathbf{A}}^{-1}:\mathbf{r}'\otimes\mathbf{r}'}{\tilde{\mathbf{A}}^{-1}:\hat{\mathbf{r}}\otimes\hat{\mathbf{r}}}\right]. \quad (35)$$

Then the binomial expansion theorem implies that

$$g_\alpha(\mathbf{r}-\mathbf{r}') = r\sqrt{\tilde{\mathbf{A}}^{-1}:\hat{\mathbf{r}}\otimes\hat{\mathbf{r}}}\left[1-\frac{1}{r}\frac{\tilde{\mathbf{A}}^{-1}:(\hat{\mathbf{r}}\otimes\mathbf{r}'+\mathbf{r}'\otimes\hat{\mathbf{r}})}{2\tilde{\mathbf{A}}^{-1}:\hat{\mathbf{r}}\otimes\hat{\mathbf{r}}}+O\left(\frac{1}{r^2}\right)\right], \quad r\to\infty \quad (36)$$

and, therefore

$$G^A(\mathbf{r},\mathbf{r}') = \exp\left\{-ik\frac{\tilde{\mathbf{A}}^{-1}:(\hat{\mathbf{r}}\otimes\mathbf{r}'+\mathbf{r}'\otimes\hat{\mathbf{r}})}{2\sqrt{\tilde{\mathbf{A}}^{-1}:\hat{\mathbf{r}}\otimes\hat{\mathbf{r}}}}\right\}\frac{\exp\left\{ikr\sqrt{\tilde{\mathbf{A}}^{-1}:\hat{\mathbf{r}}\otimes\hat{\mathbf{r}}}\right\}}{4\pi r\sqrt{\det\tilde{\mathbf{A}}}\sqrt{\tilde{\mathbf{A}}^{-1}:\hat{\mathbf{r}}\otimes\hat{\mathbf{r}}}}$$

$$+O\left(\frac{1}{r^2}\right), \quad r\to\infty. \quad (37)$$

From the above expression becomes obvious that in the far-field, the anisotropic directivity of the medium is dictated by the factor

$$\alpha(\hat{\mathbf{r}}) = \sqrt{\tilde{\mathbf{A}}^{-1}:\hat{\mathbf{r}}\otimes\hat{\mathbf{r}}} \quad (38)$$

which degenerates to unit in the isotropic case.
In terms of the isotropic factor α, the fundamental solution assumes the asymptotic form

$$G^A(\mathbf{r},\mathbf{r}') = \exp\left\{-ik\frac{\tilde{\mathbf{A}}^{-1}:(\hat{\mathbf{r}}\otimes\mathbf{r}'+\mathbf{r}'\otimes\hat{\mathbf{r}})}{2\alpha(\hat{\mathbf{r}})}\right\}\frac{\exp\{ik\alpha(\hat{\mathbf{r}})r\}}{4\pi\alpha(\hat{\mathbf{r}})r\sqrt{\det\tilde{\mathbf{A}}}}+O\left(\frac{1}{r^2}\right), \quad r\to\infty \quad (39)$$

and at the same time it provides the expansion

$$\frac{ikg_\alpha(\mathbf{r}-\mathbf{r}')-1}{g_\alpha^2(\mathbf{r}-\mathbf{r}')}\hat{\mathbf{n}}'\cdot\tilde{\mathbf{A}}^{-1}\cdot(\mathbf{r}-\mathbf{r}')=\frac{\hat{\mathbf{n}}'\cdot\tilde{\mathbf{A}}^{-1}}{\alpha^2(\hat{\mathbf{r}})}\cdot\left[ik\alpha(\hat{\mathbf{r}})\hat{\mathbf{r}}-\frac{\hat{\mathbf{r}}}{r}\right.$$

$$\left.+\frac{ik\tilde{\mathbf{A}}^{-1}:(\hat{\mathbf{r}}\otimes\mathbf{r}'+\mathbf{r}'\otimes\hat{\mathbf{r}})}{2\alpha(\hat{\mathbf{r}})r}\hat{\mathbf{r}}-\frac{ik\alpha(\hat{\mathbf{r}})}{r}\mathbf{r}'\right]+O\left(\frac{1}{r^2}\right),\qquad r\to\infty. \tag{40}$$

Inserting (39) and (40) in (31), performing the calculations and keeping only the leading asymptotic term we arrive at the far-field representation

$$u(\mathbf{r})=g^A(\hat{\mathbf{r}};\hat{\mathbf{k}})h(\alpha(\hat{\mathbf{r}})kr)+O\left(\frac{1}{r^2}\right),\qquad r\to\infty \tag{41}$$

where the anisotropic scattering amplitude enjoys the representation

$$g^A(\hat{\mathbf{r}};\hat{\mathbf{k}})=-\frac{ik}{4\pi\sqrt{\det\tilde{\mathbf{A}}}}\int_S\hat{\mathbf{n}}'\cdot\left[\frac{ik\tilde{\mathbf{A}}^{-1}\cdot\hat{\mathbf{r}}}{\alpha(\hat{\mathbf{r}})}u^+(\mathbf{r}')+\nabla_{\mathbf{r}'}u^+(\mathbf{r}')\right]\times$$

$$\times\exp\left\{-\frac{ik}{\alpha(\hat{\mathbf{r}})}\tilde{\mathbf{A}}^{-1}:\frac{\hat{\mathbf{r}}\otimes\mathbf{r}'+\mathbf{r}'\otimes\hat{\mathbf{r}}}{2}\right\}ds(\mathbf{r}') \tag{42}$$

and

$$h(\alpha(\hat{\mathbf{r}})kr)=\frac{\exp\{i\alpha(\hat{\mathbf{r}})kr\}}{i\alpha(\hat{\mathbf{r}})kr} \tag{43}$$

denotes the radiative fundamental solution for a point source at the origin.

Using the symmetry of the dyadic $\tilde{\mathbf{A}}^{-1}$, as well as the direction dependent vector

$$\mathbf{d}(\hat{\mathbf{r}})=\frac{\tilde{\mathbf{A}}^{-1}\cdot\hat{\mathbf{r}}}{\sqrt{\tilde{\mathbf{A}}^{-1}:\hat{\mathbf{r}}\otimes\hat{\mathbf{r}}}}=\frac{\tilde{\mathbf{A}}^{-1}\cdot\hat{\mathbf{r}}}{\alpha(\hat{\mathbf{r}})} \tag{44}$$

the scattering amplitude is written as

$$g^A(\mathbf{r},\mathbf{r}')=-\frac{ik}{4\pi\sqrt{\det\tilde{\mathbf{A}}}}\int_S\left[ik\mathbf{d}(\hat{\mathbf{r}})\cdot\hat{\mathbf{n}}'u^+(\mathbf{r}')+\frac{\partial}{\partial n'}u^+(\mathbf{r}')\right]\exp\{-ik\mathbf{d}(\hat{\mathbf{r}})\cdot\mathbf{r}'\}ds(\mathbf{r}')\,. \tag{45}$$

For $\hat{\mathbf{r}}=(\mathrm{O}_1,\mathrm{O}_2,\mathrm{O}_3)$ we obtain

$$\mathbf{d}(\hat{\mathbf{r}}) = \frac{\sum_{\kappa=1}^{3} \frac{\mathrm{O}_\kappa}{s_\kappa^2} \hat{\mathbf{x}}_\kappa}{\sqrt{\sum_{\kappa=1}^{3} \frac{\mathrm{O}_\kappa^2}{s_\kappa^2}}} \tag{46}$$

where s_1^2, s_2^2, s_3^2 are the three eigenvalues of $\tilde{\mathbf{A}}$.

Once again, for an isotropic medium where $\tilde{\mathbf{A}} = \tilde{\mathbf{I}}$, $\alpha(\hat{\mathbf{r}}) = 1$, $\mathbf{d} = \hat{\mathbf{r}}$, (45) degenerates to

$$g(\hat{\mathbf{r}};\hat{\mathbf{k}}) = -\frac{ik}{4\pi} \int_S \hat{\mathbf{n}}' \cdot [ik\hat{\mathbf{r}}u^+(\mathbf{r}') + \nabla_{\mathbf{r}'} u^+(\mathbf{r}')] \exp\{-ik\hat{\mathbf{r}} \cdot \mathbf{r}'\} ds(\mathbf{r}') \tag{47}$$

and, Eq. (41) is written as

$$u(\mathbf{r}) = g(\hat{\mathbf{r}};\hat{\mathbf{k}}) h(kr) + O\left(\frac{1}{r^2}\right), \quad r \to \infty. \tag{48}$$

It is of interest to observe that while, for an isotropic medium, g is given as a superposition of plane waves propagating in the direction $-\hat{\mathbf{r}}$ with a constant wave number k for the case of an anisotropic medium, the corresponding plane waves propagate in the direction $-\mathbf{d}(\hat{\mathbf{r}})$ and their wave numbers $k\,|\,\mathbf{d}(\hat{\mathbf{r}})\,|$ are directionally dependent.

The actual application of the above theory to the case of scattering by a sphere embedded within an anisotropic medium is under current investigation.

5. REFERENCES

1. Dassios, G. and Kleinman, R.: *Low Frequency Scattering*, Oxford University Press, Oxford, 2000.
2. Dassios, G. and Lindell, I.V.: Uniqueness and reconstruction for the anisotropic Helmholtz decomposition, J. of Physics A: Mathematical and General 35 (2002), 5139-5146.
3. Duff, G.F.: The Cauchy problem for elastic waves in an anisotropic medium, Philos. Trans. Roy. Soc. London Ser. A 252 (1960), 249-273.
4. Hähner, P.: On the uniqueness of the shape of a penetrable, anisotropic obstacle, J. of Computational and Applied Mathematics 116 (2000), 167-180.
5. Lindell, I.V.: Methods for Electromagnetic Field Analysis, 2nd edn, Oxford University Press, Oxford, 1995.
6. Lindell, I.V. and Dassios, G.: Generalized Helmholtz decomposition and static electromagnetics, J. of Electromagnetic Waves and Applications 14 (2000), 1415-1428.
7. Musgrave, M.J.P.: Elastic waves in anisotropic media, in Progress in Solid Mechanics II, Sneddon and Hill (eds.), North-Holland, New York, 1960.
8. Nayfeh, A.H.: Wave Propagation in Layered Anisotropic Media, Elsevier, Amesterdam, 1995.
9. Piller, N.B. and Martin, O.J.F.: Extension of the generalized multipole technique to anisotropic media, Optics Communications 150 (1998), 9-14.

10. Schleicher, J., Tygel, M., Ursin, B. and Bleistein, N.: The Kirchhoff-Helmholtz integral for anisotropic elastic media, Wave Motion 34 (2001), 353-364.
11. Stevenson R.C.: Green's function for the Helmholtz equation of a scalar wave in an anisotropic halfspace. SIAM J. Applied Mathematics 50 (1990), 199-215.
12. Thomson, C.J.: Modelling surface waves in anisotropic structures I. Theory, Physics of the Earth and Planetary Interiors 103 (1997), 195-206.

WAVE PROPAGATION IN COMPOSITES

J. T. VERBIS[a,b], S. V. TSINOPOULOS[a,b],
J. P. AGNANTIARIS[a] and D. POLYZOS[a,b]
[a]*Mechanical and Aeronautical Engineering Department*
University of Patras
[b]*Institute of Chemical Engineering and High Temperature Chemical Processes-FORTH*
GR-265 00 Patras, Greece

Abstract. The present work deals with dispersion and attenuation of elastic waves propagating in particulate and fiber reinforced composite materials. An iterative effective medium approximation based on single scattering considerations, for the quantitative estimation of wave dispersion and attenuation is applied. The single scattering problem is solved numerically by means of an advanced boundary element code valid for two and three-dimensional elastodynamic analysis. Representative numerical results concerning the frequency dependent behavior of the wave velocity and attenuation coefficient of various types of particulate and fiber reinforced composites are presented. The obtained results are compared to those taken either experimentally of numerically by other investigators.

1. INTRODUCTION

When the phase velocity of a harmonic wave, propagating in a medium, is a frequency dependent function then the medium is characterized dispersive. Similarly, when the amplitude of the wave decays in space, with a frequency dependent exponential coefficient, the medium is called attenuating. Fiber-reinforced as well as particulate composite materials are both dispersive and attenuating mediums. For a perfectly elastic and infinitely extended composite system, the manifestation of dispersion and attenuation is due to the multiple scattering of the incident wave by the randomly distributed inhomogeneities of the composite.

Within the last decades, many investigators have extensively studied wave propagation in fiber-reinforced composites. However, a general theory that can predict accurately the dynamic behavior of these materials is not yet available, because of the difficulties in simulating the multiple scattering mechanisms. The pioneering works of Foldy [1], Lax [2] and Waterman and Truell [3] can be considered as the foundations of the study of multiple scattering. Some representative works which are based on multiple scattering considerations are those by Weaver & Pao [4], which provides dispersion and attenuation curves by means of

E.E. Gdoutos and Z.P. Marioli-Riga (eds), Recent Advances in Composite Materials, 35-46

the Kramers-Kronig expressions, Bose & Mal [5], Varadan et al. [6], Kim et al. [7], Bose [8] and Liu and Kriz [9].

Another important category of works deals with self-consistent theories. According to these theories the dynamic behavior of the composite is predicted through the solution of one or more single scattering problems. In these problems the incident wave is propagated in an effective medium and is scattered by the microstructure of the composite. Here one can mention the works of Soven [10], Kim et al. [11], Sabina and Willis [12], Huang and Rohklin [13], Yang and Mal [14] and Kim [15].

In the present work, the dispersion and attenuation of elastic waves propagating in a fiber reinforced or a particulate composite material are evaluated by means of a self-consistent iterative effective medium approximation (IEMA). In the case of fiber-reinforced composites the present IEM methodology combines the dynamic self-consistent effective medium (DSCEM) approximation of Kim [15] and the quasicrystalline approximation of Waterman and Truell [3]. The solution of the single scattering problems, imposed by the IEMA model, is accomplished through a two-dimensional (2D) boundary element methodology (BEM) (Verbis et al. [16]) by taking into account that the plane longitudinal (P) or transverse (SH or SV) mean plane wave is propagating through an axis perpendicular to the axis of the fibers.

For particulate composites the IEMA used here involves, in general, non-spherical three-dimensional inclusions and is making use of the single inclusion self-consistent condition of Kim et al. [11]. As is done in the case of a single layer fiber reinforced composite, also in the case of a particulate composite the IEMA assumes that the effective stiffness of the composite is the same with the corresponding static one and evaluates iteratively the frequency dependent dynamic density of the composite. The complex value of the effective dynamic density and the static effective stiffness of the composite determine, eventually, the wave speed and the attenuation coefficient of the plane wave propagating through the composite. The single scattering problem imposed by the self-consistent condition is solved numerically via an advanced BEM code derived by Tsinopoulos et al. [17], for axisymmetric geometries, and by Kattis et al. [18] for three-dimensional (3D) applications.

In what follows, in section 2 the main steps of the present IEMA are described while in section 3 the BEM used for the solution of the scattering problems are presented in brief. In section 4 representative numerical results obtained by the use of the present IEMA are presented and they concern dispersion and attenuation curves by a Lead/Epoxy (*Epon 828z*) particulate composite (26,1% and 52% inclusions) and a Boron/Epoxy (54% fibers) fiber reinforced composite. All the results are compared to those obtained by other investigators either numerically or experimentally. Finally, in section 5 a multiple scattering problem dealing with a propagating plane wave in a slab containing 108 randomly distributed fibers is solved numerically through the 2D boundary element method and the results are compared with those obtained by solving the single scattering problem in the case when the slab is homogeneous with dynamic effective material properties predicted by the use of the present IEMA.

2. ITERATIVE EFFECTIVE MEDIUM APPROXIMATION

An elastic wave propagating in a composite material can be considered as a sum of a mean wave and a number of fluctuating waves. The mean wave exists in a homogeneous medium having the effective properties of the composite, while the fluctuating waves are the result of the multiple scattering of the mean wave by the randomly distributed material variations with respect to those of the effective medium [10]. Under this consideration, the average of fluctuating fields should be vanish, at any direction within the effective medium:

$$\left\langle \hat{\boldsymbol{k}} \cdot \tilde{\boldsymbol{T}} \cdot \hat{\boldsymbol{k}} \right\rangle = 0 \tag{1}$$

where < > denotes average over the composition and the shape of the scatterers, $\tilde{\boldsymbol{T}}$ is a matrix corresponding to the total multiple scattering operator for the fluctuating waves and $\hat{\boldsymbol{k}}$ is the propagation direction of the mean wave. Eq. (1) is well known as self-consistent condition and can be used to determine the dynamic effective properties of the composite material. However, due to the prohibitive computational cost of the evaluation of the operator $\tilde{\boldsymbol{T}}$, Soven [10] proposed a simplified self-consistent condition, which reads

$$\left\langle \hat{\boldsymbol{k}} \cdot \tilde{\boldsymbol{t}} \cdot \hat{\boldsymbol{k}} \right\rangle = 0 \tag{2}$$

where $\tilde{\boldsymbol{t}}$ is a single scattering operator coming from the diffraction of the mean wave by each composition, i.e. matrix and particles/fibers, embedded in an infinitely extended effective medium. Devaney [19] proved that the condition (2) could also be written as a function of the far-field scattering amplitudes in forward direction. Thus, for the case of either a particulate or a single layer fiber reinforced composite, where identical homogeneous inclusions or elastic fibers are embedded in a homogeneous elastic matrix, condition (2) assumes the following form:

$$n_1 g^{(1)}(\hat{\boldsymbol{k}},\hat{\boldsymbol{k}}) + (1-n_1) g^{(2)}(\hat{\boldsymbol{k}},\hat{\boldsymbol{k}}) = 0 \tag{3}$$

where n_1 represents the volume fraction of the particles/fibers and $g^{(1)}(\hat{\boldsymbol{k}},\hat{\boldsymbol{k}})$, $g^{(2)}(\hat{\boldsymbol{k}},\hat{\boldsymbol{k}})$ are the forward scattering amplitudes taken by the solution of the two single wave scattering problems illustrated in Fig. 1.

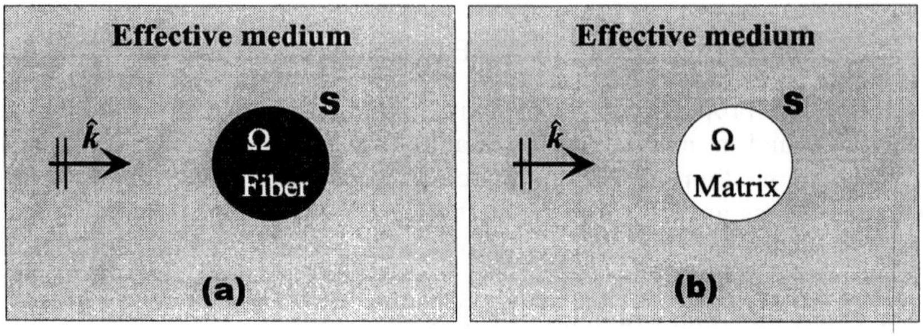

Figure1. Single wave scattering problems of a mean wave propagating in the effective medium; a) the scatterer is the fiber (problem I) and b) the scatterer is a matrix inclusion with identical to the fiber geometrical properties (problem II).

In the present work, the self-consistent condition (3) is satisfied numerically following an iterative procedure. This iterative effective medium procedure can be summarized as follows: Consider a single layer fiber reinforced or particulate composite material consisting of a homogenous, isotropic matrix medium with density, shear and Young moduli, ρ_2, μ_2 and E_2, respectively. In the matrix randomly distributed, long parallel elastic fibers or elastic 3D particles, of identical properties (ρ_1, μ_1 and E_1) in a volume fraction equal to n_1 are embedded. For the case of a fiber-reinforced composite a harmonic elastic wave either longitudinal (P) or transverse (SH or SV) is propagated through the composite with a propagation vector being perpendicular to the fibers. Due to the presence of the particles/fibers multiple scattering occurs and thereby the mean wave is both dispersive and attenuated and its complex wavenumber $k_d^{eff}(\omega)$ can be written as

$$k_d^{eff}(\omega) = \frac{\omega}{C_d^{eff}(\omega)} + i\alpha_d^{eff}(\omega) \tag{4}$$

with $C_d^{eff}(\omega)$ and $\alpha_d^{eff}(\omega)$ being the frequency dependent phase velocity and attenuation coefficient, respectively of a mean plane wave. The subscript d denotes either longitudinal ($d \equiv P$) or transverse ($d \equiv$ SH, SV) wave.

Next, the composite material is replaced by an elastic homogeneous and isotropic medium with effective shear and Young moduli μ^{eff} and E^{eff}, respectively given by the static mixture models of Christensen [20] for particulate composites or Halpin and Tsai (Ashton & Halpin [21]) for fiber-reinforced composites.

At the first step of the IEMA the effective density is assumed to be:

$$\left(\rho^{eff}\right)_{step\,1} = n_1\rho_1 + (1 - n_1)\rho_2 \tag{5}$$

Then, the real effective wave number $\left(k_d^{eff}\right)_{step1}$ is straightforward evaluated through (4), using the material properties μ^{eff} and E^{eff}.

At the second step of the IEMA, utilizing the material properties obtained from the first step, the two single wave scattering problems, illustrated in fig. 1, are solved. The solution of the scattering problems is accomplished numerically by means of a boundary element code valid for 2D or 3D elastodynamic analysis, which is described in the next section. Combining, as in relation (3), the evaluated forward scattering amplitudes, i.e.,

$$g_d(\hat{\boldsymbol{k}},\hat{\boldsymbol{k}})=n_1 g_d^{(1)}(\hat{\boldsymbol{k}},\hat{\boldsymbol{k}})+(1-n_1)g_d^{(2)}(\hat{\boldsymbol{k}},\hat{\boldsymbol{k}}) \tag{6}$$

and making use of the dispersion relation proposed by Foldy [1], the new effective wave number of the mean wave is computed, which is given by

$$\left[\left(k_d^{eff}\right)_{step2}\right]^2=\left[\left(k_d^{eff}\right)_{step1}\right]^2+\frac{4n_1}{a^2}g_d(\hat{\boldsymbol{k}},\hat{\boldsymbol{k}}) \tag{7}$$

for fiber-reinforced composites or by

$$\left[\left(k_d^{eff}\right)_{step2}\right]^2=\left[\left(k_d^{eff}\right)_{step1}\right]^2+\frac{3n_1}{a^3}g_d(\hat{\boldsymbol{k}},\hat{\boldsymbol{k}}) \tag{8}$$

for particulate composites, where a is the radius of the particles/fibers. The new, complex now, density $\left(\rho^{eff}\right)_{step2}$ is evaluated from the $\left(k_d^{eff}\right)_{step2}$ and μ^{eff}, E^{eff}. Then the second step is repeated with the same material properties and the new density $\left(\rho^{eff}\right)_{step2}$ until the self-consistent condition (3) is satisfied. Finally, from the relation (4) the frequency dependent, effective phase velocity $C_d^{eff}(\omega)$ and the attenuation coefficient $\alpha_d^{eff}(\omega)$, are computed.

3. BEM CODE FOR THE SOLUTION OF WAVE SCATTERING PROBLEMS

In this section, the boundary element methodology, used for the solution of the elastic wave scattering problems required in the above-mentioned IEM procedure, is presented.

Consider a long homogeneous elastic fiber of arbitrary cross-section embedded into an infinitely extended, homogeneous isotropic elastic matrix and Cartesian coordinate system $Ox_1x_2x_3$ with the x_1x_2 plane being perpendicular to the fiber. Alternatively one can consider a 3D arbitrarily shaped particle embedded into an infinitely extended, homogeneous isotropic elastic matrix and Cartesian coordinate system $Ox_1x_2x_3$. A time harmonic plane wave $\boldsymbol{u}^I$ (with $e^{-i\omega t}$ suppressed) impinges upon the particle/fiber with a circular frequency ω. The incident wave is propagating across the direction $\hat{\boldsymbol{k}}$ and is polarized in $\hat{\boldsymbol{d}}$ direction. The solution of this scattering problem can be obtained by the solution of a combined system of integral equations written as

$$\tilde{c}(\boldsymbol{r})\boldsymbol{u}^{ext}(\boldsymbol{r})+\int_S \tilde{\boldsymbol{T}}^{ext}(\boldsymbol{r},\boldsymbol{r}')\boldsymbol{u}^{ext}(\boldsymbol{r}')dS=\int_S \tilde{\boldsymbol{U}}^{ext}(\boldsymbol{r},\boldsymbol{r}')\boldsymbol{t}^{ext}(\boldsymbol{r}')dS+\boldsymbol{u}^{I}(\boldsymbol{r}) \quad (9)$$

$$\left(\tilde{\boldsymbol{I}}-\tilde{c}(\boldsymbol{r})\right)\boldsymbol{u}^{int}(\boldsymbol{r})+\int_S \tilde{\boldsymbol{T}}^{int}(\boldsymbol{r},\boldsymbol{r}')\boldsymbol{u}^{int}(\boldsymbol{r}')dS=\int_S \tilde{\boldsymbol{U}}^{int}(\boldsymbol{r},\boldsymbol{r}')\boldsymbol{t}^{int}(\boldsymbol{r}')dS \quad (10)$$

where S is the boundary of the space Ω occupied by one particle/fiber, $\boldsymbol{r}$ is a field point, $\boldsymbol{r}'$ is a source point laying on the boundary S, $\boldsymbol{u}$, $\boldsymbol{t}$ are the displacements and traction vectors, respectively, $\tilde{\boldsymbol{U}}$ and $\tilde{\boldsymbol{T}}$ represent the frequency domain fundamental solutions for 2D or 3D elastodynamic analysis, respectively, the indices "ext" and "int" indicate exterior and interior to the particle/fiber quantities, respectively, and $\tilde{c}(\boldsymbol{r})$ is the usual jump tensor. More details about the BEM can be found in the book of Manolis and Beskos [22].

Far away from the scatterer, the scattered field satisfies the radiation conditions and admits asymptotic representations as a function of the scattering amplitudes g_r, g_φ and g_θ across the directions of a polar coordinate system, having its origin interior to the scatterer. Those representations can be found in the work of Dassios and Kiriaki [23] when P, SH and SV waves in the 2D case, or P and SH waves in the 3D case, are considered.

In the present work, the boundary value problem and the scattering amplitudes are calculated by means of a frequency domain BEM code, valid for 2D and 3D elastodynamic analysis, developed by the authors. According to this code, the boundary of the scatterer S is discretized into three-noded or eight-noded quadratic line or surface elements. Collocating the integral equations (9) and (10) at all nodes and satisfying the continuity conditions at the interface of the particle/fiber and the matrix, a linear system of equations is obtained with unknowns all the nodal components of the displacement and traction. As soon as the boundary problem is solved, the evaluation of the scattering amplitudes is straightforward, through their integral representations [23].

4. NUMERICAL RESULTS

In the present section the prediction capability of the proposed IEMA is examined. The first problem deals with the dispersion curves prediction of a Lead/Epoxy (Epon 828z) particle/matrix composite with equally sized solid spherical particles, while the second one concerns the prediction of the dispersion and attenuation of a

Boron/Epoxy fiber/matrix composite. In both cases the results of the present IEMA are compared with the ones obtained by other numerical or experimental methods.

4.1. Lead/Epoxy (Epon 828z) particulate composite

The composite material under consideration contains spherical particles with radius a=660 μm and volume fractions 26.1% and 52%. The material properties of the composite constituents are summarised in Table 1.

Table 1. Material properties of particles and matrix

Phase (Material)	C_P (m/s)	C_S (m/s)	λ (GPa)	μ (GPa)	ρ (Kgr/m^3)
Particles (Lead)	2210	860	38.475	8.357	11300
Matrix (Epoxy)	2640	1200	4.915	1.731	1202

The frequency dependent phase velocity of a longitudinal wave propagating in the composite is evaluated via the IEMA and is shown in fig.2 for the two volume fractions of lead particles. The obtained results are presented versus the dimensionless frequency $k_P^m a$, with k_P^m being the wave number of a longitudinal wave propagating in the matrix medium, and they are compared to those taken experimentally by Kinra and Rousseau [24] and numerically by Kim et al.[11].

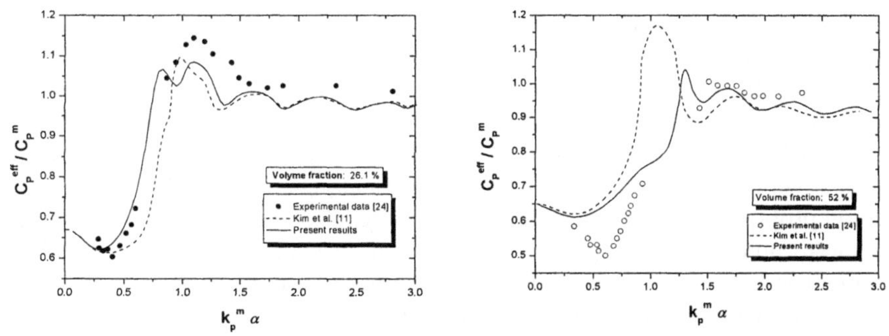

(a) *(b)*

Figure 2. Effective longitudinal wave speed versus dimensionless frequency.
a) 26.1% particles, b) 52% particles
° Experimental data by Kinra et al. [24], ---- Numerical results by Kim et al. [11], – Present results

Although the discrepancy between numerical results, obtained by the IEM approximation, and the experimental ones provided by Kinra and Rousseau increases as the volume fraction increases, one can say that, quantitatively, the results are in a very good agreement. The most important conclusion extracted by these two diagrams is that the IEM approximation clearly predicts the shift of the lowest and higher resonance frequencies to higher values of $k_p^m a$ as the volume fraction of the particles increases. The position of these frequencies can be easily found in the diagrams of Fig. 2 since at these resonance frequencies the effective velocity of the propagating wave takes the highest values. Comparing the prediction efficiency between the IEM approximation and that of Kim et al. one can say that for high concentrations the prediction of the IEMA are more close to the experimental observations.

4.2. Boron/Epoxy fiber-reinforced composite

Here the fiber-reinforced composite contains fibers of Boron at 54% volume fraction embedded in Epoxy matrix. Both fibers and matrix are homogeneous and isotropic and their properties are given in Table 2. The diameter of the fibers is equal to 100 μm.

Table 2. Material properties of fibers and matrix

Material	E(GPa)	μ (GPa)	v	ρ (Kgr/m^3)
Fibers (Boron)	379.2	160.678	0.18	2682
Matrix (Epoxy)	5.033	1.797	0.4	1261

Figures *3a* and *3b* portray the effective velocity and the attenuation coefficient versus frequency of a transverse SH wave propagating in the considered here composite with propagation direction and polarization vector being perpendicular and parallel to the fibers, respectively. The wave dispersion prediction, made by the present IEMA, is compared to the experimental one of Tauchert and Guzelsu [25] and also with a prediction made by the numerical generalized self-consistent (GSC) method proposed by Yang and Mal [14]. Finally, also shown in the following graphs are curves obtained by the present IEMA when, instead of the dispersion relation of Eq. (7), use is made of a dispersion relation proposed in the work of Waterman and Truell [3] involving not only the scattering amplitudes of not only the forward direction, but also the ones of the backward one.

Observing the results of Fig. *3a* one can say that, both the IEMA (Foldy) and the IEMA (Waterman and Truell) are in very good agreement with the measured results, while the GSC approximation as a nearly constant deviation from the experimental results at all frequencies. Unfortunately, Tauchert and Guzelsu do not provide, in their work, experimental results dealing with attenuation of the propagating waves. Nevertheless, the main conclusion extracted from figure *3b* is that as the frequency increases te attenuation coefficient, predicted by the present IEMA, increases more smoothly than that provided by the GSC approximation.

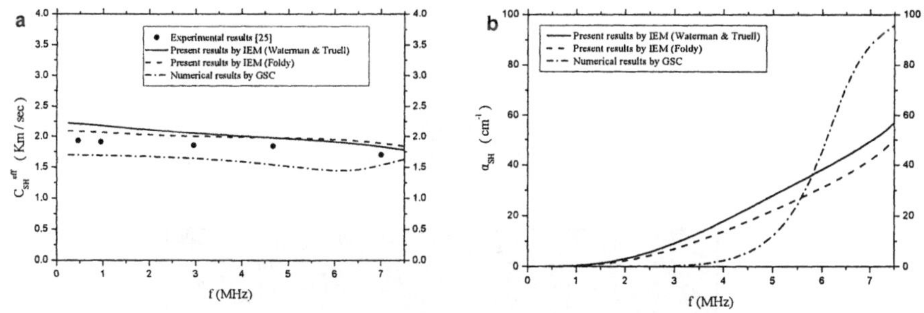

Figure 3. a) Phase velocity and b) attenuation coefficient of a transverse SH wave propagating in a Boron/Epoxy Fiber/Matrix composite with fiber volume fraction 0.54

5. A MULTIPLE SCATTERING PROBLEM

In order to demonstrate the efficiency of the present IEMA procedure, a large scale multiple scattering problem of a slab containing 108 randomly distributed fibers, is solved. The matrix is AA520 Aluminum and the inclusions are Alumina Al2O3 ceramic fibers of a diameter equal to 20 μm, in a volume fraction $n_I = 0.32$. Both matrix and fibers are homogeneous and isotropic and their properties are given in Table 2.

Table 2: Material properties of each phase in Al2O3/ AA520 fiber/matrix composite.

Phase (Material)	E (Gpa)	μ (GPa)	ν	ρ (kgr/m^3)
Fiber (Alumina Al2O3)	360	144	0.250	3700
Matrix (AA520 Aluminum)	66	25	0.31	2600

The problem is solved numerically by use of the present BEM code for 2D elastodynamic analysis. The dimensions of the slab are taken equal to 180x540 μm and the fibers are placed numerically through a random number generator function [26] as shown in Fig. 4.

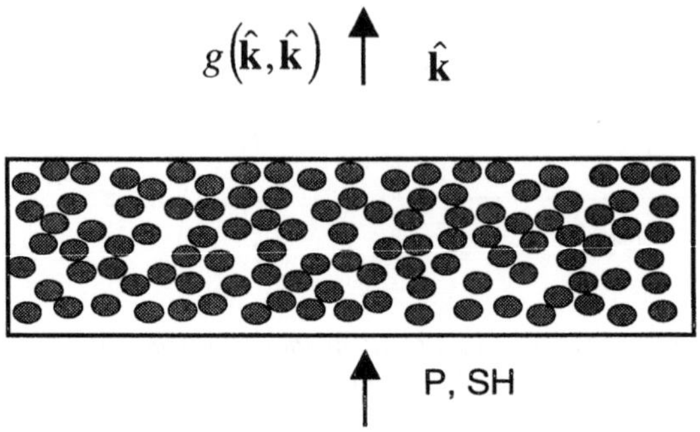

Figure 4. Multiple wave scattering by a slab of fiber-reinforced composite material

Both longitudinal (P) and transverse horizontal (SH) waves, propagating perpendicular to the fibers (Fig. 4) are considered. The magnitude of the forward scattering $g(\hat{k},\hat{k})$ as a function of the dimensionless frequency $k^m a$, with k^m being the wave number of the matrix medium is calculated. The obtained results are displayed in Figs 5 and 6 and compared to those taken from the solution of the single scattering problem where the slab is homogeneous with dynamic effective material properties predicted by the present IEMA.

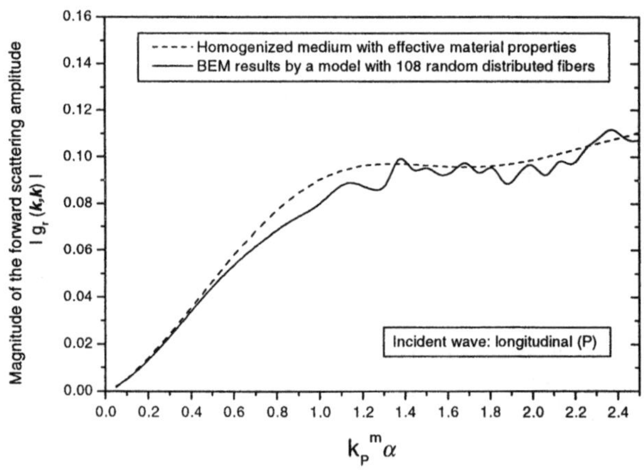

Figure 5. Magnitude of the forward scattering amplitude as a function of the dimensionless frequency of a longitudinal wave propagating in a Al2O3/ AA520 fiber/matrix composite material with fiber volume fraction n_f=0.35.

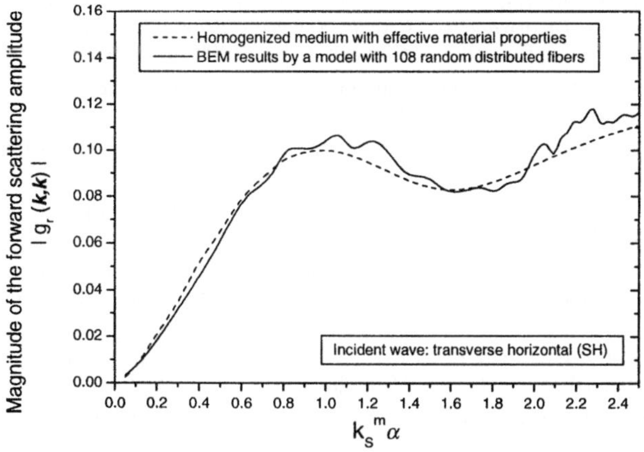

Figure 6. Magnitude of the forward scattering amplitude as a function of the dimensionless frequency of a transverse horizontal (SH) wave propagating in a Al2O3/ AA520 fiber/matrix composite material with fiber volume fraction n_1=0.35.

As it is evident, the proposed iterative effective medium methodology can predict efficiently the dynamic effective material properties of a fiber reinforced composite material for both P and SH incident waves.

6. REFERENCES

1. Foldy, L.L.: The multiple scattering of waves, *Phys. Rev.* **67** (1945), 107-119.
2. Lax, M.: The effective field in dense systems, *Phys. Rev,* **88** (1952), 621-629.
3. Waterman, P. C., and Truell, R. : Multiple scattering of waves, *J. Math. Phys.* **2** (1961), 512-537.
4. Weaver, R.L. and Pao, Y.H.: Dispersion relations for linear wave propagation in homogeneous and inhomogeneous media, *J. Math. Phys.* **22**/9 (1981), 1909-1918.
5. Bose, S.K. and Mal, A.K.: Elastic waves in a fiber-reinforced composite, *J. Mech. Phys. Solids* **22** (1974), 217-229.
6. Varadan, V.K., Varadan, V.V. and Pao, Y.H.: Multiple scattering of elastic waves by cylinders of arbitrary cross section II. Pair-correlated cylinders, *J. Acoust. Soc. Am.* **78**/5 (1995), 1874-1878.
7. Kim, J.Y., Ih, J.G. and Lee, B.H.: Dispersive wave propagation in viscoelastic matrix reinforced by elastic fibers, *J. Acoust. Soc. Am.* **95** (1994), 1213-1222.
8. Bose, S.K.: Ultrasonic plane SH wave reflection from a uni-directional fibrous composite slab, *J. Sound Vibr.* **193**/5 (1996), 1069-1078.
9. Liu, W. and Kriz, R.D.: Axial shear waves in fiber-reinforced composites with multiple interfacial layers between fiber core and matrix, *Mech. Mat.* **31** (1999), 117-129.
10. Soven, P.: Coherent-potential model of substitutional disordered alloys, *Phys. Rev.* **156** (1967), 809-813.
11. Kim, J.Y., Ih J.G. and Lee, B.H.: Dispersion of elastic waves in random particulate composites, *J. Acoust. Soc. Am.* **97**/3 (1995), 1380-1388.
12. Sabina, F.J. and Willis, J.R.: A simple self-consistent analysis of wave propagation in particulate composites, *Wave Motion* **10** (1988), 127-142.

13. Huang, W. and Rohklin, S.I.: *Frequency dependences of ultrasonic wave velocity and attenuation in fiber composites. Theory and experiments, Review of Progress in Quantitative Nondestructive Evaluation*, D.O. Thompson and D.E. Chimenti (eds), **14**, Plenum Press, New York, 1995.

14. Yang, R.B. and Mal, A.K.: Multiple scattering of elastic waves in a fiber-reinforced composite, *J. Mech. Phys. Solids* **42** (1994), 1945-1968.

15. Kim, J.Y.: Dynamic self-consistent analysis for elastic wave propagation in fiber composites, *J. Acoust. Soc. Am.* **100** (1996), 2002-2010.

16. Verbis, J.T., Kattis, S.E., Tsinopoulos, S.V. and Polyzos, D.: Wave dispersion and attenuation in fiber composites, *Computational Mechanics* **27** (2001), 244-255.

17. Tsinopoulos, S.V., Kattis, S.E., Polyzos, D. and Beskos, D.E.: An advanced boundary element method for axisymmetric elastodynamic analysis, *Comp. Meth. Appl. Mech. Engng* **175** (1999), 53-70.

18. Kattis, S.E., Polyzos, D. and Beskos, D.E.: Vibration isolation by a row of piles using 3-D frequency domain BEM, *Int. J. Num. Meth. Engng* **46** (1999), 713-728.

19. Devaney, A.J.: Multiple scattering theory for discrete, elastic, random media, *J. Math. Phys.* **21**/11 (1980), 2603-2611.

20. Christensen, R.M.: A critical evaluation for a class of micromechanics models, *J. Mech Phys. Solids* **38**/3 (1990), 379-404.

21. Ashton, J.E. and Halpin, J.C., *Primer on Composite Materials: Analysis,* Progress in materials Science Series, Volume III, Technomic, USA, 1969.

22. Manolis G. and Beskos D.E.: *Boundary element methods in elastodynamics*, Unwin-Hyman, London, 1988.

23. Dassios, G. and Kiriaki, K.: The low-frequency theory of elastic wave scattering, *Quart. Appl. Math.* **42** (1984), 225-248.

24. Kinra, V.K. and Rousseau, C.: Acoustical and Optical Branches of wave propagation, *J. Wave Mater. Interaction* **2** (1987), 141-152.

25. Tauchert, T.R. and Guzelsu, A.N.: An experimental study of dispersion of stress waves in a fiber-reinforced composite, *J. Appl. Mech.* (1972), 98-102.

26. IMSL *Math/Library User's Manual*, Version 3.0, Visual Numerics Inc., Houston, Texas, 1994.

THE VALIDITY OF THE VARIABLE SEPARABLE METHOD IN SPECIAL LOADING CASES IN COMPOSITE MULTI-MATERIAL WEDGES

J.H. STABOULOGLOU and E.E. THEOTOKOGLOU
Faculty of Applied Sciences,
Laboratory of Strength of Materials
The National Technical University of Athens, Zographou Campus,
Theocaris Bld., GR-157 73, Athens, Greece

Abstract. The problem of a multi-material composite wedge under a normal and sheer loading at its external faces is considered with a variable separable solution. The special cases of the infinite isotropic wedge, the two materials wedge and the multiple-materials wedge under shear and normal loading of the order r^{-1} and r^{-2} of the radial coordinate r, are studied.

1. INTRODUCTION

The presence of wedges involving different materials is appeared in adhesively bonded joints between metallic and composite adherents .The original work of Williams [1] makes use of the Airy stress function to obtain solutions separable in the radial coordinate r, and the angular coordinate θ. Numerous studies addressing material and geometric complications have followed [2-6].

In our previous studies [7-8] we have studied the infinite isotropic wedge under linearly distributed loading and the multi-material wedge under normal and shear loading using a variable separable solution. The solution of the above problems is proved to be self-similar for $n = 0,1,2,\pm 3,\ldots$, where n is the order of the radial coordinate r.But in the special cases of $n=-1$, for the shear loading and $n=-1,-2$ for the normal loading, we can not find a solution because the system of equations can not be solved.

In this paper we try to confront the special cases $n=-1$ for the shear loading and $n=-1,-2$ for the normal loading of a single material wedge under the validity of the variable separable solution, and to determine the angle of the wedge θ_0. In the case of a two-materials or a multi-materials wedge, it is examined under which circumstances it is avoided the loading from an enforced displacement at the interfaces between the dissimilar materials when the shear and normal distributed

E.E. Gdoutos and Z.P. Marioli-Riga (eds), Recent Advances in Composite Materials, 47-60

loading at the external faces of the wedge are of the order $1/r$ and $1/r^2$. In every case, in the proposed study the validity of the variable separable solution is verified.

2. GENERALITIES

If the distributed loads along the faces of an infinite isotropic multimaterial wedge fulfills the self similarity condition

$$F(ar) = aF(r) \tag{1}$$

then the elastostatic wedge problem is reduced to a self similar problem where the stress field is given by the variable separable form

$$\sigma_{ij}(r,\theta) = h_{ij}(r) g_{ij}(\theta)\,, \quad i,j = r,\theta \tag{2}$$

In the case of a shear distributed loading [8], $T_\alpha(r) = C_\alpha r^n$ and $T_\beta(r) = C_\beta r^n$ at the faces $\theta = a$ and $\theta = \beta$ respectively, the system of equations derived from the equilibrium of forces is

$$rh_{rr}(r)\int_\alpha^\beta g_{rr}(\theta)\cos\theta d\theta - r^{n+1}\int_\alpha^\beta g_{r\theta}(\theta)\sin\theta d\theta - F(r)\left(C_\alpha\cos\alpha - C_\beta\cos\beta\right) =$$
$$= \int_\alpha^\beta g_{rr}(\theta)\cos\theta d\theta - \int_\alpha^\beta g_{r\theta}(\theta)\sin\theta d\theta - \left(C_\alpha\cos\alpha - C_\beta\cos\beta\right)F(1) \tag{3}$$

$$rh_{rr}(r)\int_\alpha^\beta g_{rr}(\theta)\sin\theta d\theta + r^{n+1}\int_\alpha^\beta g_{r\theta}(\theta)\cos\theta d\theta - F(r)\left(C_\alpha\sin\alpha - C_\beta\sin\beta\right) =$$
$$= \int_\alpha^\beta g_{rr}(\theta)\sin\theta d\theta + \int_\alpha^\beta g_{r\theta}(\theta)\cos\theta d\theta - \left(C_\alpha\sin\alpha - C_\beta\sin\beta\right)F(1)$$

where C_α, C_β constants and $F(s) = \int s^n ds$; $n = \pm1, \pm2, \pm3, \ldots$
or if $n \neq -1$

$$rh_{rr}(r)\int_{\alpha}^{\beta} g_{rr}(\theta)\cos\theta d\theta - r^{n+1}\left[\frac{C_\alpha \cos\alpha - C_\beta \cos\beta}{\eta+1} + \int_{\alpha}^{\beta} g_{r\theta}(\theta)\sin\theta d\theta\right] =$$

$$= \int_{\alpha}^{\beta} g_{rr}(\theta)\cos\theta d\theta - \left[\frac{C_\alpha \cos\alpha - C_\beta \cos\beta}{\eta+1} + \int_{\alpha}^{\beta} g_{r\theta}(\theta)\sin\theta d\theta\right]$$

$$(4)$$

$$rh_{rr}(r)\int_{\alpha}^{\beta} g_{rr}(\theta)\sin\theta d\theta - r^{n+1}\left[\frac{C_\alpha \sin\alpha - C_\beta \sin\beta}{\eta+1} - \int_{\alpha}^{\beta} g_{r\theta}(\theta)\cos\theta d\theta\right] =$$

$$= \int_{\alpha}^{\beta} g_{rr}(\theta)\sin\theta d\theta - \left[\frac{C_\alpha \sin\alpha - C_\beta \sin\beta}{\eta+1} - \int_{\alpha}^{\beta} g_{r\theta}(\theta)\cos\theta d\theta\right]$$

In the case of a normal distributed loading [8], $N_\alpha(r) = C_\alpha r^n$ and $N_\beta(r) = C_\beta r^n$ at the faces $\theta = a$ and $\theta = \beta$ respectively, from the equilibrium of moments, we have

$$r^2 h_{r\theta}(r)\int_{\alpha}^{\beta} g_{r\theta}(\theta)d\theta - (C_a - C_b)(F_1(r) - F_1(1)) = \int_{\alpha}^{\beta} g_{r\theta}(\theta)d\theta\,; \quad F_1(r) = \ell nr \tag{5}$$

Relation (5) if $n \neq -2$ becomes

$$r^2 h_{r\theta}(r) = r^{\eta+2} \Leftrightarrow h_{r\theta}(r) = r^n\,; \quad \int_{\alpha}^{\beta} g_{r\theta}(\theta)d\theta = \frac{C_\alpha - C_\beta}{\eta+2} \tag{6}$$

Similarly, the system of equations derived from the equilibrium of forces if $n \neq -1,-2$, takes the form

$$rh_{rr}(r)\int_{\alpha}^{\beta} g_{rr}(\theta)\cos\theta d\theta - r^{n+1}\left\{\int_{\alpha}^{\beta} g_{r\theta}(\theta)\sin\theta d\theta - \frac{C_\alpha \sin\alpha - C_\beta \sin\beta}{\eta+1}\right\} =$$

$$= \int_{\alpha}^{\beta} g_{rr}(\theta)\cos\theta d\theta - \int_{\alpha}^{\beta} g_{r\theta}(\theta)\sin\theta d\theta + \frac{C_\alpha \sin\alpha - C_\beta \sin\beta}{\eta+1}; \quad n = 1,2,\pm 3,\ldots$$

$$rh_{rr}(r)\int_{\alpha}^{\beta} g_{rr}(\theta)\sin\theta d\theta + r^{n+1}\left\{\int_{\alpha}^{\beta} g_{r\theta}(\theta)\cos\theta d\theta - \frac{C_\alpha \cos\alpha - C_\beta \cos\beta}{\eta+1}\right\} = \tag{7}$$

$$= \int_{\alpha}^{\beta} g_{rr}(\theta)\sin\theta d\theta + \int_{\alpha}^{\beta} g_{r\theta}(\theta)\cos\theta d\theta - \frac{C_\alpha \cos\alpha - C_\beta \cos\beta}{\eta+1}; \quad \eta = 1,2,\pm 3,\ldots$$

It is obvious from relations (4), (6) and (7) that the orders of r, $n=-1$ in the shear loading case and $n=-1,-2$, in the normal loading case, constitute special cases.

3. SPECIAL CASES

In the case of shear loading of order $n=-1$, we have

$$T_a(r) = {C_\alpha}/{r}, \quad T_\beta(r) = {C_\beta}/{r}, \quad h_{r\theta}(r) = {1}/{r}, \quad F(s) = \int \frac{ds}{s} = \ell ns$$

thus the system (3), becomes

$$\left[rh_{rr}(r)-1\right]\int_{\alpha}^{\beta} g_{rr}(\theta)\cos\theta d\theta - \left(C_\alpha \cos\alpha - C_\beta \cos\beta\right)\ell nr = 0$$
$$\left[rh_{rr}(r)-1\right]\int_{\alpha}^{\beta} g_{rr}(\theta)\sin\theta d\theta - \left(C_\alpha \sin\alpha - C_\beta \sin\beta\right)\ell nr = 0 \tag{8}$$

the homogeneous system (8) is valid, only in the cases (Figs. 1,2)

$$C_\alpha = -C_\beta = T = ct, \quad \beta = \alpha + \pi, \quad \theta_0 = \pi \tag{9}$$

or

$$C_\alpha = C_\beta = T = ct, \quad \beta = \alpha + 2\pi, \quad \theta_0 = 2\pi \tag{10}$$

In the case of normal loading of order $n=-1$, we have $N_\alpha(r)=C_\alpha/r$, $N_\beta(r)=C_\beta/r$, $h_{r\theta}(r)=1/r$, $F(s)=\ell ns$, thus the system arised from the equilibrium of forces, finally takes the form

$$\begin{aligned}&[1-rh_{rr}(r)]\int_\alpha^\beta g_{rr}(\theta)\cos\theta d\theta-\left(C_\alpha\sin\alpha-C_\beta\sin\beta\right)\ell nr=0\\&[1-rh_{rr}(r)]\int_\alpha^\beta g_{rr}(\theta)\sin\theta d\theta+\left(C_\alpha\cos\alpha-C_\beta\cos\beta\right)\ell nr=0\end{aligned}\tag{11}$$

The homogeneous system (11) is valid only if (Figs. 3,4)

$$C_\alpha=-C_\beta=N=ct\,,\quad \beta=a+\pi\,,\quad \theta_0=\pi \tag{12}$$

or

$$C_\alpha=C_\beta=N=ct\,,\quad \beta=a+2\pi\,,\quad \theta_0=2\pi \tag{13}$$

In the case of normal loading of order $n=-2$, from relation (5), it is obtained

$$h_{r\theta}(r)=1/r^2\,,\quad C_\alpha=C_\beta=N=ct \tag{14}$$

From relation (14) and the equilibrium of forces, it is obtained the self similar problem

$$\begin{aligned}&N_\alpha(r)=N_\beta(r)=N/r^2\,,\quad N=ct\\&h_{ij}(r)=1/r^2\,,\quad i,j=r,\theta\\&\int_\alpha^\beta g_{rr}(\theta)\cos\theta d\theta-\int_\alpha^\beta g_{r\theta}(\theta)\sin\theta d\theta-N\left(\sin\alpha-\sin\beta\right)=0\\&\int_\alpha^\beta g_{rr}(\theta)\sin\theta d\theta+\int_\alpha^\beta g_{r\theta}(\theta)\cos\theta d\theta+N\left(\cos\alpha-\cos\beta\right)=0\end{aligned}\tag{15}$$

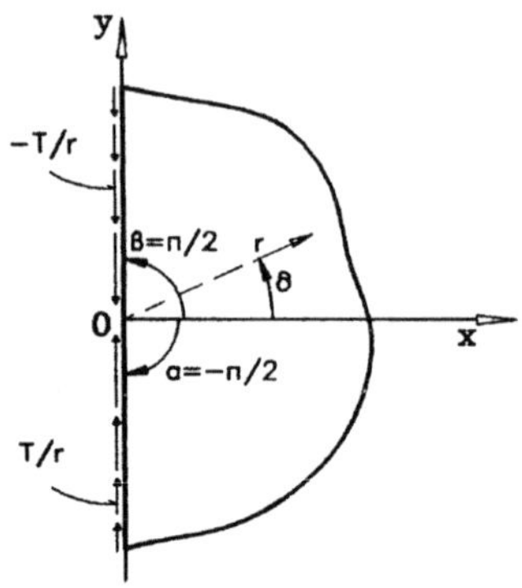

Figure 1.

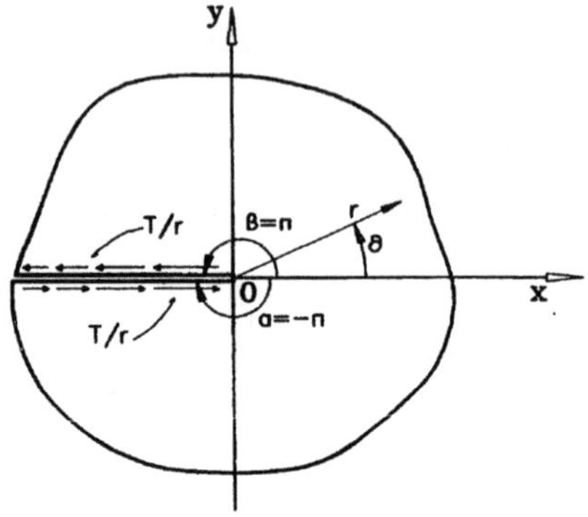

Figure 2.

4. FORMULATION OF THE SINGLE MATERIAL WEDGE

The Airy stress function $M(r,\theta)$, the stress field $\sigma_{ij}(r,\theta)$ and the displacement field $u_i(r,\theta)$ $(i=r,\theta)$, in the case of a single material isotropic wedge under distributed shear $T_i(r)$ $(i=\alpha,\beta)$ of order $n=-1$, or normal loads $N_i(r)$ $(i=\alpha,\beta)$ of order $n=-1,-2$, when the self similarity property is valid, are given by

(i) Shear load, $n=-1$, $\theta_0=\pi$ (Fig. 1)

$$T_\alpha(r)=T/r\,,\quad T_\beta(r)=-T/r\,,\quad M(r,\theta)=-Tr\ell nr\cos\theta$$

$$\sigma_{rr}=\sigma_{\theta\theta}=-\frac{T}{r}\cos\theta\,,\quad \sigma_{r\theta}=-\frac{T}{r}\sin\theta \tag{16}$$

$$2\mu u_r=-\frac{T}{2}\left[(\kappa+1)\theta\sin\theta-\cos\theta+(\kappa-1)\ell nr\cos\theta\right]$$

$$2\mu u_\theta=-\frac{T}{2}\left[(\kappa+1)\theta\cos\theta-\sin\theta-(\kappa-1)\ell nr\sin\theta\right]$$

where μ is the shear modulus, $u=(3-4\nu)$ for plane strain, $\kappa=(3-\nu)/(1+\nu)$ for generalized plane stress, ν being the Poisson's ratio.

(ii) Shear load, $n=-1$, $\theta_0=2\pi$ (Fig. 2)

$$T_\alpha(r)=T_\beta(r)=T/r\,,\quad M(r,\theta)=Tr\ell nr\sin\theta$$

$$\sigma_{rr} = \sigma_{\theta\theta} = \frac{T}{r}\sin\theta \ , \quad \sigma_{r\theta} = -\frac{T}{r}\cos\theta \tag{17}$$

$$2\mu u_r = \frac{T}{2}\left[-(\kappa+1)\theta\cos\theta - \sin\theta + (\kappa-1)\ell nr\sin\theta\right]$$

$$2\mu u_\theta = \frac{T}{2}\left[(\kappa+1)\theta\sin\theta + \cos\theta + (\kappa-1)\ell nr\cos\theta\right]$$

(iii) Normal load, $n=-1$, $\theta_0=\pi$ (Fig. 3)

$$N_\alpha(r) = N/r\,, \quad N_\beta(r) = -N/r\,, \quad M(r,\theta) = -Nr\ell nr\sin\theta$$

$$\sigma_{rr} = \sigma_{\theta\theta} = -\frac{N}{r}\sin\theta \ , \quad \sigma_{r\theta} = \frac{N}{r}\cos\theta \tag{18}$$

$$2\mu u_r = -\frac{N}{2}\left[-(\kappa+1)\theta\cos\theta - \sin\theta + (\kappa-1)\ell nr\sin\theta\right]$$

$$2\mu u_\theta = -\frac{N}{2}\left[(\kappa+1)\theta\sin\theta + \cos\theta + (\kappa-1)\ell nr\cos\theta\right]$$

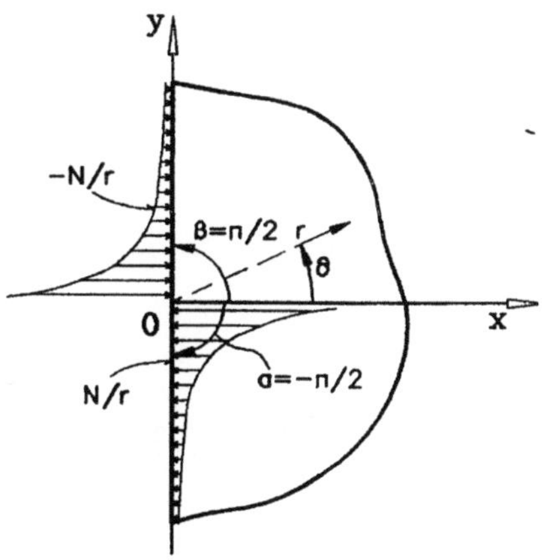

Figure 3.

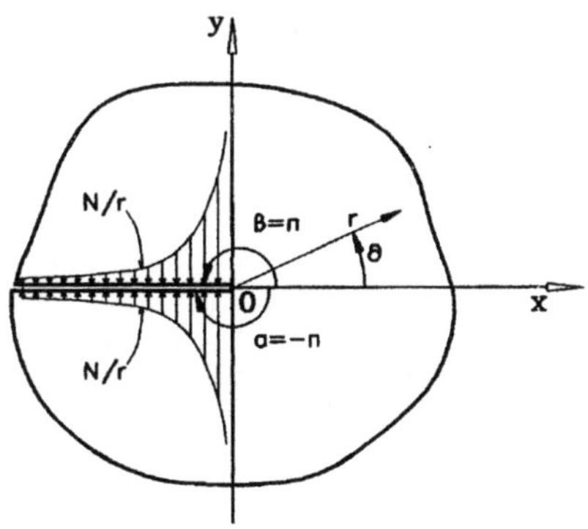

Figure 4.

(iv) Normal load, $n=-1$, $\theta_0=\pi$ (Fig. 4)

$$N_\alpha(r) = N_\beta(r) = N/r, \quad M(r,\theta) = -Nr\ell nr\cos\theta$$
$$\sigma_{rr} = \sigma_{\theta\theta} = -\frac{N}{r}\cos\theta, \; \sigma_{r\theta} = -\frac{N}{r}\sin\theta \tag{19}$$
$$2\mu u_r = -\frac{N}{2}[(\kappa+1)\theta\sin\theta - \cos\theta + (\kappa-1)\ell nr\cos\theta]$$
$$2\mu u_\theta = -\frac{N}{2}[(\kappa+1)\theta\cos\theta - \sin\theta - (\kappa-1)\ell nr\sin\theta]$$

(v) Normal load, $n = -2$ (Fig. 5)

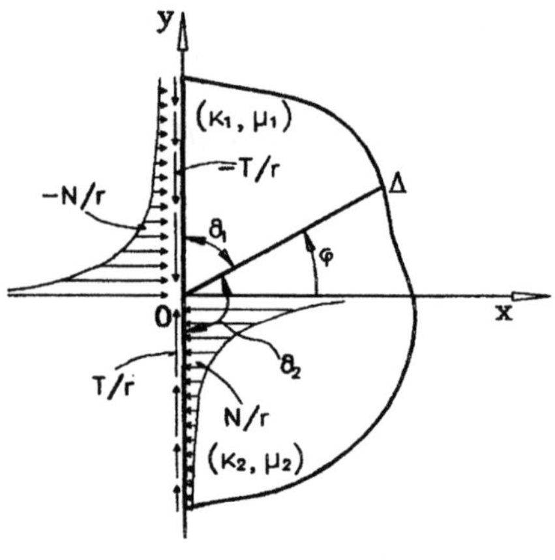

Figure 5.

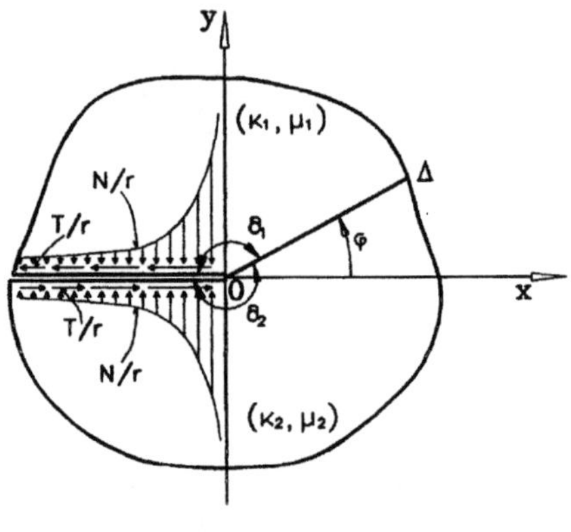

Figure 6

$$M(r,\theta) = -C_\alpha \ell nr$$
$$\sigma_{rr}(r,\theta) = -C_\alpha/r^2, \quad \sigma_{r\theta}(r,\theta) = 0, \quad \sigma_{\theta\theta}(r,\theta) = C_\alpha/r^2$$
$$2\mu u_r(r,\theta) = C_\alpha/r, \quad u_\theta(r,\theta) = 0 \tag{20}$$

where

$$C_\alpha = C_\beta = N, \quad \theta_0 \neq \pi \;\text{ or }\; 2\pi, \; \sin\theta_0 \neq \theta_0\cos\theta_0$$

5. THE BIMATERIAL WEDGE UNDER SHEAR AND NORMAL DISTRIBUTED LOADS OF ORDER $n=-1$

In this section it will be investigated the possibility of existence of a wedge of angle $\theta_0=\theta_1+\theta_2=\pi$ or $\theta_0=\theta_1+\theta_2=2\pi$ (Figs. 7,8), where the self-similarity property and the separation of variables will valid. For this reason at the interface between the dissimilar materials with material constants (κ_1,μ_1) and (κ_2,μ_2), it must be ensured the continuity of stress field and the "cooperation" of the displacement field. With the term "cooperation" we mean that the displacements at the interface for both materials, must be equal and must be identical with the displacements appeared at the radius $\theta=\varphi$ (which coincides with the interface) in a single material wedge (κ_1,μ_1) (or (κ_2,μ_2)) of angle $\theta_0=\pi$ or $\theta_0=2\pi_0$. In this way it doesn't appear any stress due to imposed displacement, at the interface between the two materials and hence the stress field of the bimaterial wedge is identical with the stress field of a single material wedge (κ_1,μ_1) (or (κ_2,μ_2)), loaded with the same distributed loads at its external faces as the bimaterial wedge At the interface $\mathrm{O}\Delta$ (Figs. 6,7), it is valid

$$4u_{i_T}^{(1)}(r,\varphi)+4u_{i_N}^{(1)}(r,\varphi)=4u_{i_T}^{(2)}(r,\varphi)+4u_{i_N}^{(2)}(r,\varphi)\,;\quad i=r,\theta \tag{21}$$

(i) $\theta_0=\theta_1+\theta_2=\pi$

From relations (16), (18) and (21), we finally obtain the homogeneous system of equations

$$\begin{aligned} T(2\varphi\sin\varphi-\cos\varphi)-N(2\varphi\cos\varphi+\sin\varphi)&=0\\ T(2\varphi\cos\varphi-\sin\varphi)+N(2\varphi\sin\varphi+\cos\varphi)&=0 \end{aligned} \tag{22}$$

under the condition

$$\frac{\kappa_1-1}{\mu_1}=\frac{\kappa_2-1}{\mu_2} \tag{23}$$

The system (22) is valid only in the case that

$$4\varphi^2=1 \tag{24}$$

$$\varphi = \frac{1}{2} rad; \quad \frac{T}{N} = \frac{\cos(1/2) + \sin(1/2)}{\sin(1/2) - \cos(1/2)} = -3{,}4082\,, \quad T = (-1)^{\frac{1}{2}+\varphi}\left(\tan\frac{\pi+2}{4}\right)N$$

$$\varphi = -\frac{1}{2} rad; \quad \frac{T}{N} = -\frac{\sin(-1/2) - \cos(-1/2)}{\sin(-1/2) + \cos(-1/2)} = 3{,}4082\,, \quad T = (-1)^{\frac{1}{2}+\varphi}\left(\tan\frac{\pi+2}{4}\right)N$$

The stress and the displacement fields of the bimaterial wedge, are given by

$$\sigma_{rr} = \sigma_{\theta\theta} = -\frac{N}{r}\left[(-1)^{\frac{1}{2}+\varphi}\tan\frac{\pi+2}{4}\cos\theta + \sin\theta\right]$$
$$\sigma_{r\theta} = -\frac{N}{r}\left[(-1)^{\frac{1}{2}+\varphi}\tan\frac{\pi+2}{4}\sin\theta - \cos\theta\right] \qquad (25)$$

For the subwedge (1) : $\varphi \leq \theta \leq \frac{\pi}{2}$

$$2\mu_1 u_r = -\frac{N}{2}\left\{(\kappa_1+1)\theta\left[(-1)^{\frac{1}{2}+\varphi}\tan\frac{\pi+2}{4}\sin\theta - \cos\theta\right] + \right.$$
$$\left. + \left((\kappa_1-1)\ell nr - 1\right)\left[(-1)^{\frac{1}{2}+\varphi}\tan\frac{\pi+2}{4}\cos\theta + \sin\theta\right]\right\}$$
$$2\mu_1 u_\theta = -\frac{N}{2}\left\{(\kappa_1+1)\theta\left[(-1)^{\frac{1}{2}+\varphi}\tan\frac{\pi+2}{4}\cos\theta + \sin\theta\right] - \right. \qquad (26)$$
$$\left. - \left((\kappa_1-1)\ell nr + 1\right)\left[(-1)^{\frac{1}{2}+\varphi}\tan\frac{\pi+2}{4}\sin\theta - \cos\theta\right]\right\}$$

For the subwedge (2) : $-\frac{\pi}{2} \leq \theta \leq \varphi$

$$2\mu_2 u_r = -\frac{N}{2}\left\{(\kappa_2+1)\theta\left[(-1)^{\frac{1}{2}+\varphi}\tan\frac{\pi+2}{4}\sin\theta - \cos\theta\right] + \right.$$
$$\left. + \left((\kappa_2-1)\ell nr - 1\right)\left[(-1)^{\frac{1}{2}+\varphi}\tan\frac{\pi+2}{4}\cos\theta + \sin\theta\right]\right\} \qquad (27)$$

$$2\mu_2 u_\theta = -\frac{N}{2}\left\{(\kappa_2+1)\theta\left[(-1)^{\frac{1}{2}+\varphi}\tan\frac{\pi+2}{4}\cos\theta+\sin\theta\right]-\right.$$
$$\left.+\left((\kappa_2-1)\ell nr-1\right)\left[(-1)^{\frac{1}{2}+\varphi}\tan\frac{\pi+2}{4}\cos\theta+\sin\theta\right]\right\} \quad (27)$$

where

$$\frac{\kappa_1-1}{\mu_1}=\frac{\kappa_2-1}{\mu_2}, \quad \varphi=\pm\frac{1}{2}rad$$

(ii) $\theta_0=\theta_1+\theta_2=2\pi$ (Fig. 7)

From relations (17), (19) and (21), we finally get the following homogeneous system of equations

$$\begin{aligned} T(2\varphi\cos\varphi+\sin\varphi)+N(2\varphi\sin\varphi-\cos\varphi)&=0\\ T(2\varphi\sin\varphi+\cos\varphi)-N(2\varphi\cos\varphi-\sin\varphi)&=0 \end{aligned} \quad (28)$$

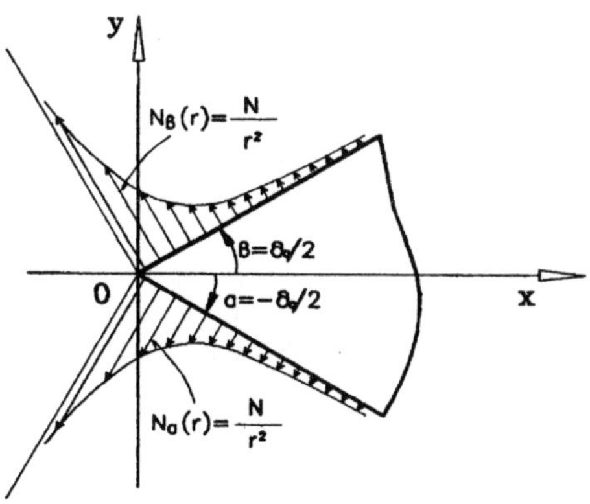

Figure 7.

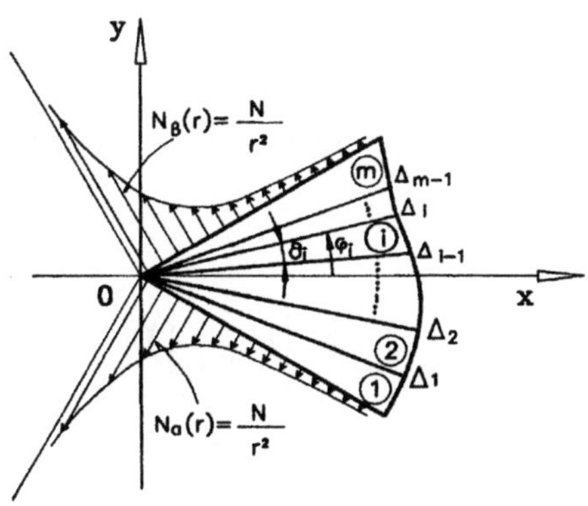

Figure 8.

The system (28) is also valid if $\varphi = \pm\frac{1}{2} rad$, where the ratio T/N is

$$\frac{T}{N} = (-1)^{\varphi-\frac{1}{2}} \tan\frac{\pi-2}{4} \tag{29}$$

The stress and the displacement field of the bimaterial wedge $\theta_0 = \theta_1 + \theta_2 = 2\pi$, are given by

$$\begin{aligned} \sigma_{rr} = \sigma_{\theta\theta} &= \frac{N}{r}\left[(-1)^{-\frac{1}{2}} \tan\frac{\pi-2}{4}\sin\theta - \cos\theta\right] \\ \sigma_{r\theta} &= -\frac{N}{r}\left[(-1)^{-\frac{1}{2}} \tan\frac{\pi-2}{4}\cos\theta + \sin\theta\right] \end{aligned} \tag{30}$$

For the sub-wedge (1); $\varphi \le \theta \le \pi$

$$\begin{aligned} 2\mu_1 u_r &= \frac{N}{2}\left\{-(\kappa_1+1)\theta\left[(-1)^{-\frac{1}{2}}\cos\theta\tan\frac{\pi-2}{4} + \sin\theta\right] + \right. \\ &\left. + ((\kappa_1-1)\ell nr - 1)\left[(-1)^{-\frac{1}{2}}\sin\theta\tan\frac{\pi-2}{4} - \cos\theta\right]\right\} \\ 2\mu_1 u_\theta &= \frac{N}{2}\left\{(\kappa_1+1)\theta\left[(-1)^{-\frac{1}{2}}\sin\theta\tan\frac{\pi-2}{4} - \cos\theta\right] + \right. \\ &\left. + ((\kappa_1-1)\ell nr + 1)\left[(-1)^{-\frac{1}{2}}\cos\theta\tan\frac{\pi-2}{4} - \sin\theta\right]\right\} \end{aligned} \tag{31}$$

For the sub-wedge (2); $-\pi \le \theta \le \varphi$

$$2\mu_2 u_r = \frac{N}{2}\left\{-(\kappa_2+1)\left[(-1)^{-\frac{1}{2}}\cos\theta\tan\frac{\pi-2}{4}+\sin\theta\right]+\right.$$

$$\left.+\left((\kappa_2-1)\ell nr-1\right)\left[(-1)^{-\frac{1}{2}}\sin\theta\tan\frac{\pi-2}{4}-\cos\theta\right]\right\}$$

(32)

$$2\mu_2 u_\theta = \frac{N}{2}\left\{(\kappa_2+1)\left[(-1)^{-\frac{1}{2}}\sin\theta\tan\frac{\pi-2}{4}-\cos\theta\right]+\right.$$

$$\left.+\left((\kappa_2-1)\ell nr+1\right)\left[(-1)^{-\frac{1}{2}}\cos\theta\tan\frac{\pi-2}{4}-\sin\theta\right]\right\}$$

where

$$\frac{\kappa_1-1}{\mu_1}=\frac{\kappa_2-1}{\mu_2}, \quad \varphi=\pm\frac{1}{2}rad$$

6. THE MULTIMATERIAL WEDGE UNDER NORMAL DISTRIBUTED LOAD, OF ORDER $n=-2$

The existence of a multi-material wedge elastostatic self similar problem (Fig. 8) under normal distributed loads of order $n=-2$, is possible if the displacements along any interface of the wedge are the same for both materials so that the stress field of the multi-material wedge coincides with the stress field of a single material wedge of equal angle θ_0 and the same distributed loads along its external faces. In this case the stress field of the multi-material wedge is given by relations (20).

Relations (20) are enforced in order to ensure the displacement's equality along the interface, thus

$$u_r^{(i)}(r,\varphi_i)=u_r^{(i+1)}(r,\varphi_i)\Leftrightarrow\frac{C_\alpha}{2\mu_{i+1}r}=\frac{C_a}{2\mu_{i+1}r}\Leftrightarrow\mu_i=\mu_{i+1}\,;\quad i=1,2,\ldots,m$$

Hence, it is resulted the equality of all the shear modulus of the multi-material wedge

$$\mu_1=\mu_2=\ldots=\mu_m=\mu=ct \tag{33}$$

7. CONCLUSIONS

The multi-material composite wedge is a very interesting and difficult problem appeared in composite materials. During the last years, comprehensive analyses have been carried out, and different problems have been studied by several investigations.

The variable-separable forms in the solutions of several elastostatic wedge problems, are proved very efficient. These forms of solutions can also be applied to more complex cases such as the multi-material wedge elastostatic problems, provided that there isn't any stress due to enforced displacements at the interfaces between the dissimilar materials.

The special cases of shear distributed loads (order $1/r$) and normal distributed loads (order $1/r$ or $1/r^2$) enforce relation (23) among the material constants and some particular angles (θ_0, φ) and ratios T/N, as the elastostatic problem of the multi-material wedge to be reduced to a self similar problem.

Although the proposed solutions are very particular, the appearance of a special inclination at the interface (half-radious interface $\varphi = \pm\frac{1}{2}rad$), is very interesting in engineering problems and needs further research.

8. REFERENCES

1. Williams, M.L.: Stress singularities resulting from various boundary- conditions in angular corners of plates in extension, *ASME, J. Applied Mechanics*, **19** (1952), 526-528.
2. Boyy, D. B.: Edge -bonded dissimilar orthogonal elastic wedges under normal and sheer loading, *ASME J. Applied Mechanics*, **35** (1968), 460-466.
3. Gdoutos, E.E. and Theocaris, P.S.: Stress concentrations at the a apex of a plane indenter acting on an elastic half plane, *ASME, J. Applied Mechanics*, **42** (1975), 688-691.
4. Dundurs, J. and Mar Kenscoff, X.: The Stenberg-Koiter conclusion and other anomalies of the concentrated couple, *ASME J. Applied Mechanics*, **56** (1989), 240-243.
5. Pageau, S.S., Joseph, P.F. and Biggers, S.B., Jr.: The order of stress singularities for bonded and disbonded three-material junctions, *Int. J. Solids Structures*, **31** (1994), 2979-2997.
6. Ting, T.C.J.: *Anisotropic Elasticity: Theory and applications*, Oxford Science Publications, New York, 1996.
7. Staboulogiou, J.H. and Theotokoglou, E.E.: The infinite isotropic wedge under linearly distributed loading, *6th National Congress of Mechanics*, Thessaloniki, July 19-21, 2001, *Aifantis E.C. and Kounadis, A.N. editors*, Vol. II (2001), 133-139.
8. Theotokoglou, E.E. and Staboulogiou, J.H.: The multi-material wedge under normal and shear distributed loading-Application in the case of a multi-material problem, *Tenth European Conference on Composite Materials (ECCM-10)*, June 3-7, 2002, Brugge, Belgium, 1-8.

THREE-DIMENSIONAL FINITE ELEMENT ANALYSIS AROUND BROKEN FIBERS IN FRPs

P.A. KAKAVAS[1], C.C. KIRITSI[2] and N.K. ANIFANTIS[3]
[1]*Assistant Professor, TEI of Patras*
[2]*Graduate student, University of Patras*
[3]*Associate Professor, University of Patras.*
GR- 263 34 Patras, Greece

Abstract. A finite element computational procedure is developed for the linear elastic analysis of a single broken fiber in unidirectional fiber-reinforced composites. Depending on the packing sequence and symmetry conditions, a slice is modeled including the segments of the broken fiber and unbroken, along with the epoxy matrix material surrounding the fibers. Considering the volumetric composition as independent variable, the computational model may treat hexagonal or square packing sequences, and any material combination. Numerical results are presented and compared with experimental and numerical ones available in the literature. Stress response predicted by the present method seems to behave in accordance with the available solutions.

1. INTRODUCTION

Recently, there has been a rapid growth in the use of advanced materials in engineering applications, owing to their versatility and high performance. The success of composites in displacing traditional materials relates directly to their high levels of specific properties, namely stiffness and strength. Among such high performance materials, the fiber-reinforced composites are made of fibers and some polymeric matrix material with fiber axis generally oriented in the direction of the applied load. One of the major aspects of these materials is the existence of interfaces separating constituent material phases. An interfacial layer [1] is defined as the inhomogeneous space region intermediate between two bulk phases in bond, and where properties are significantly different from, but related to, the properties of the bulk phases. The stiff fibers mainly carry the tensile load in the axial direction of unidirectional composites. It is well understood that the tensile strength of materials is controlled by fiber failure. A fiber fracture generally causes a strain concentration to exist in the surrounding fiber, matrix, and interphase regions. Therefore, to improve reliability standards and failure resistance of composite materials, the stress concentration in the neighborhood of this break must studied in detail. The axial failure of the fibers is controlled by fiber properties. An appropriate description of this influence is given by the Weibull distribution of strength for brittle materials. Single, isolated fiber breaks will occur in the composite according to the statistical distribution of the fiber strengths. Important parameters in the calculation of strength

E.E. Gdoutos and Z.P. Marioli-Riga (eds), Recent Advances in Composite Materials, 61-70

of the composite are the stress concentration in the next neighboring fibers around the broken fibers and the ineffective length of the broken fiber. Both of these properties can be calculated analytically, but little finite element analysis has been done to investigate the influence of the assumptions that have to be made. The neighboring surviving fibers adjacent to a break will be subject to a stress concentration over a positively affected length. This stress overload increases the probability that a neighboring fiber will fail. The stress concentration factor is related to the interfiber spacing, the material properties, the number of broken fibers (cluster size) and the ineffective length.

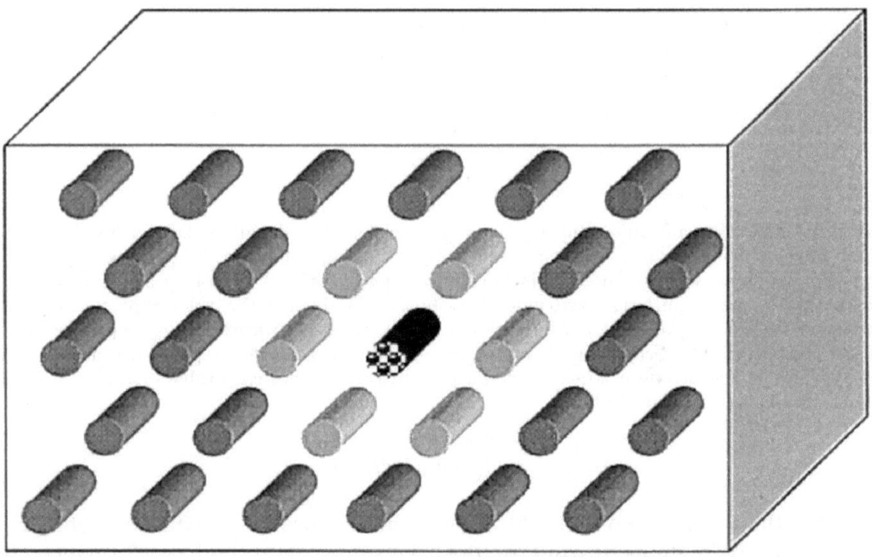

Figure 1. A sample of fiber-reinforced composite containing a broken fiber

The failure of a fiber-reinforced composite material is complicated. Initially, all the fibers are intact and able to carry the same load. When an increasing load is applied to the composite the weakest fibers will eventually break and the shed loads will be transferred either to intact fibers or to matrix. The amount of the load, which is transferred, depends on several parameters, e.g. the sliding resistance and the strength of the fiber-matrix interface, material properties, etc. After the first break occurrence, the load is redistributed and this can cause other fibers to fail and thus shed further load to intact fibers. At some point the composite as a whole will be unable to carry additional load and failure will ensue. According to the work of Landis *et al.* [2], the composite can fail between the extremes of global load sharing on the one hand and nearest neighbor local load sharing on the other. The global load sharing is associated with a material where fiber breaks do not cause stress concentrations in the other intact fibers and the stress along a broken fiber recovers to the applied stress linearly from the break. This type of load transfer requires sliding along the interface of fiber and matrix and a relative stiff matrix.

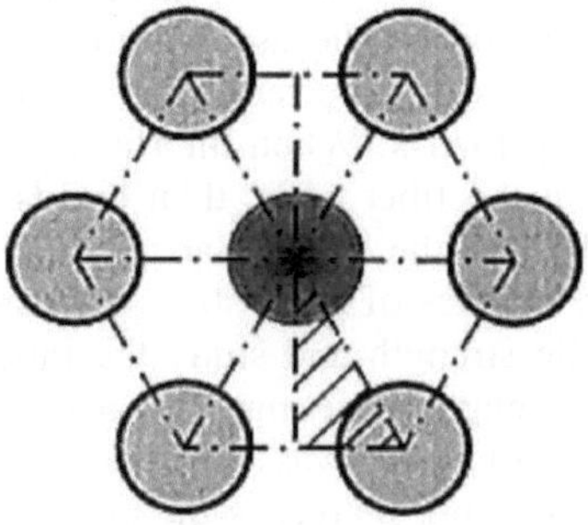

Figure 2. Hexagonal fiber arrangement with the central fiber broken.

Most of the composites materials involve complicated the three dimensional domain of attraction. Hence, the best way to model such materials is to use the finite element technique to investigate the stress distributions in the constitutive materials i.e. fiber and matrix. Shear lag models have become a common feature in micro-stress analysis of composite materials. This type of analysis is used to describe a model that presents a three-dimensional structure, such as a fiber, as a one-dimensional axial load carrying entity. The shear lag models assume that the matrix cannot carry any axial loads, the fibers are well bonded to the matrix, the fibers are arranged in a uniformly spaced array, and the fibers are only allowed to displace along their axis. Cox firstly introduced the analysis based on the shear lag concept [3], and his analysis considered the stresses along a single broken fiber in an elastic matrix and other neighboring fibers were not directly included in the model. The stress concentrations near a single fiber break in a unidirectionally reinforced fiber composite have been investigated using a shear lag theory within the framework of finite elements by Landis *et al.* [2]. These researchers for uniformly spaced, well-bonded fibers embedded in a matrix that cannot carry axial loads have formulated a model. A new formulation to represent sliding in any shear lag system was presented by Landis *et al.* [2] and used to relax the zero matrix axial stiffness and uniform fiber spacing assumptions and to investigate the effects on the stress concentrations near a single broken fiber. For the case of the perfect composite the finite element model have shown to be in excellent agreement with the exact solution of the influence function model.

The Micro Raman Spectroscopy has been employed as a tool to measure the micro mechanical behavior of composites. Galiotis and his coworkers [4-6] have experimentally examined the micro mechanical behavior of single filament model composites including residual stresses caused by processing, stress transfer from the matrix to fiber during loading, and the interfacial behavior as a function of tensile and compressive loading, interface surface treatment and environmental exposure.

They reported the first application of the micro Raman spectroscopy experimental technique as a strain monitoring method in model composites. The axial strain was found to build from the ends of the sort fiber and to reach a maximum value at the center of the fiber, exactly as Cox predicted it theoretically. The computation of the interfacial shear stress was based on the assumption that in single fiber composite geometries the tensile load is transferred to the fibers solely by a shearing mechanism between fiber and matrix. When interfacial failure occurs the interfacial shear stress drops to zero at the fiber ends, then builds to a maximum and finally decays to zero at the middle of the fiber. Gao *et al.* [7] combined the theory of bundle strength with the mechanics of local stress concentration to develop a micro mechanical model for tensile strength and study the influence of micro mechanical properties on the tensile strength of composite and optimal design for tensile-controlled loading applications. It was shown that the existence of stress concentrations due to fiber fracture increases the chance of local crack propagation and, therefore, decreases the strength values predicted by the bundle strength theory. Using the above formulation Gao et al have studied how the tensile strength influenced by the changes in the Weibull shape factor, the stress concentration factor, the volume fraction and the ineffective length as well as the associated material parameters of the composite, i.e. the modulus of elasticity of the fiber and the shear modulus of the resin. Hikami and Chou [8, 9] have examined the two-dimensional multi-filament failure problem of unidirectional fiber composites, focusing specifically on the stress concentration factors of fibers adjacent to the cracks. The physical problems are analyzed by the two-dimensional shear lag analysis under two different loading conditions. The general solution of the governing equations of the elastic field has been obtained in terms of the Legendre polynomials. Based on this solution, closed form expressions of stress concentration factors in all fibers have been derived.

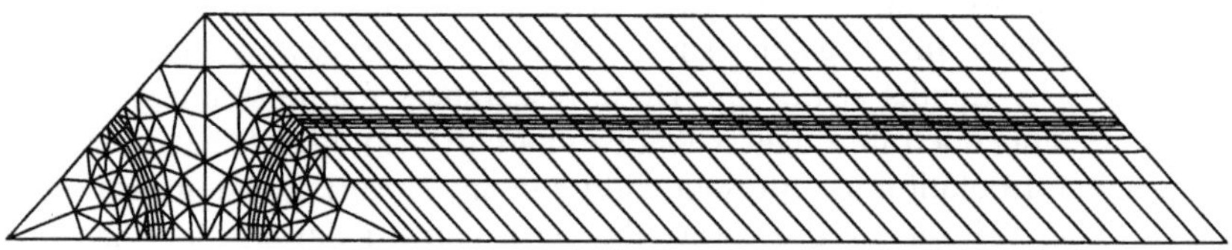

Figure 3. Finite element mesh of a square fiber arrangement.

Anifantis [1] has numerically predicted the micro mechanical stress states developed within fibrous composites that contain a heterogeneous interphase region, by considering variations in topology, materials properties and adhesion characteristics. Square and hexagonal arrays of fibers were used in order to develop the finite element grids and his numerical results have illustrate the influence of material parameters on local critical stress states and indicate regions of high stress concentration. Recent studies [10-16], present a model for predicting the stress distribution along a broken fiber in a unidirectional composite. These studies have suggested that interface debonding may occur as a result of local matrix failure, resulting in the well-known splitting phenomenon observed in tensile tests. A decreasing interfacial shear stress was derived from the Poisson contractions and Coulomb friction, along the debonded length. As is was pointed out, by Morais [17], the debond is followed by a matrix yielding zone, where the interfacial shear stress was assumed to be equal to the matrix yield stress and in addition there is an elastic zone, where the interfacial shear stress follows a classical exponential law. The scope of the present study is to evaluate the stress distribution around a broken fiber in a unidirectional reinforced polymer matrix composite. The finite element analysis was applied using three-dimensional elements for the generation of the finite element grid. A hexagonal array of fibers was assumed in order to formulate the geometry of the testing sample.

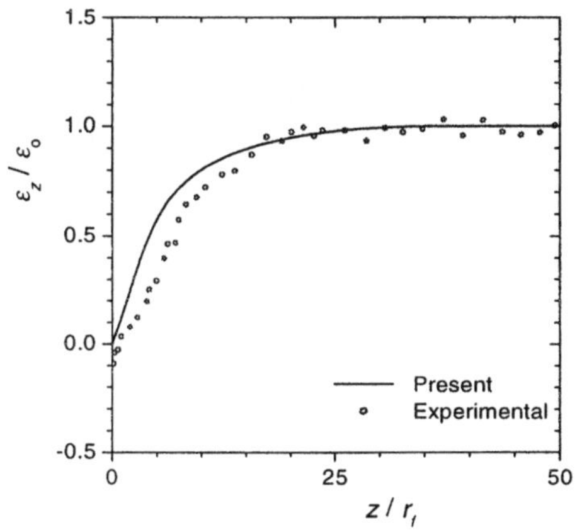

Figure 4. Longitudinal strain along the broken fiber.

2. MODELING ISSUES

The three-dimensional domain, the type of loading and boundary conditions, along with the presence of inhomogeneities render the problem unresolved by analytical procedures, leading to numerical approximations. Among these, the finite element method is adopted for its simplicity and generality. This approach is greatly simplified by assuming the fibers are arranged in some type of regular and periodic array, which allows the analysis to be performed on a representative unit cell rather than the entire composite. The two fiber arrangements assumed for this study were the square and the hexagonal. Figure 1 illustrates a simplified cross-section of a hexagonal fiber arrangement, in which an internal fiber is broken. Figure 2 shows the cross-section of the hexagonal fiber arrangement with the broken one located at the centre. The hatched triangle depicts the cross-section of the corresponding unit cell selected for stress analysis. For the square packing arrangement, a similar situation exists and a corresponding unit cell may be defined. Independently of the packing sequence, the geometry of the unit cell may be written in terms of fiber volumetric composition υ_f, the fiber radius r_f, and the length L of the unit cell.

Both unit cells consist from three separate subdomains in which the fiber and matrix materials are defined. Additional simplifications to the unit cell can be obtained by utilizing the symmetry of its geometry and applied loading conditions. These conditions induce development of a computational simulation for the elastic analysis of closely packed fibrous composites utilizing the finite element approach. The proposed model is uniquely defined by the geometry of the unit cell and the material properties. The type of elements being used in this analysis was six or eight-noded prismatic elements. These types of elements were chosen due to their capabilities on efficient modeling of curved surfaces and their inter-compatibility. The six-noded elements were used to optimize the meshes. Figure 3 illustrates a part of the finite element grid developed to model the square fiber array. In the regions of the interfaces and near the broken surface, appropriate mesh refinements were applied. These typical unit cell configurations consist from 11515 elements and 8106 nodes, in total. The unit cells in which fibers are nearly in contact (very high values of fiber volume fraction) are modeled separately.

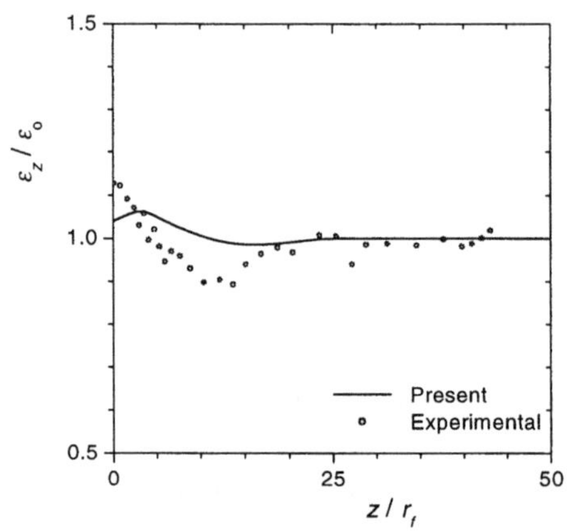

Figure 5. Longitudinal strain along the unbroken fiber.

To keep the outer boundaries their flat shape, proper constraints are imposed over them in the form of unit outward normal displacement. Hence, the vertical component of the displacement on the upper surface is $v = 1$, and the horizontal component of the displacement on the right surface is $u' = 1$, where u' is the local horizontal component of the displacement. To study the stress states developed within such composites, a generalized computational simulation was developed utilizing the previously described finite element model, which employs parametric definition of geometry and material properties. For a given array of fibers, utilizing the fiber radius r_f, and the fiber volume fraction υ_f, as global design variables, the topology and the morphology characteristics of the finite element model are completely described. Then, considering material properties for both the fiber and the matrix, the material properties of the model are properly assigned. This simulation employs automatic mesh generation, assignment of properties and constrains, setup algorithms, which use geometric and material global design variables described above to define a particular case. Furthermore, this simulation enables analysis of any particular fiber-reinforced composite, including geometry variations and material imperfections.

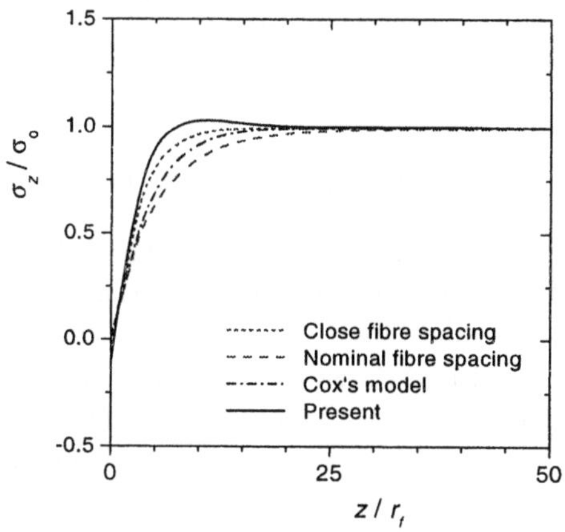

Figure 6. Longitudinal stress along the broken fiber.

Recently, has been proposed a model [11], which accounts for the effect of fiber breakage on the tensile strength of unidirectional composites. In the present study, it will be used similar approach in three-dimensional space. The tensile strength, in continuous fiber-reinforced composites, is controlled by the stress distributions surrounding fiber fractures, i.e. the stress concentrations in fibers near the fibers that have fractured. The redistribution of stress at the neighbourhood of a broken fiber is analyzed on the assumptions that the fibers support only tensile loading, the matrix supports and transmits shear loading, and the matrix between the broken fiber and the adjacent fibers is responsible for the shear transfer. The structure and properties of the fiber/matrix interface play a major role in the mechanical and physical properties of the composites. The analytical and experimental analysis of this problem is usually based on the following assumptions:

(a) perfect bonding exist between the fiber and the matrix i.e. no strain discontinuity exists at the interface fiber/matrix (b) the fibers are arranged in a regular or repeating array. When a tensile loading is applied on the sample, both parts of a broken fiber move away from each other. A matrix cracking, a partial interface debonding or both phenomena can occur. Under perfect bonding conditions at the fiber/matrix interface, and before the fiber failure, on the matrix acts a significant tensile load and one can expect a matrix cracking that is the first local failure mechanism. Near the ends of the broken fiber exist high values of matrix shear strains, which cause yielding.

Table 2. Material properties

Material	*E* (GPa)	v_{xy}
Carbon fiber	396	0.2
Epoxy matrix	3.4	0.36

3. NUMERICAL RESULTS

The composite is consistent from carbon fibers emended into epoxy resin. Both materials are assumed being isotropic and their material parameters are depicted in Table 1. The normalized axial strain, $\varepsilon_z / \varepsilon_o$ along the broken carbon fiber is shown in Figure 4. The solid line shows the numerical solution based on the three-dimensional finite element analysis while the open circles indicate the experimental data published by Heuvel *et al.* [15]. The experimental points at zero strain shows negative values, which can be attributed to prestretching of the material during the curing of the specimen. The normalized axial strain along the broken fiber was level off at about twelve fiber diameters. The finite element analysis based on the 3-D approach follows closely the experimental data extracted from such system using the Raman spectroscopy technique. The axial strain, $\varepsilon_z / \varepsilon_o$ within the broken fiber as a function of the distance along the fiber is shown in Figure 5. The half-length of the fiber is 200 μm, and its diameter is 6 μm, hence the aspect ratio, α, i.e. the ratio of the loading surface to the unloading area of the fiber is approximately 0.375 10^{-2}. The interfiber spacing between the broken fiber and its nearest adjacent fiber is 12 μm, i.e. twice the fiber diameter. The open circles in Figure 5 show the experimental data published by *Heuvel at al.* [15] for the strain profile in a broken Tenax HMS-40 carbon fiber. This type of fibers were embedded into epoxy resin, where its characteristics are given in reference [16]. The experimental results of the strain build near the fiber break were obtained using the Raman spectroscopy technique. In the original paper by Heuvel *et al.* [15] the strain profile near the fiber break was taken negative values that were attributed to prestretching of fibers during the preparation of the composite system. The finite element method gave the right profile of the strain at the nearest unbroken carbon fiber. A strong peak value of the axial strain was observed at the first unbroken fibre. The axial stress normalized upon the stress applied at large distance from the break of the fiber is plotted in Figure 6. A comparison is made between the Cox's theory [3] and the finite element solution proposed by Nedele et al. [11] for two different interfiber spacing. The finite element solution based on the three-dimensional analysis proposed in the present article is above the other curves. All the curves were level off at about twelve fiber diameters from the fiber break. The axial stress in the adjacent fiber is shown in Figure 7. The stress was computed at the middle of the adjacent fiber. A strong peak was observed at the unbroken fiber above the break of the fiber. A comparison of the proposed finite element model with the finite element solution of Nedele et al [11] is shown in the figure for two different case (1) close fiber spacing and (2) nominal fibre spacing ($4\,r_f$). The plots of the axial stress and strain in the broken and adjacent fiber give a smooth variation of stress and strain, indicating that the finite element mesh is fine enough. It was observed that the overall strain in the composite is not altered by the single fiber break; hence a drop in the local fiber strain has to occur to fulfil the requirement of axial strain compatibility.

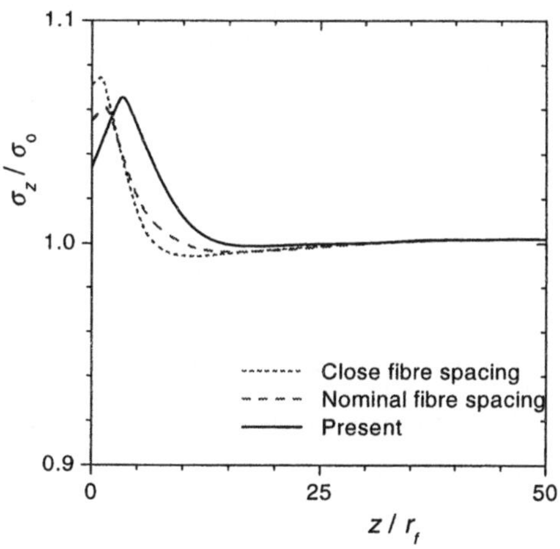

Figure 7. Longitudinal stress along the unbroken fiber.

4. CONCLUSIONS

The three-dimensional finite element analysis has predicted very well the experimental data for the stress and strain distribution around broken fibers in fiber-reinforced polymers. Also the stress and strain distribution in the adjacent fibers was fitted well using the finite element analysis and the experimental data evaluated by

Raman spectroscopy. The hexagonal packing array of fibers was proved being simple model and a reasonable representation of the real composites. According to the proposed analysis with finite elements it is possible to predict the enhanced probability of failure for the adjacent fibers.

5. REFERENCES

1. Anifantis, N.K.: Micromechanical stress analysis of closely packed fibrous composites, *Composites Science and Technology* **60** (2000), 1241-1248.
2. Landis, C.M. and McMeeking, R.M.: Stress concentrations in composites with interface sliding matrix stiffness and uneven fiber spacing using shear lag theory, *International Journal of Solids and Structures* **36** (1999), 4333-4361.
3. Cox, H.L.: The Elasticity and strength of paper and other fibrous materials, *British Journal of Applied Physics* **3** (1952), 72-79.
4. Mastron, C., Gabbitas, B., Adams, J., Marshall, P. and Galiotis, C.: Measurement of stress concentration around fibre breaks in carbon-fibre/epoxy tows, *Composites Science and Technology* **57** (1997), 913-923.
5. Chohan, V. and Galiotis, C.: Effects of interface, volume fraction and geometry on stress redistribution in polymer composites under tension, *Composites Science and Technology* **57** (1997), 1089-1101.
6. Schadler, L.S. and Galiotis, C.: Fundamentals and applications of micro Raman spectroscopy to strain measurements in fiber reinforced composites, *International Materials Reviews* **40** (1995), 116-134.
7. Gao, Z., Reifsnider and Carman, G.: Strength prediction and optimization of composites with statistical fiber flaw distributions, *Journal of Composite Materials* **26** (1992), 1678-1705.
8. Hikami, F. and Chou, T.W.: Statistical treatment of transverse crack propagation in aligned composites, *AIAA* **22** (1984), 1485 – 1495.
9. Hikami, F. and Chou, T.W.: Explicit crack problem solutions of unidirectional composites: Elastic stress concentrations, *AIAA* **28** (1990), 499-505.
10. Andrews, M.C. and Young, R.J.: Fragmentation of aramid fibres in single-fibre model composites, *Journal of Materials Science* **30** (1995), 5607-5616.
11. Nedele, M.R. and Wisnom, M.R.: Stress concentration factors around a broken fibre in a unidirectional carbon fibre-reinforced epoxy, *Composites* Part A **25A** (1994), 549-557.
12. Nedele, M.R. and Wisnom, M.R.: Three-dimensional finite element analysis of the stress concentration at a single fibre break, *Composites Science and Technology* **51** (1994), 517-524.
13. Carman, G.P., Lesko, J.J. and Reifsnider, K.L.: Micromechanical analysis of fiber fracture, *Composite Materials: Fatigue and Fracture*, 4th Volume, ASTM STP 1156, Stinchcomb, W.W. and Ashbaugh, N.E., Editors, American Society for Testing and Materials, Philadelphia, (1993) 430-452.
14. Abu-Farsakh, G.A., Abdel-Jawad, Y.A. and Abu-Laila, Kh.M.: Micromechanical characterization of tensile strength of fiber composite materials, *Mechanics of Composite Materials and Structures* **7** (2000), 105-122.
15. Heuvel, P.W.J., Peijs, T. and Young, R.J.: Failure phenomena in two-dimensional multi-fibre microcomposites. Part 4: A Raman spectroscopic study on the influence of the matrix yield stress on stress concentrations, *Composites* Part A **31** (2000), 165-171.
16. Hui, C.Y., Phoenix S.L. and Ibnabdeljalil M.: An exact close form solution for fragmentation of Weibull fibers in a single filament composite with applications to fiber- reinforced ceramics, *Journal of Mechanics and Physics of Solids*, **43** (1995), 1551-1585.
17. Morais, A. B.: Stress distribution along broken fibres in polymer-matrix composites, *Composite Science and Technology* **61** (2001), 1571-1580.

IN SITU ASSESSMENT OF THE MICROMECHANICS OF LARGE SCALE BRIDGING IN CERAMIC COMPOSITES

K. DASSIOS[1,2,3], C. GALIOTIS[1,4],
V. KOSTOPOULOS[2,1] and M. STEEN[3]
[1]Institute of Chemical Engineering and High Temperature Processes, Stadiou, Foundation for Research and Technology Hellas, Stadiou str., Platani, Patras GR-265 00, Greece
[2]University of Patras, Department of Mechanical Engineering and Aeronautics, Rio GR-265 00, Greece
[3]Institute for Energy, Joint Research Centre, European Commission, Westerduinweg 4, 1755 LE Petten, Netherlands
[4]University of Patras, Department of Materials, Rio GR-265 00, Greece

Abstract. The macro- and micro-mechanics of Large Scale Bridging in SiC/MAS-L composites were assessed by means of tensile testing with in situ LRM probing of Double Edge Notch specimens. The macromechanical behaviour was analyzed in terms of bridging stresses. A dedicated Raman calibration curve was established for a fibre extracted from the composite; the curve served in the transformation of wavenumber shifts collected from bridging fibres in the composite to bridging strain. Bridging strain profiles were established along the notched ligament of the specimen and their shape was discussed in conjunction with local notch effects. The failure strain was not attained within the exposed length of the fibres but within the embedded length. The individual contributions of intact fibres and pull-out fibres were distinguished and discussed in the text.

1. INTRODUCTION

Mechanical testing of full composites provides statistical data on their reliability and information on their macroscopical properties. On the other hand, diffraction and spectroscopic techniques can provide information on the micromechanics that govern the fracture behaviour [1]. The high spatial resolutions and large penetration depths associated with Laser Raman Microscopy (LRM) have established this technique as the only experimental method to date for assessing the deformation of ceramic fibres at the microscale.

The LRM technique is applied in this study for the *in situ* quantification of strain in SiC fibres within the bridging zones developing during testing of Double-Edge-Notched (DEN) SiC/MAS-L composites. Initially, the magnitude of Raman wavenumber shifts of the fibres from their stress-free position was calibrated by loading individual SiC fibres extracted from the composite in axial tension. The inverse methodology was applied in the ceramic composite: the frequency shift in the Raman spectra collected from bridging fibres that were present between matrix

E.E. Gdoutos and Z.P. Marioli-Riga (eds), Recent Advances in Composite Materials, 71-79

macrocracks during testing was converted to strain via the established calibration curve. This study highlights the potential of the LRM technique when applied in full CFCCs, especially for the *in situ* assessment of deformation in the reinforcing phase. It also aims at providing a real measure of strains and stresses within the bridging fibres of SiC/MAS-L composites and to demonstrate differences between real, experimental bridging values in the microscale and values derived from macromechanical models.

2. MATERIAL

The material used in thitudy is a laminated cross -ply SiC/MAS-L composite processed by EADS (France). The glass-ceramic matrix contains MgO, Al_2O_3, SiO_2 and LiO_2 and is made via the sol-gel route. The reinforcing fibres are SiC Nicalon NL202 with a chemical composition in weight concentration of 56.6% Si, 31.7% C and 11.7% O. Plates of 2.0 and 3.0 mm thickness with 8 and 12 plies respectively were produced via hot pressing. The laminae were stacked together in a symmetric $[0\text{-}90^{o}]_{2s}$ and $[0\text{-}90^{o}]_{3s}$ orientation for the 2.0 and 3.0 mm thick plate respectively. The effective volume fraction of the fibres in the loading direction is 0.17 [2] whereas the matrix stiffness (75 GPa) and failure strain are lower than the corresponding values for the fibre and cracks appear first in the matrix.

Double Edge Notch specimens were machined from the composite plates using a 150 μm thick diamond disk with the external plies were oriented parallel to the tensile direction (0°). Specimens with 0.4, 0.5 and 0.6 initial notch-to-width ratios were produced using a diamond disk of the same thickness while the specimen width, *W*, was 12 mm for the 2 mm thick specimens and 10 mm for the 3 mm thick specimens.

Mechanical testing under monotonic tensile loading conditions was performed on a MTS® 858 servo-hydraulic tabletop equipped with a 25 kN load cell under a constant crosshead velocity of 0.01 mm/min. Displacement was measured directly on the specimens by fixing a MTS® axial extensometer (gauge length 25 mm) to one side.

3. MACROMECHANICAL ANALYSIS

3.1 Calculation of Crack Opening

The non-linear part of the total displacement of the system corresponds exclusively to crack opening, i.e. separation of the opposite sides of the macrocrack which, in turn, is always associated with the presence of a bridging zone [3-7]. Hence, the crack opening can be calculated by subtracting the elastic displacement of the system from the total recorded displacement. If D_0 is the initial slope of the measured load-displacement curve, the total crack opening, *e*, is obtained from the instantaneous load-displacement pairs (*F*,*d*) through:

$$e(d) = d - \frac{F(d)}{D_0} \tag{1}$$

3.2 Bridging Stress Evaluation Methodology

Given that non-linear contributions to the measured displacement stem directly from the presence of the bridging zone, the bridging work that contributes to the increase of crack opening resistance can be calculated as the area under the load versus crack opening curve:

$$R = \Delta G = \int_0^e \frac{F(e)}{t \cdot (W - 2\alpha_0)} \cdot de \tag{2}$$

In equation (2), t is the specimen thickness, α_0 is the notch length and $t \cdot (W - 2\alpha_0)$ represents the area of the bridging zone, i.e. the cross section of the bridged ligament of the specimen. Accordingly, the integrand in equation (2) represents the nominal bridging stress, $\sigma_{br}(e)$.

$$\sigma_{br}(e) = \frac{F(e)}{t \cdot (W - 2\alpha_0)} \tag{3}$$

3.2 Macromechanical Bridging Stresses

Following the elastic displacement correction of the measured load-displacement curves through equation (1), the bridging stress field is calculated as a function of crack opening directly from the bridging load using equation (3). This experimental bridging law is presented in Fig. 1 for DEN specimens with three different notch-to-width ratios and different thickness. The tail behaviour is a result of friction due to fibre pull-out.

4. *IN SITU* STRAIN MEASUREMENTS AT THE MICROSCALE

4.1. Raman Calibration Curve for the SiC-Nicalon fibre

The calibration was performed on SiC mono-filaments extracted from the composite using an alcohol dispersal technique. Single fibres of a length of ca. 30 mm were bonded to conventional paper "window" tabs that provided an effective gauge length of 10 mm and were tested in monotonic axial tension on a custom-made micro-extension base. The fibres were subjected to a uniform displacement that was applied sequentially with a step of 20 μm or an equivalent strain step of 0.2 %. Three Raman spectra were collected at each extension level on the same position on the fibre at the middle of the gauge length. The dependence of the 1350 cm^{-1} band position on tensile strain is presented in Fig. 2. A linear relation between the 1350 cm^{-1} band position and fibre strain is observed and the shift rate is calculated as 7.1 cm^{-1} per percent of applied strain with a standard deviation of 0.33 cm^{-1}. Fibre failure occurred at approximately 1.3 % applied strain.

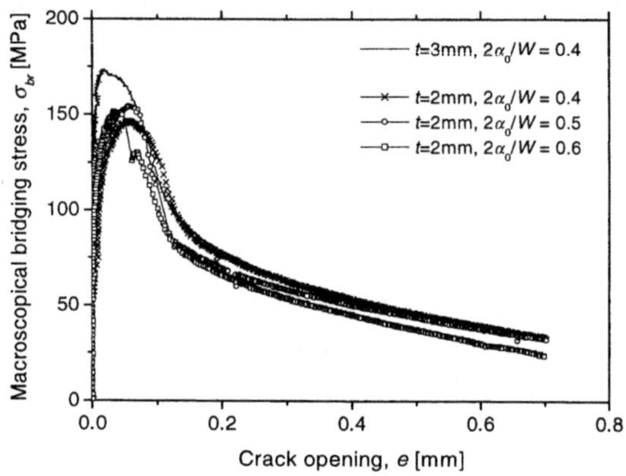

Fig. 1. SiC/MAS-L composite: Directly measured bridging laws for different notch dimensions.

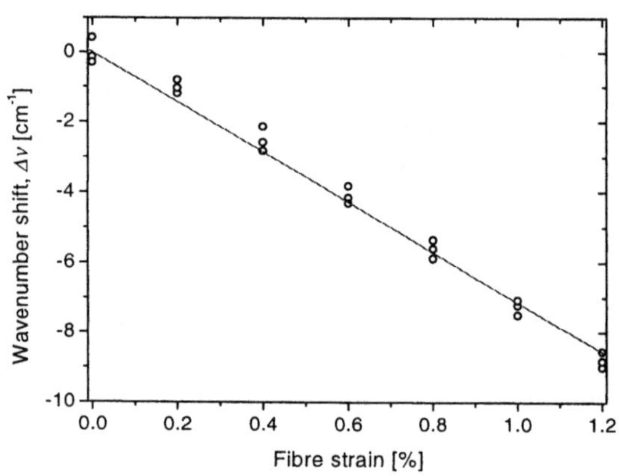

Fig. 2. The strain-based Raman calibration curve obtained for a graphitized SiC-Nicalon fibre extracted from the composite.

4.2 In situ Raman Measurements

After the development of the dominant macrocrack a vast amount of bridging fibres was revealed on the external surface of the composite; these fibres served as collection sites for the Raman microprobe. A microscope picture of the bridging zone obtained during testing of a SiC/MAS-L double edge notched composite gives an impression of the extent of the bridging effect (Fig. 3). The extension-controlled procedure was systematically paused (extension hold) and Raman spectra were collected from individual fibres within the bridging zone. At each extension level, spectra were collected on 100 bridging fibres at equally spaced locations along the ligament. For each fibre, three spectra were collected along its length at different positions near the middle of the exposed length in order to obtain a statistical measure of the validity of the measurement.

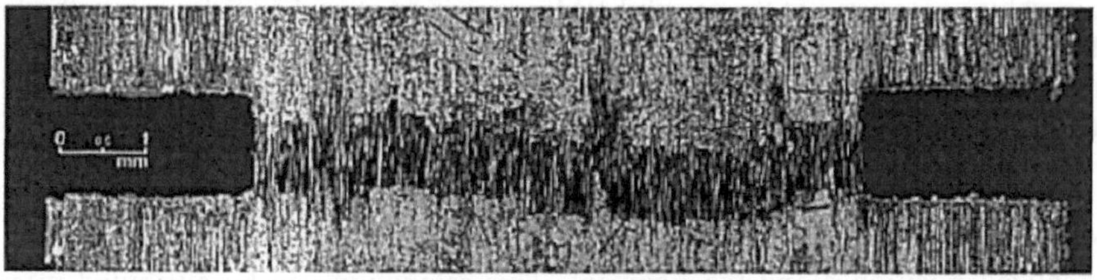

Fig. 3. Large Scale Bridging in SiC/MAS-L composites.

The spectra were analyzed with respect to wavenumber shifts of the 1350 cm^{-1} band from its stress-free position. The exact stress-free position of the band was determined after the end of the test while the specimen was stress-free by collecting Raman spectra on the free standing tips of 50 pulled-out fibres. The standard deviation of the stress-free position of the 1350 cm^{-1} band, 0.38 cm^{-1}, represents the accuracy of the Raman measurements. The established wavenumber shifts were converted to fibre strain using the dedicated Raman calibration curve (Fig. 2).

The nominal macromechanical bridging zone strain was obtained through the elastic correction of the recorded strain (calculated as recorded displacement divided by extensometer gauge length), in a manner similar to that followed for the calculation of crack opening (equation 1). The established nominal bridging strain serves in the comparison of the macromechanical behaviour to the actual bridging zone strain obtained via LRM probing of single bridging fibres.

Fig. 4 represents the macromechanically-derived bridging stress as a function of nominal bridging strain obtained for a SiC/MAS-L DEN specimen of t=2 mm and $2\alpha_0/W$=0.5. The testing procedure was interrupted at four different displacement levels to allow for in-situ LRM scanning along of the bridging zone; the loci of the LRM measurements are demonstrated by the transient load relaxations due to the crosshead extension-hold procedure. The strain values obtained at each step via the Raman microprobe are presented in Fig. 5.

The first LRM-acquisition step (Fig. 5a) was performed at the onset of the fully developed macrocrack where the separation of the two crack flanks was still small and the majority of fibres were still intact. At this stage, a few negative strain values were captured by the Raman microprobe which stem from fibres that have not yet assumed the externally applied deformation and still carry the compressive strain due to fibre/matrix thermal expansion mismatch effects during composite fabrication. As the composite is loaded further, all fibres assume the externally applied deformation.

The second LRM-acquisition step (Fig. 5b) corresponds to a composite state where a large amount of fibres fail within the matrix resulting in an increase of the pull-out contribution to the total behaviour. At this stage no information on the integrity of the fibres is available and the captured Raman strains are a combined result from intact and failed (pull-out) fibres. Most of the fibres have failed at the third acquisition step (Fig. 5c) where the pull-out regime commences and the measured strains have a small contribution from intact fibres. Equivalently, the strain values measured at the fourth LRM-acquisition step (Fig. 5d) are purely due to pull-out fibres. Moreover, at this stage, pull-out has progressed to the extent that many fibres have separated completely from the matrix with their tips protruding outside the matrix. This phenomenon is manifested by the zero-strain values in Fig. 5d and, in much smaller scale even in earlier stages, as for example in Fig. 5c.

By examination of Fig. 5a-d it is observed that fibre strain relaxes in the vicinity of the notch roots, more specifically within the regions of 0-0.1 and 0.9-1.0 normalized position along the ligament. This relaxation can be attributed to stress concentration gradients that develop locally around the notch due to continuity singularities and lead to pre-mature fibre failures hence creating a "strain build-up" effect in the measured profiles.

While the trend of the LRM strain profiles is demonstrated explicitly by Fig. 5a-d, the considerable scatter in the magnitude of the profiles is a direct result of the statistical nature of the failure strain of the fibres and the different fibre failure locations within the composite. Adequate quantification parameters are provided in the form of *average* and *maximum* measured strains with the difference in their values being, in every case, much higher than the standard deviation of the measurements. The average fibre strain corresponds to the plateau of the spline regression whereas an accurate upper limit for the maximum strain is provided by subtraction of the stress-free standard deviation ($0.38cm^{-1}$) from the maximum measured wavenumber shift. On the other hand, the small scatter of multiple measurements performed along each fibre is within the standard deviation margins; an observation indicating that strains are uniform along the scanned length of the fibres.

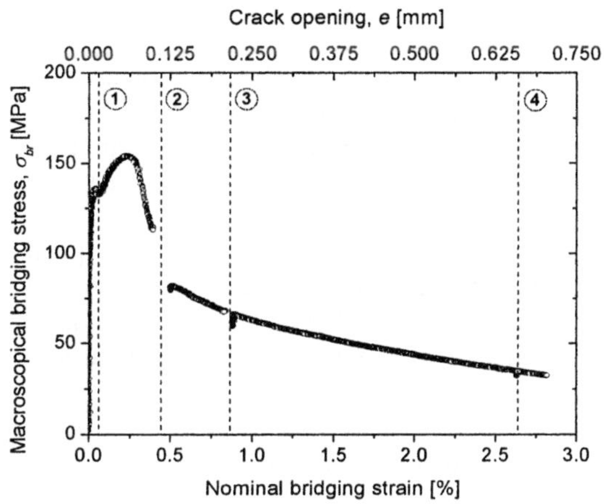

Fig. 4. Mechanical testing of SiC/MAS-L composite: Typical macromechanical behaviour and loci of in situ LRM measurements (dashed lines, steps 1-4).

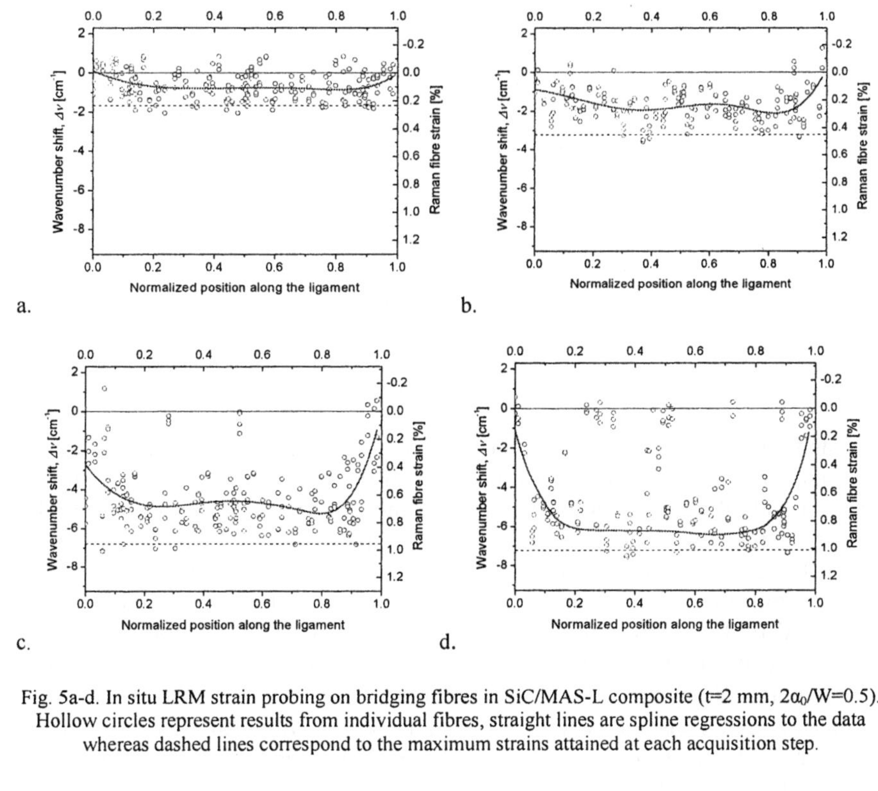

Fig. 5a-d. In situ LRM strain probing on bridging fibres in SiC/MAS-L composite (t=2 mm, $2\alpha_0/W=0.5$). Hollow circles represent results from individual fibres, straight lines are spline regressions to the data whereas dashed lines correspond to the maximum strains attained at each acquisition step.

5. DISCUSSION

A comparison between the macromechanical behaviour of the SiC/MAS-L composite and the local behaviour of fibres within the bridging zone is presented in Fig. 6 where the average (hollow circles) and maximum (solid squares) strain on bridging fibres are plotted as a function of nominal macromechanical bridging zone strain for each LRM-acquisition step. The solid line represents the one-to-one behaviour whereas the dotted lines correspond to the typical error associated with the LRM measurements on the specific fibre, equal to the standard deviation of the stress-free position of the 1350 cm^{-1} band. The x-axis values of Fig. 6 were obtained from the macromechanical bridging stress versus nominal bridging strain curve (Fig. 4) at each LRM-acquisition step. Also plotted in both axes of Fig. 6 is the fibre failure strain, $\varepsilon_{failure}$=1.3%, as established during the calibration of the 1350 cm^{-1} band position with respect to applied strain.

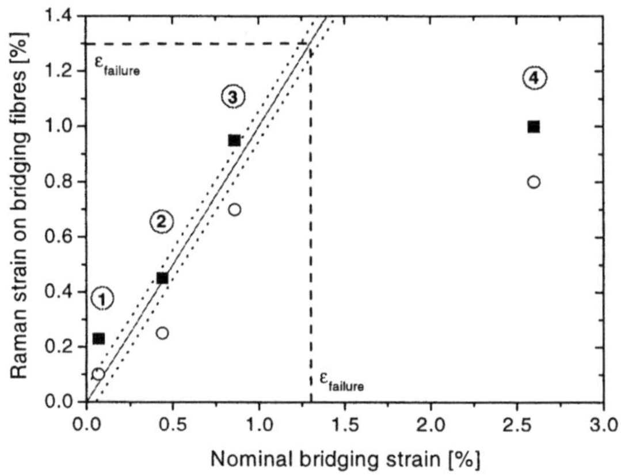

Fig. 6. Large Scale Bridging: Comparison of actual behaviour of fibres in the microscale to macromechanically-derived results.

A sound agreement between the macromechanical and Raman results is apparent in the first three acquisition steps in contrast to the case of bridging by pure pull-out where a wide divergence in noted between the two strains. It must be reminded that the first pair-value in Fig. 6 corresponds to pure strain of intact bridging fibres. Given the fact that at the third acquisition step a great number of fibres already undergo pull-out, it is observed that the maximum strain measured on combined intact and pull-out fibres reaches a plateau value of ca. 1% while the fibre failure strain is never attained within the macrocrack. In conjunction with the small strain scatter noted along the fibre length this observation implies that, while axial fibre strain remains constant within the length of the fibre that is exposed by the macrocrack, it increases along the embedded length and attains its critical value at the fibre failure location.

6. CONCLUSIONS

In this study, the LRM technique was applied for the direct in situ quantification of strains acting on distinct bridging fibres during monotonic tensile loading of a SiC/MAS-L composite with LSB characteristics. Bridging strain profiles were obtained by scanning the bridging zone along the ligament of the specimen with the Raman microprobe. The established profiles exhibited a strain build-up effect spanning 10% of the ligament symmetrically at each notch. The magnitude of measured local strains indicated that the vast majority of fibres fail along their embedded length. The pull-out contribution to the total composite behaviour was considerable owing to a weak fibre/matrix interphase and was analysed separately from the intact fibre behaviour.

7. REFERENCES

1. Galiotis, C.: Interfacial studies on model composites using laser Raman spectroscopy, *Composites Science and Technology* **42** (1991), 125-150
2. Brenet, P., Conchin, F., Fantozzi, G., Reynaud, P., Rouby, D. and Tallaron, C.: Direct measurements of crack-face bridging tractions: a new approach to the fracture behaviour of ceramic/ceramic composites, *Composites Science and Technology* **84** (1996), 817-823
3. Kostopoulos, V. and Pappas, Y. Z.: Toughening Mechanisms in Long Fibre Ceramic Matrix Composites, *Comprehensive Composite Materials* **4** (2000), ed. Kelly, A. & Zweben, C., Elsevier publ., 95-114
4. Marshall, D.B. and Cox, B. N.: The Mechanics of Matrix Cracking in Brittle-Matrix Fiber Composites, *Acta Metall.* **33** (1985), 2013-2021
5. Cox, B. N., Marshall, D.B. and Thouless M.D.: Influence of Statistical Fiber Strength Distribution on Matrix Cracking in Fiber Composites, *Acta Metall.* **37** (1989), 1933-1943
6. Sutcu, M.: Weibull Statistics Applied to Fiber Failure in Ceramic Composites and Work of Fracture, *Acta Metall.* **37** (1989), pp 651-661
7. Curtin, W. A.: Theory of Mechanical Properties of Ceramic-Matrix Composites, *J. Am. Ceram. Soc.* **74** (1991), 2837-2845

2. *Stress and Failure Analysis*

DURABILITY ANALYSIS OF POLYMER MATRIX COMPOSITES: 1978-2003 – RESULTS AND NEEDS

A.H. CARDON
Dept. Mechanics of Materials and Constructions (MEMC)
Faculty of Applied Sciences and Engineering (FTW)
Free University of Brussels (V.U.B.)
Pleinlaan 2, B-1050 Brussels, Belgium

Abstract. Polymer matrix composites exhibit some local and global time dependent behaviour due to the viscoelastic nature of the matrix in interaction with the elastic nature of the reinforcement. As usual this viscoelastic behaviour is also influenced by the environmental conditions: especially temperature changes and moisture uptake and diffusion. Internal transformations in the polymer matrix, known as physical and/or chemical ageing have also to be taken into account. For reliable applications of polymer matrix composites in structural components it is necessary to have a prediction of the stiffness and strength evolutions function of time, environmental changes and a complex loading history. The analysis of the stress-strain relations allows us to obtain the stiffness evolution. For the strength evolution we can use a limit criterion with time and environmental dependent characteristics, but it is also possible to predict damage initiation and development related to the loading history. After an overview of different methods we introduce a nonlinear viscoelastic-viscoplastic model for stiffness prediction under quasi static loading conditions with corrections for interactions between different loading modes. Finally we will discuss the prediction of the strength evolution and we suggest a follow-up of the damage development by a full field optical method able to observe the heterogeneisation and localisation of the strain field. Predictions of stiffness and strength evolutions give us the basis of a complete durability analysis for those polymer matrix composite based structural components.

1. INTRODUCTION

When we have to design a structural component fabricated from a time-dependent material system such as a polymer matrix composite, we need not only the instantaneous hygrothermomechanical properties, but also some reliable predictions of those properties with time, environmental conditions and stress history.

We define durability analysis as the prediction of the changes of the properties of a material system subjected to an estimated complex mechanical loading history interacting with changing environmental conditions over an imposed life time. The results of this durability analysis must give us the basic information in order to evaluate the residual structural integrity and the reliability of the construction component.

Different methodologies have been proposed over the last 25 years:

- a micromechanical method as proposed by C.Chamis [1] ;

E.E. Gdoutos and Z.P. Marioli-Riga (eds), Recent Advances in Composite Materials, 83-94

- a "material" method based on the construction of time-temperature-stress property maps as proposed by H.F. Brinson [2-3-4];
- a "structural " method based on a failure mode analysis leading to the identification of "critical" elements controlling the ultimate failure, as proposed by K. Reifsnider et al. [5-6-7]
- a direct damage analysis starting on micro- or meso-level, and/or based on irreversible thermodynamics as proposed and developed by R. Talreja [8-9], P. Ladevèze [10-11], and Y.J. Weitsman [12-13].

It is also important to refer to the large number of papers on the analysis of fatigue behaviour of composite systems and related life predictions. We mention here explicitly only the work of B. Harris [14], F.K. Gamstedt [15] and G.Caprino et al. [16] but many international conferences on fatigue behaviour of composites were organised over the last decade.

The micromechanics method starts from the properties of the fibres and the matrix computing the properties of the lamina using the assumption of perfect adhesion between fibres and matrix. The laminate properties are obtained using a laminate theory and by integrating over lamina and laminates predictions are formulated for a complete macroscopic structural component using simplified micro-macro relations. In this approach the basic test specimen are defined on the physical level comparable to the fibre diameter.

In the structural method the analysis of the failure mode of a representative specimen of the composite material system introduces critical and sub-critical elements. Failure of sub-critical elements does not lead total failure but the stresses of those elements are transferred to the other elements increasing the stress level in the critical element until total failure occurs. This method can be applied for any type of loading and any type of material.

In the damage methods, damage parameters are defined from experimental observation and damage evolution equations are used in order to compute the damage development and accumulation. Damage parameters can be introduced as internal variables in a thermodynamic based formulation. After identification of the associated generalised forces a simulation of the damage development can be computed. Specific time dependent behaviour is not taken into account by those damage methods.

The fatigue life analysis methods are using linear and nonlinear damage rules, generalisations of Miner's rule and the generalised S/N curves analysis in order to evaluate the residual strength. Generally no specific time dependent behaviour is included in those fatigue analysis approaches.

Our research axis is related to the development and extension of the "material" approach proposed by H.F .Brinson, including the work of C.C. Hiel [17], H.R. Brouwer [18], X. Xiao [19], Y. Qin [20] and P. Bouquet [21].

We also have to mention the results of cooperation on this subject with the group of Antonio Torres-Marques (University of Porto), e.g. R. Miranda-Guedes [22], and with the group of George Papanicolaou (University of Patras), e.g. G. Papanicolaou and S. Zaoutsos [23].

2. CHARACTERISTIC TEST SPECIMEN AND METHOD OF ANALYSIS

A composite material system is a structured continuum made of different materials in interaction. The interaction levels between fibres and matrix in a lamina, and between lamina in a laminate, are not easy to measure. A direct measure is practically impossible and we have to use some mixed numerical-experimental method in order to quantify those interactions after using the concept of imperfect interface conditions as proposed by J. Achenbach and Zhu [24], and developed by Z. Hashin [25].

Even if it is possible to obtain the thermomechanical characteristics of the fibres and the bulk resin we have to assume some specific level of interaction between matrix and fibres in order to compute the characteristics of the composite lamina. Considering also the fact that the interaction between fibres and resin will be strongly influenced by the curing process we have chosen to start from the unidirectional reinforced lamina as basic test specimen for the instantaneous properties and the prediction of the time and temperature dependent evolution of those properties.

This unidirectional reinforced lamina, or laminate, is strongly anisotropic in constitutive behaviour changing from elasto-plastic in the fibre direction, to viscoelastic-viscoplastic in the transverse direction. The basic test specimen under loading conditions in the fibre direction, different off-axis directions and the transverse direction for different environmental conditions is used to predict the long-term stiffness and strength variations on the basis of short term test-results using different acceleration factors.

For those predictions we have applied the method proposed by H. Brinson [2], generalisation of the time-temperature superposition principle for simple thermorheological materials, including the stress level, or some stress invariant, as accelerating factor. This generalisation is based on the nonlinear viscoelastic constitutive equations developed by R. Schapery [26]. Based on the analysis of the creep recovery residual strains a viscoplastic correction has to be applied as proposed by Y. Qin [20]. Such a viscoplastic correction was also introduced by M. Tuttle et al. [27] , and discussed by X. Xiao et al. [28].

The prediction of the changes of the characteristics of the polymer matrix composite under quasi static loading conditions is not complete enough for more general applications. It is also necessary to consider the fatigue behaviour, e.g. B. Harris [14], and the result of interactions between static and dynamic loadings as was shown by J. Bax [29].

3. NONLINEAR VISCOELASTIC-VISCOPLASTIC ANALYSIS

The one-dimensional Schapery model can be seen as a nonlinear extension of the 3-parameter Burgers model for standard viscoelastic solids,

$$\varepsilon(t) = g_o S_o \sigma(t) + g_1 \int \Delta S \frac{d[g_2 \sigma(\psi')]}{d\psi'} d\psi' \quad (1)$$

$$d\psi = \frac{dt}{a_s} \quad ; \quad a_\sigma = a_D a_G^{-1}$$

ΔS is the transient compliance

g_o, g_1, g_2 and a_σ are nonlinearising functions.

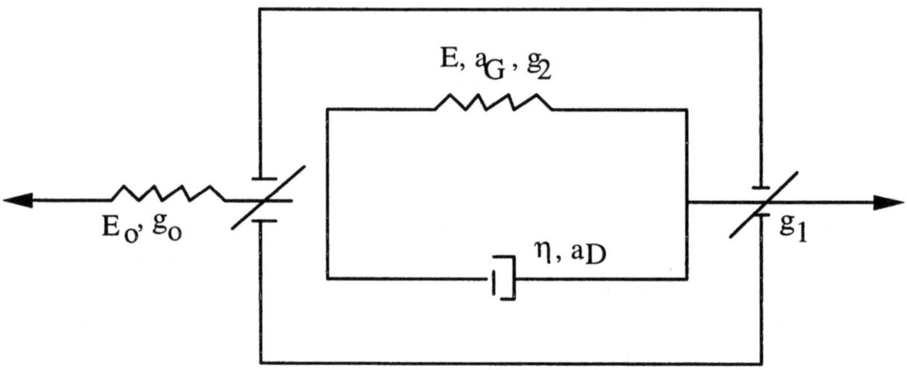

Figure 1. One-dimensional nonlinear Schapery model

If $g_o = g_1 = g_2 = a_\sigma = 1$, equation (1) reduces to the classical linear constitutive equation for non ageing viscoelastic solids.

g_i and a_σ are functions of stress level, temperature and other environmental conditions.

The prediction of the stiffness evolution starting from creep and creep recovery results using equation (1) was successfully applied by Brinson et al. [2], Tuttle et al. [30], Hiel [17] and Xiao [31], for different graphite reinforced polymer composites with thermoset and thermoplastic matrices.

When we mention that the application was successful we mean that the collection of short term creep and creep recovery experimental results can be handled by a master curve treatment allowing to obtain predictions for the stiffness on a long term basis.

In the master curve procedure curve fitting is used and this can lead to growing cumulative errors. Qin [20] has shown that curve fitting can be replaced by the application of some neural network approach. A limited part of the complete creep and creep recovery results for different stress levels are used as "learning" elements for the neural network-based fuzzy inference system (NNFIS), Qin et al. [32].

Applied so a specific graphite-epoxy system (Fibredux, Ciba-Geigy, Hexcel, 920C-TS-5-42) the obtained output was in excellent agreement with the direct "long" term experimental results.

When we consider the 4-parameter Burgers model for a viscoelastic continuum under creep and creep recovery the residual strain after recovery, related to the flow occurring during the creep period characterised by an oblique asymptote has to be proportional to the period of this creep flow.

If we observe a residual strain not corresponding to this time period, or if we have a residual strain after a creep period with an horizontal asymptote, we have to conclude that our material system exhibit some plastic behaviour. Such behaviour was observed by Tuttle et al. [27], Xiao et al. [28] and by Qin et al. [32].

The viscoplastic part of the observed behaviour can be modelled by a Zapas-Crissman functional:

$$\varepsilon_{VP} = \phi\left\{\int g_3[\sigma(t)]\right\}dt . \tag{2}$$

With a specific expression

$$\varepsilon_{VP}(t) = C\sigma^{Nm}t^m \tag{3}$$

we have for the creep period $[0,t_1]$ with a constant stress level σ_o :

$$\varepsilon_C = g_o S_o \sigma_o + g_1 g_2 \sigma_o \frac{S_\infty - S_o}{1+\left(\frac{\tau_o a_\sigma}{t}\right)^n} + C\sigma_o^{Nm} t^m \quad , \tag{4.1}$$

S : compliances for t=0 and $t \to \infty$

τ_o, n, N, m : parameters.

For the recovery ($t>t_1$) we obtain:

$$\varepsilon_r = g_2(S_\infty - S_o)\left\{\left[1+\left(\frac{\tau_o}{\frac{t_1}{a_\sigma}+t-t_1}\right)^n\right]^{-1} - \left[1+\left(\frac{\tau_o}{t-t_1}\right)^n\right]^{-1}\right\}\sigma_o + C\sigma_o^{Nm} t_1^n \ . \tag{4.2}$$

The results shown in Figure 2, demonstrate that the creep recovery evolution given by the VE-VP model is in much better agreement with the experimental data than those given by the simple nonlinear VE-model.

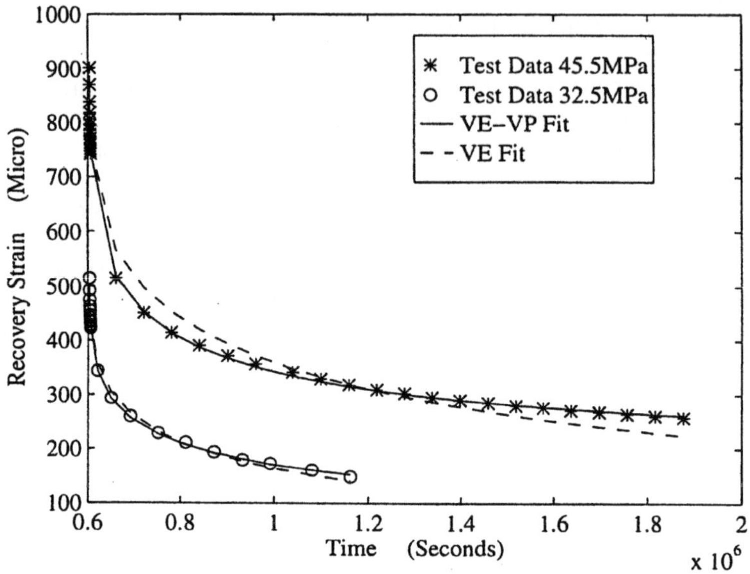

Figure 2. Predictions of $\varepsilon_2(t)$ under σ_2 = cte from VE and VE-VP models compared to the experimental data.

3. INTERACTIONS BETWEEN LOADING MODES

Any durability analysis is related to the specific material system and to the estimated loading conditions for a given application over an imposed time period.

In some problems we have only to consider static or quasi-static loadings. For other applications without permanent loadings we have only to consider fatigue behaviour. Not only mechanical loadings occur but also temperature changes and moisture uptake and diffusion are important for polymer matrix composites.

If we are able to predict the responses to quasi static loading and to fatigue loading we have also to analyse the possible interactions between those two single loading modes.

As early as 1970, Bax [29], observed experimental evidence on glass fibre reinforced polyester and epoxy filament-wounded tubes under internal pressure that a limited number of low frequency oscillations of the pressure accelerates the creep behaviour and significantly reduces the lifetime. Only 20 loading cycles may result in a reduction of the lifetime of some 80%. In order to verify this behaviour, some cyclic loadings were applied to the UD-reinforced Fibredux 920-C-TS-5 followed by a static loading. As can be seen on Figure 3 for a given loading level, we observe an acceleration of the creep behaviour and we have an important difference with the predictions obtained by the VE-VP model.

The acceleration of the creep behaviour is not observed for all loading levels and at lower levels the dynamic loading does not seem to modify the time dependent evolution due to the quasi static loading. The acceleration is probably related to some damage introduced by the dynamic loading and it is important to identify the limit loading levels safe for those interaction effects.

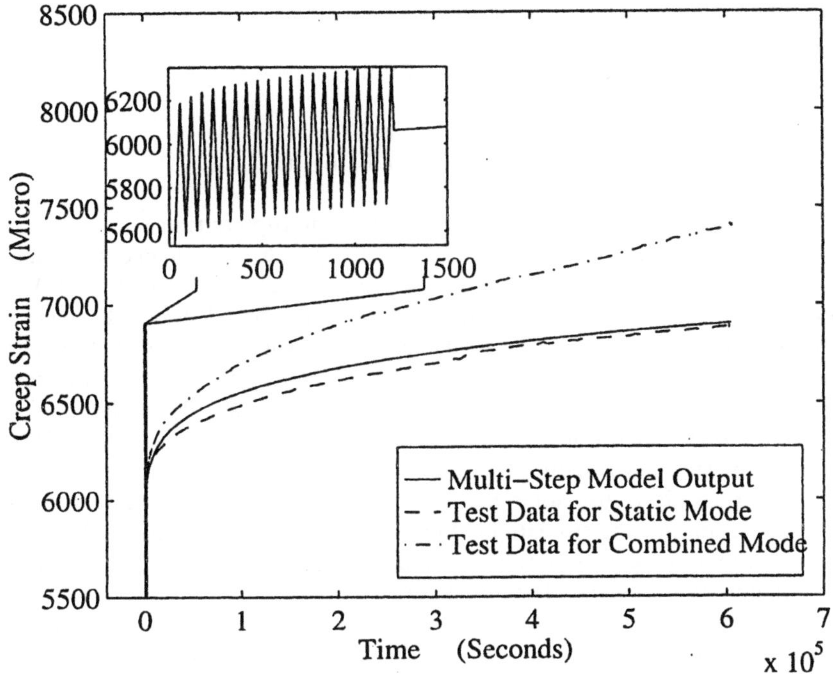

Figure 3. Comparison between the creep evolution in the transverse direction of a unidirectional reinforced composite without and with a first cyclic loading.

5. STRENGTH - DAMAGE

If we want to predict the life time under a specific loading programme, or if we want to identify the maximum admissible loadings for an imposed life time, we have not only to predict the stiffness behaviour but we must also predict the time dependent strength behaviour or the damage initiation and development.

This can be modelled by using a Tsai-Hill failure criterion applied at the ply level with time and temperature dependent strength levels followed by a ply-discount scheme as described by M. Tuttle [33].

$$\frac{\sigma_{11}^2}{X^2(T)} - \frac{\sigma_{11}\sigma_{22}}{X^2(T)} + \frac{\sigma_{22}^2}{[Y(t,T)]^2} + \frac{\tau_{12}^2}{[S(t,T)]^2} \geq 1 \quad , \tag{5}$$

with
X (T) : temperature dependent strength in the fibre direction ;
Y (t,T) : time and temperature dependent strength in the transverse direction;
S (t,T) : time and temperature dependent shear strength.

Y and S are assumed to be of the following form :

Y (t,T) = A(T) – B(T) log (t)
S (t,T) = C(T) – D(T) log (t) .

A, B, C and D are constants to be obtained by creep to rupture tests.
Cumulative damage has to be taken into account if ply stresses change with time and this may happen even if the external load remains constant due to the VE-VP behaviour. This is also what we consider as local time dependent effects where fibre stresses increase and matrix stresses decrease as result of the relaxation effects. Those stress transfers are dependent from the interaction level between fibres and matrix and the amount of stress transfers can be measured by the moduli of imperfect interface conditions as introduced by Achenbach et al. [24]. When a lamina fails the stiffness is reduced following the classical ply discount scheme.

It is also possible to use some master curve scheme similar to that used for the stiffness evolution, for the strength evolution as proposed by Y. Miyano [34]. In our modelling scheme it is also possible to add a specific damage model to the initial VE-VP model. We consider damage as an irreversible rupture of interactions at a certain material scale level. As consequence a certain number of planes with orientations n, in a point are unable to build up a stress vector and stress transfers become impossible for those orientations.
A damage model can be introduced similar to the viscoplastic model with some yield stress element decreasing function of the number of cycles, of time, of environmental changes and of some internal time factor if we have to take into account ageing and/or degradation.
The combined VE-VP damage model gives us an expression for the total strain :

$$\varepsilon = \varepsilon_e + \varepsilon_{ve} + \varepsilon_{vp} + \varepsilon_d \quad , \tag{6}$$

$\varepsilon_e + \varepsilon_{ve}$, given by the Schapery equations,

ε_{vp} , described by the Zapas-Crissman functional,

$\varepsilon_d = \Delta\lambda_d \dfrac{\partial f}{\partial \sigma}$, where f is a potential function similar to those used in plasticity.

This damage extension of our basic VE-VP model is similar to the approach proposed by Tamusz et al. [35]. A number of failure tests on lamina have to be

performed in order to identify the significant damage parameters and the potential function. Comparison of the predictions with real time experimental results as obtained by Tuttle and Brinson [30], show that the predictions are too optimistic.
Those discrepancies are probably due to the applied laminate theory for the introduction of the results obtained on unidirectional reinforced lamina. Laminate theories suppose perfect interface conditions between lamina even if in the physical reality we observe resin rich regions between lamina.
For the long term predictions it is necessary to introduce imperfect interface conditions between lamina in order to take into account the initiation and development of delaminations during the loading cycle.

6. EXPERIMENTAL DAMAGE CONTROL

Modulus degradation in the stress-strain behaviour coupled with permanent deformation after unloading are indicators of damage development in material systems. An integrated indication is obtained from cyclic loading. This is easy for an elastic material system but if the material has some time dependent behaviour it is much more time consuming to obtain a damage indication by cyclic creep and creep recovery tests. The final failure as result of some damage development is largely due to the presence of a "weak" link, some initial defect followed by time, temperature, moisture and local stress concentration evolutions. The transformation of the homogeneous stress-strain behaviour, through heterogeneisation and localisation is the way to arrive at the matrix dominated failure.

The initiation of the final failure is local and we have to use experimental methods allowing us to obtain strain values commensurate with the size of the inferred damage region.
Traditional strain measures obtained by electrical strain gages are not able to give a good overview of the damage initiation and development. We need full field optical, non contact, methods such as photoelastic coatings, Moiré interferometry, shearography, stress pattern analysis by thermoelastic emission (SPATE), thermography and the most recent one, electronic speckle pattern interferometry (ESPI). Some possible applications of SPATE were studied by D.Van Hemelrijck [36], and damage initiation analysis using ESPI was analysed by P.Bouquet [21]. In the ESPI-method the out of plane displacements give us a clear indication of some localisation and heterogeneisation of the strain field as result of some damage initiation in a volume under the observed surface.

All those full field optical methods are based on the observation of phenomena on the outer surface of the test specimen or the real construction component. It will be necessary to apply some mixed numerical-experimental methods to quantify the relations between events in the volume and surface related manifestations.
The ESPI, especially by the out of plane effects, is able to give us the necessary information about the damage initiation, development and localisation.

7. CONCLUSIONS – FUTURE DEVELOPMENTS

On the basis of short-term creep and creep recovery tests, completed by fatigue tests and cyclic loadings at different constant and changing environmental conditions we are able to predict long term stiffness evolution by application of a NLVE-VP model.

Short term failure analysis allows us to obtain the time and environment dependent strength characteristics, basis for the definition of admissible stress states corresponding to an imposed life time.

Full field optical methods allow us to follow the localisation of the strain field, giving us a good indicator of the damage initiation, development and failure. Those optical information give us the necessary elements for the definition of a damage component to be added to the basic NLVE-VP model so that we are able to predict residual structural integrity conditions for a construction under a combined mechanical and environmental loading history. We have to develop mixed numerical-experimental methods in order to arrive at the measure of an in volume event starting from the observation of a surface event. A better understanding of the ageing phenomena and degradation of matrix, fibre and interaction levels in between is a basic condition for more reliable structural applications of fibre reinforced polymer matrix composite construction components.

8. ACKNOWLEDGEMENTS

This research was made possible by financial support from the Science Foundation of the Flemish Region of Belgium (FWO-Vlaanderen) and from the Research Council of the Free University Brussels (OZR-VUB). For parts of this research we have also received financial support from the International Association for Promotion of the Cooperation with Scientists from the New Independent States of the former Soviet Union (INTAS-96-2113).

Many thanks are due to the technical staff of the department and especially to Frans Boulpaep. Without the permanent support and assistance of Myriam Bourlau this work could never have been achieved.

9. REFERENCES

1. Chamis, C.: Simplified composite micromechanics equations for strength, fracture, toughness and environmental effects, *Sampe Quarterly* 15, (1984), 41-51;
2. Brinson, H.F., Morris, D.H. and Yeow, Y.T.: A new experimental method for the accelerated characterisation of composite materials, 6th International Conference on Experimental Stress Analysis, Th. Gaymann (Ed.), *VDI Berichte*, 313, (1978), 395-400.
3. Brinson, H.F.: A nonlinear viscoelastic approach to durability predictions for polymer based composite structures, Durability Analysis of Polymer Based Composite Systems for Structural Applications (Eds. A. Cardon - G. Verchery), Elsevier Applied Science, (1991), pp. 46-64.
4. Brinson, H.F.: Matrix dominated time dependent failure predictions in polymeric matrix composites", *Composite Structures* 47, (1999), 445-456.

5. Reifsnider, K.L. and Stinchcomb, W.W. : A critical element model for residual strength of fatigue-loaded composite coupons, *Composite Materials Fatigue and Fracture,* ASTM STP 907, H.T. Hahn (Ed.), (1986), 298-313.
6. Reifsnider, K.L.:The critical element model : a modeling philosophy, *Eng. Fracture Mechanics*, vol. 25, nos. 5-6, (1986), 739-749.
7. Reifsnider, K.L., Iyengar, H., Case, S. and Xu, Y.L.: Damage tolerance and durability of fibrous material systems : A micro kinetic approch, *Durability Analysis of Structural Composite Systems* (Ed. A.H. Cardon), Balkema, (1996), pp. 123-144
8. Talreja, R.: Damage mechanics of composite materials based on thermodynamics with internal variables. Durability of Polymer Based Composite Systems for Structural Applications (Eds. A.H. Cardon - G.Verchery), Elsevier Applied Science, (1991), pp. 65-79.
9. Talreja, R.: A synergistic damage mechanics approach to durability of composite material systems, Progress in Durability Analysis of Composite Systems, A.H.Cardon, H.Fukuda, K.Reifsnider (Eds.), Balkema, (1996), 117-129.
10. Ladevèze, P., Allix O., Daudeville, L.: Mesomodelling of damage for laminate composites - Application to delamination, Inelastic Deformation of Composite Materials (Ed. G. Dvorak), Springer, Berlin, (1991), 607-622.
11. Ladevèze, P.: A damage computational approach for composites : Basic aspects and micromechanics strength/stiffness indicators", Progress in Durability Analysis of Composite Systems, A.H.Cardon, H.Fukuda, K.Reifsnider (Eds.), Balkema, (1996), 131-138.
12. Weitsman Y.J. and Elahi, M.: Effects of fluids on the deformation, strength and durability of polymeric composites – An overview, *Mechanics of Time Dependent Materials*, vol. 4, n°2, (2000), 107-126.
13. Weitsman, Y.J.: Effects of fluids on polymeric composites", Durability Analysis of Composite Systems 2001, Miyano, Cardon, Reifsnider, Fukuda and Ogihara (Eds.), Swets & Zeitlinger, Lisse, (2002), 123-128.
14. Harris, B.: Fatigue behaviour of polymer-based composites and life prediction methods, Durability Analysis of Structural Composite Systems, A.H. Cardon (Ed.), Balkema, Rotterdam, (1996), 49-84.
15. Gamstedt, E.K.: Fatigue in composite lamiminates – A qualitative link from micromechanisms to fatigue life performance, Recent Developments in Durability Analysis of Composite Systems, Cardon, Fukuda, Reifsnider and Verchery (Eds.), Balkema, Rotterdam, (2000), 87-100.
16. Caprino, G. and D'Amore, A.: Fatigue life of graphite/epoxy laminates subjected to tension-compression loadings, *Mechanics of Time Dependent Materials* 4, 2, (2000), 139-154.
17. Hiel, C.C.: The nonlinear viscoelastic response of resin matrix composites, PhD-thesis, Vrije Universiteit Brussel (VUB), 1983.
18. Brouwer, H.R.: Nonlinear viscoelastic characterisation of transversely isotropic fibrous composites under biaxial loadings, PhD-thesis, Vrije Universiteit Brussel (VUB), 1986.
19. Xiao, X.R.: Viscoelastic characterisation of thermoplastic matrix composites - APC2, PhD-thesis, Vrije Universiteit Brussel, (VUB), 1987.
20. Qin, Y.: Nonlinear viscoelastic-viscoplastic characterisation of polymer matrix composites, PhD-thesis, Vrije Universiteit Brussel (VUB), 1996.
21. Bouquet, P.: Critical analysis of the long-term behaviour of polymer matrix composites under a complex mechanical loading history, PhD-thesis, Vrije Universiteit Brussel (VUB), 2002.
22. Miranda-Guedes, R. : On creep rupture prediction of polymer and FRP laminates, Durability Analysis of Composite Systems, Miyano, Cardon, Reifsnider, Fukuda and Ogihara (Eds.), Swets and Zeitlinger, Lisse, (2002), 175-181.
23. Papanicolaou, G. and Zaoutsos, S.: On the prediction of the nonlinear viscoelastic behaviour of polymer matrix composites", *Composites Science and Technology*, 59, (1999), 1311-1319.
24. Achenbach, J.D. and Zhu, H.: Effect of the interfacial zone on mechanical behaviour and failure of fibre reinforced composites, *Journal of Mechanics and Physics of Solids*, vol. 37, (1989), 381-393.
25. Hashin, Z.: Composite materials with interphase : Thermoelastic and inelastic effects, Inelastic Deformation of Composite Materials (Ed. G.J. Dvorak), Springer, Berlin, (1991), 3-34.
26. Schapery, R.A.: A theory of non-linear thermoviscoelasticity based on irreversible thermodynamics, Proceedings 5th US National Congress on Applied Mechanics, ASME, (1996), 511-530.
27. Tuttle, M.E., Pasricha, A. And Emery, A.F.: Time dependent behaviour of IM7/5620 composites subjected to cyclic loads and temperatures, in Mechanics of Composite Materials – Nonlinear effects, AMD, vol. 159, ASME, New York, (1993), 343-357.

28. Xiao, X.R., Hiel, C.C. and Cardon, A.H.: Characterisation and modelling of nonlinear viscoelastic response of peek resin and peek composites", *Comp. Engrg.* 4(7), (1994), 681-702.
29. Bax, J.: Deformation behaviour and failure of glass fibre reinforced resin materials, *Plastics and Polymers*, (1970), 27-30.
30. Tuttle, M.E. and Brinson, H.F.: Predictions of the long-term creep compliance of general composite laminates, *Experimental mechanics*, 26, (1986), 89-102.
31. Xiao, X.R.: Studies of the viscoelastic behaviour of a thermoplastic resin composite, *Composite Science and Technology*, 34, (1989), 163-182.
32. Qin, Y., Guedes, R. and Cardon, A.H.: Characterisation and prediction of nonlinear viscoelastic-viscoplastic behaviour of CRP under cyclic loadings using a neuro-fuzzy model, Progress in Durability Analysis of Composite Systems (Eds. K.Reifsnider, D.Dillard and A.Cardon), Balkema, (1998), 249-253.
33. Tuttle, M.E.: A framework for long-term durability predictions of polymeric composites. Progress in Durability Analysis of Composite Systems, Cardon, Fukuda & Reifsnider (Eds.), Balkema, Rotterdam, (1996), 169-176.
34. Miyano, Y.: Long term prediction method for static, creep and fatigue strengths of CFRP composites, Progress in Durability Analysis of Composite Systems, Cardon, Fukuda & Reifsnider (Eds.), Balkema, Rotterdam, (1996), 177-188.
35. Tamusz, V., Andersons, J., Aniskevich, K. and Korsgaard, J.: Creep damage interaction in composites, Progress in Durability Analysis of Composite Systems, Cardon, Fukuda & Reifsnider (Eds.), Balkema, Rotterdam, (1996), 143-149.
36. Van Hemelrijck, D.: Contribution to the development of thermoelasticity for the experimental mechanics of composite systems, PhD-thesis, Vrije Universiteit Brussel (VUB), 1992.

EXPERIMENTAL AND NUMERICAL ANALYSIS OF MATRIX CRACKING IN BRITTLE COMPOSITES

D. VAN HEMELRIJCK, H. CUYPERS, J. WASTIELS, G. KALOGIANNAKIS and W.P. DE WILDE
Vrije Universiteit Brussel, Faculteit Toegepaste Wetenschappen, Dienst Mechanica van Materialen en Constructies (MEMC), Pleinlaan 2, B–1050 Brussel, Belgium

Abstract. Inorganic Phosphate cement (IPC) is a cementitious material developed at the "Vrije Universiteit Brussel" which can be reinforced with E-glass fibres. In order to improve its application in civil engineering constructions like wall and roof panels, the constitutive behaviour of IPC should be fully understood and accurately modelled. Using a stochastic cracking theory, the averaged matrix crack distance can be calculated as a function of the applied stress. This theory is experimentally verified using a stereomicroscope equipped with a video camera and image acquisition board. It is shown that the experimentally obtained value of the average crack distance as a function the applied stress correlates quite well with the theoretically obtained value.

1. INTRODUCTION

Inorganic Phosphate cement (IPC) is a cementitious material developed at the "Vrije Universiteit Brussel, Department Mechanics of Materials and Constructions (MEMC)". IPC is a two-component system, consisting of a calcium silicate powder and a phosphate acid based solution of metal oxides. Like most cementitious materials, IPC has high compression strength but its tensile strength is quite low. Fortunately, after hardening IPC provides a non-alkaline environment, allowing E-glass fibres to be introduced for added stiffness, strength and toughness. Other advantages are the high fire resistance and the fact that in case of fire no toxic gasses are released (all components are inorganic). One application currently under study is the use of E-glass fibre reinforced IPC as face material in sandwich wall and roof panels for buildings. However, for an economical and safe design without excessive safety factors the constitutive behaviour of IPC must be accurately modelled. In compression, the behaviour of IPC is nearly linear elastic up to failure [6]. Unfortunately, in tension the behaviour of IPC is completely different. Since cracks initiate and propagate at very low stress levels, the stress-strain relation is linear only in a very small region. A first attempt to model this behaviour (the well-known ACK-theory) was done by Aveston *et al.* [1] and Aveston and Kelly [2]. However, according to Bauweraerts [3] and Gu *et al.* [7], the main shortcoming of this theory, when applied to IPC, is the fact that in the ACK theory the matrix strength is

E.E. Gdoutos and Z.P. Marioli-Riga (eds), Recent Advances in Composite Materials, 95-108

assumed to be a deterministic variable. Based on the work of Curtin *et al.* [4,5], who presented a statistical treatment of matrix crack evolution in UD-reinforced ceramic micro-composites H. Cuypers [6] presented a stochastic cracking theory for E-glass reinforced IPC. Two Weibull parameters (σ_r and m) account for the stochastic cracking behaviour of the matrix, which is highly dependent on the presence of fibres. Another parameter introduced into the model is the shear stress (τ_0) along the matrix-fibre interface (depends on the impregnation of the fibre bundle by the matrix). These three material properties are obtained by finding a 'best fit' between the theoretical and experimental stress-strain curves. Once the Weibull-parameters are known, the average crack spacing can be predicted theoretically as a function of the applied stress.

2. THEORETICAL BACKGROUND

2.1 Introduction

Figure 1 shows a typical stress-strain curve of a brittle matrix composite. Based on the ACK-theory we may divide this curve into three zones: the pre-cracking zone (zone I), the multiple cracking zone (zone II) and the post-cracking zone (zone III).

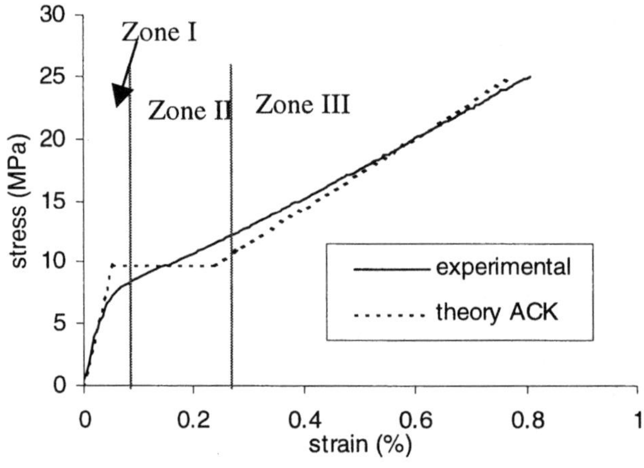

Figure 1. Stress-strain behaviour of IPC composite (2D-random oriented fibre reinforcement, 10%) in tension.

Basic assumptions, used in the development of the ACK theory as well as in the development of the stochastic cracking theory, are:

- The fibres are only capable of carrying load along their longitudinal axis. They provide no bending stiffness.

- The matrix-fibre bond is weak. The adhesion shear bond strength τ_{au} is low. Propagation of matrix cracks leads to matrix-fibre debonding along a certain length (slip length, δ_0) in the vicinity of a matrix crack (see fig. 2).
- Once the matrix and the fibre are debonded, a pure frictional shear stress τ_0 replaces the previously existing adhesion shear stress τ_a .
- At the debonded matrix-fibre interface, this sliding shear stress τ_0 is constant with slip.
- The frictional interface shear stress τ_0 is constant along the debonded interface.
- Poisson effects of the fibre and matrix are neglected.
- Global load sharing is used for the fibres.
- The matrix normal stresses are assumed to be uniform in a section, transverse to the loading direction.

2.2 Pre-cracking zone (zone I)

Zone I is often also called the 'linear elastic zone'. The composite obeys the law of mixtures. The composite stiffness E_{c1} is thus function of the fibre and matrix stiffness, E_f and E_m and the fibre and matrix volume fractions, V_f and V_m.

$$E_{c1} = E_f V_f^* + E_m V_m \tag{1}$$

with :

$$V_f^* = \eta_l \eta_\theta V_f \tag{2}$$

η_l and η_θ are factors containing the influence of the fibre length and orientation respectively.

2.3 Multiple cracking zone (zone II)

2.3.1 Introduction

As the tensile strength of IPC is rather low, matrix cracks will appear in the composite at relatively low tensile stress levels. When a first matrix crack initiates and reaches a fibre, debonding of the matrix-fibre interface occurs, since it is assumed that the matrix-fibre interface is weak. At the vicinity of a crack face, the matrix is stress-free. Along the debonded interface, the assumption of the existence of a constant frictional matrix-fibre interface shear stress τ_0 is used. Away from the crack face, this frictional interface shear stress provides stress transfer from fibres to the matrix. Figure 2 illustrates the evolution of the normal stresses in matrix and fibres as a function of the distance from a crack face, provided the existence of other cracks is neglected. x is the distance from the crack face. Slip length δ_0 is the x-distance from a crack face where the matrix stress reaches the maximum value σ_m^{ff}. σ_m^{ff} is the far field matrix stress (stress in the matrix at an infinite distance from the crack face):

$$\sigma_m^{ff} = \frac{E_m \sigma_c}{E_{c1}} \tag{3}$$

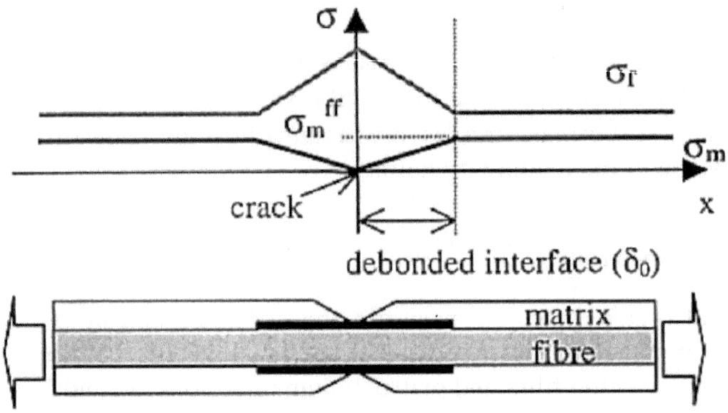

Figure 2. Evolution of the matrix (σ_m) and fibre (σ_f) normal stresses as function of the distance from the crack face, x.

It is shown in figure 2 that the matrix stresses are lower at the vicinity of a crack face, within distance δ_0 to the left and the right of a crack. This means it becomes very unlikely that new matrix cracking will occur within distance δ_0 to the left and the right of this matrix crack. This zone is therefore often called exclusion zone. Debonding length δ_0 can be determined from equilibrium of forces around a fibre and is:

$$\delta_0 = \frac{V_m r \sigma_m^{ff}}{V_f^* 2\tau_0} \tag{4a}$$

where r is the fibre radius.

Widom [9] determined that the average final crack distance $\langle cs \rangle_f$ was equal to:

$$\langle c_S \rangle_f = 1.337\delta_0 \tag{4b}$$

2.3.2 ACK-theory

In the ACK-theory it is assumed that all the matrix cracking occurs at the same stress level.

2.3.3 Stochastic cracking theory

In the stochastic cracking theory the nature of the matrix failure strength is introduced through a probability distribution. Cracks can now occur gradually at various stress levels. Once the first matrix crack is initiated we have fibres bridging this matrix crack and consequently the external load can be increased, leading to the initiation and propagation of new matrix cracks at higher stress levels outside the already existing exclusion zones. From three-point bending tests on pure IPC matrix specimens [6] it was concluded that the two-parameter Weibull probability distribution seems to be the most appropriate choice to model the stochastic nature of the IPC tensile strength. The shape of the Weibull [8] cumulative distribution, describing the failure of specimens in uniform tension, is:

$$P = 1 - \exp\left[-\left(\frac{\sigma}{\sigma_R} \right)^m \right] \tag{5}$$

where:

P = failure probability
σ = uniform tensile stress in the material
σ_R = reference failure stress
m = Weibull modulus

According to the stochastic cracking theory, the average crack spacing <cs> is a function of the applied composite stress. Equation (5) provides information about the percentage of inherent matrix flaws, which are able to propagate at a certain stress level. If multiple cracking occurred completely, the average crack spacing <cs> equals $<cs>_f$, being the final crack spacing. If only partial multiple cracking occurred, <cs> is still larger than the final crack spacing. The average crack spacing <cs> at a certain composite stress is found by dividing the final crack spacing $<cs>_f$ by the percentage of matrix cracks that already propagated, which is expressed by equation (5):

$$\langle cs \rangle = \langle cs \rangle_f \left[1 - \exp\left[-\left(\frac{\sigma_c}{\sigma_R} \right)^m \right] \right]^{-1} \tag{6}$$

with: σ_c being the composite stress

In case the stochastic cracking theory is used, the formulation of the stress-strain relationship should be done for two cases, as shown in figure 3.

Figure 3a shows the matrix stress (σ_m) and fibre stress (σ_f) in the composite if the average crack spacing <cs> is larger than $2\delta_0$ (lower stress levels, occurrence of first matrix cracks). Figure 3b shows these stresses if <cs> is smaller than $2\delta_0$ (higher stress levels, end of multiple cracking zone).

From figure 3a, the composite stress-strain formulation can be determined in case <cs> > $2\delta_0$:

$$\varepsilon_c = \frac{\sigma_c}{E_{c1}}\left(1+\frac{\alpha\delta_0}{\langle cs\rangle}\right) \quad \text{with} \quad \alpha = \frac{E_m V_m}{E_f V_f^*} \tag{7}$$

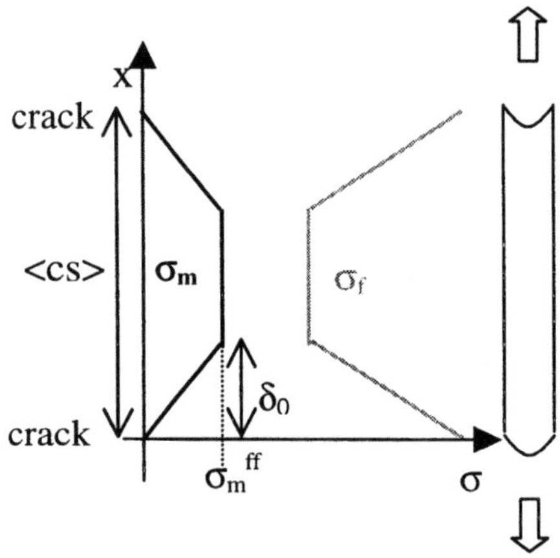

Figure 3a. Stresses in matrix and fibre along a cracked composite, <cs> > $2\delta_0$.

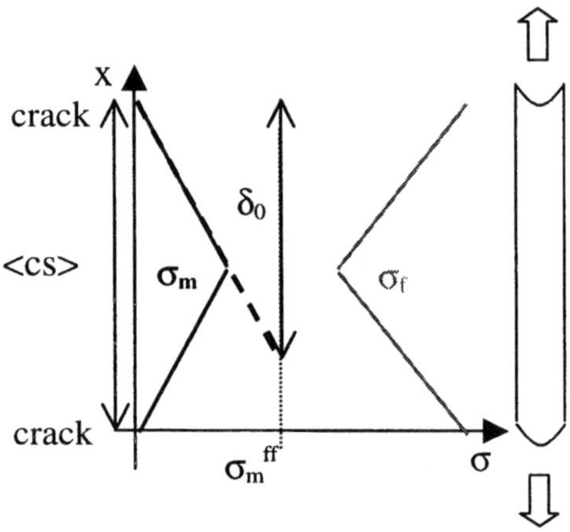

Figure 3b. Stresses in matrix and fibre along a cracked composite, <cs> < $2\delta_0$.

From figure 3b, the composite stress-strain formulation can be determined in case <cs> < $2\delta_0$:

$$\varepsilon_c = \sigma_c\left(\frac{1}{E_f V_f^*} - \frac{\alpha\langle cs\rangle}{4\delta_0 E_{c1}}\right) \tag{8}$$

In the stochastic cracking theory the following material properties: (i) the composite reference cracking stress, σ_R ; (ii) the Weibull modulus, m and (iii) the frictional interface shear stress, τ_0 will be determined through minimisation of the difference between the theoretical and experimental stress-strain curves. Once these three parameters are determined, the averaged crack spacing <cs> can be calculated using equation 6.

2.4 Post-cracking zone (zone III)

Once full multiple cracking occurred, the fibres only provide further stiffness. The stiffness in zone III can therefore be expressed as:

$$E_{c3} = E_f V_f^* \tag{9}$$

3. EXPERIMENTAL VERIFICATION

3.1 Optical stereo microscope

The use of an optical stereo microscope is an alternative method for the determination of the Weibull parameters related to the number of cracks and averaged distance between the cracks both as a function of the tensile load. Figure 4 shows the setup used.

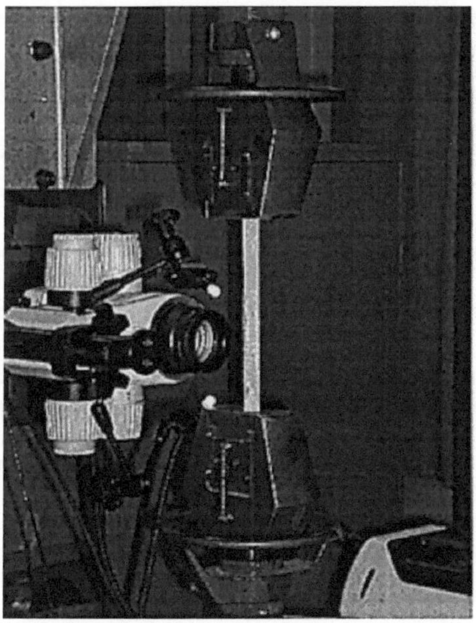

Figure 4 Test setup with stereo microscope.

The stereo-microscope (Leica MZ12.5, total magnification up to 150x) and analog video camera are mountable on the test frame and can be moved in both horizontal and vertical direction. A National Instruments image acquisition board (Imaq-PCI-1411) is used to digitise the images (8 bit, greyscale). With the aid of a special software (Crack-D) written in the Labview version 6 programming environment and using the IMAC Vision toolbox, the cracks are detected. In order to improve the detectability (contrast) of the cracks an ink solution is used after each loading step. Once all crack edges are detected, the picture is calibrated to obtain the distances between the detected cracks and the average crack distance can then be determined as a function of the applied stress.

4. EXPERIMENTS

4.1 Specimens

Two IPC laminates (UD1 & UD2) were fabricated using hand lay-up technique. The IPC matrix was reinforced with quasi-unidirectional E-glass fibres (Syncoglas Roviglas R17/141, 158 g/m^2). We had 141g/m^2 glass fibres in the longitudinal direction and only 17 g/m^2 glass fibres in the transverse direction. The fibre diameter was 14μm. Each bundle contained about 1500 single fibres. The IPC laminates were cured at ambient conditions for 24 hours and afterwards post-cured at 60°C for 24 hours. The average thickness of the laminates was about 2mm. From these plates, specimens with a length of 250mm and a width of 18mm were cut. The obtained fibre volume fractions of the plates are listed in table 1.

Table 1. Plate properties.

plate	*reinforcement*	*average* V_f *(%)*	*stdev* V_f *(%)*
UD1	unidirectional	10.5	0.3
UD2	unidirectional	10.4	0.6

The properties of the pure IPC matrix are [6]: E-modulus = 18GPa and the average failure strength σ_{mu} = 7.8MPa. Initial values of the Weibull parameters have also been determined on these pure IPC specimens. In case the specimens have a length of 250mm, a width of 18mm and a thickness of 2mm, the matrix reference cracking stress of pure matrix is 8.2MPa. With a fibre volume fraction of about 10% the reference cracking stress of the composite will be 10.5 MPa. The obtained value of the Weibull modulus 'm' is 9.3.

4.2 Experimental results

From each laminate (UD1 and UD2), three specimens are now loaded in tension up to fracture. Figure 5 shows the stress-strain curves, obtained from the specimens of the laminate UD1. The material properties determined from these stress-stain curves are respectively the ultimate stress σ_{cu} and the ultimate strain ε_{cu}. The results are given in table 2.

Table 2 Ultimate composite stress and strain.

Specimen	σ_{cu} *(MPa)*	ε_{cu} *(%)*
A1	122	1.948
A10	112	1.615
A19	104	1.831

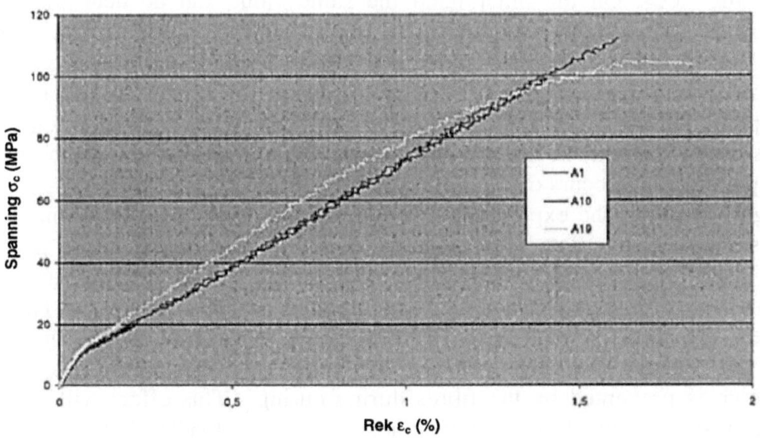

Figure 5. Stress-strain curve specimens plate UD1

According to P. Purnell [10] the matrix-fibre interface shear stress τ_0 can be calculated using equation 10, which was established using the ACK-theory. The final average crack distance was experimentally determined. For both laminate UD1 and UD2 $\langle cs \rangle_f$ =0.93mm. For the laminate UD1 the matrix-fibre interface shear stress τ_0 is then equal to 0.35 MPa.

$$\tau_0 = \frac{1.337 r \sigma_{mu} V_m}{2 V_f \langle cs \rangle_f} \tag{10}$$

The material properties (see table 3) determined using the stochastic theory are the Weibull modulus (m), reference cracking stress (σ_R) and the average matrix-fibre interface shear stress (τ_0). The initial values used for the best fit-procedure between the experimental and theoretical curve were: m = 9.3, σ_R = 10.5MPa and τ_0 = 0.35MPa. The final values are given in table 3.

Table 3. Determination of the laminate properties with the stochastic theory

	τ_0 *(MPa)*	m *(-)*	σ_R *(MPa)*
initial	0.35	9.3	10.5
UD1	0.43	2.6	16.9
UD2	0.52	1.4	14.8

Using equation 6 we are now able to define a theoretical curve for the averaged crack spacing <cs>. On the other hand the same value can be determined as a function of the applied tensile loading using the stereo-microscope and the software Crack-D. Ten specimens from the laminate UD1 and 8 specimens from the laminate UD2 were tested in tension up to 60 MPa. Along the length of the specimen different pictures were taken and the average crack spacing calculated. Each picture represents about 100mm² of the specimen. From figure 6 it can already be noticed that multiple cracking occurs over a large stress interval.

Figure 7 shows the experimental results of the average crack distance of the specimens tested from plate UD1 together with the theoretical curve given by equation 7 and for the values listed in table 3. As may be observed in figure 7 the correlation between the experimental and theoretical results is quite good. However, at relatively low stress levels (up to 10 MPa), the experimental curves show lower values of the average cracking distance. This is probably due to shrinkage of the matrix which is prevented by the fibres during curing. This effect will introduce local internal stresses which will initiate cracks at very low loading levels.

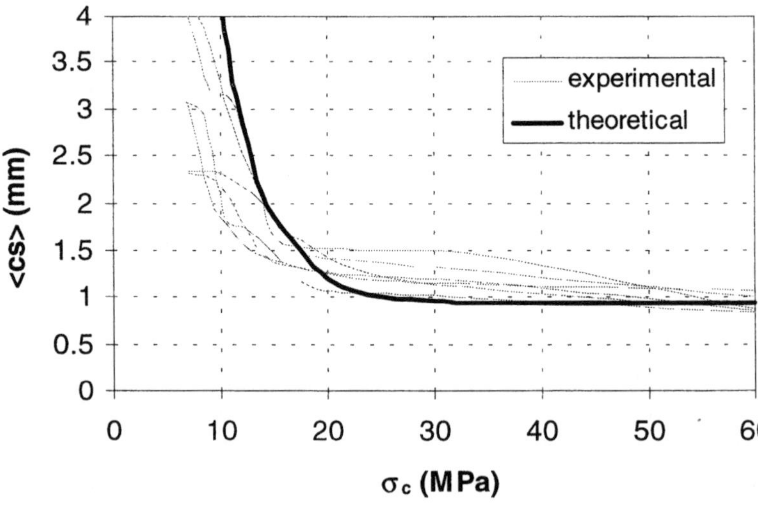

Figure 7. Average crack distance as a function of the applied load of the ten specimens of the plate UD1.

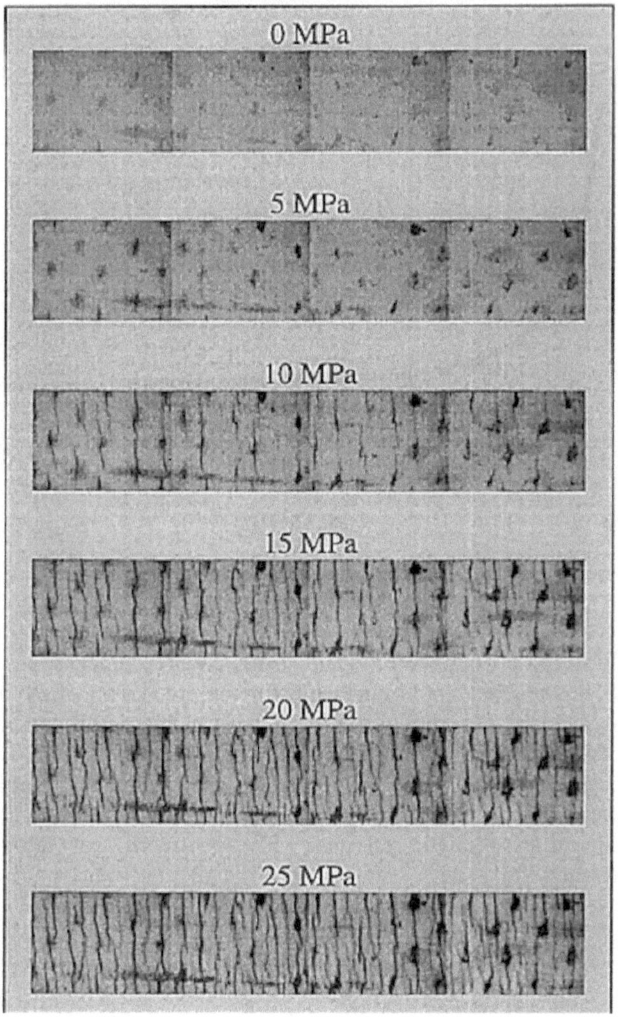

Figure 6. Matrix cracking as a function of the tensile load.

The experimental and theoretical results of plate UD2 are shown in figure 8. The correlation between theory and experiment is even better, also at low stress levels.

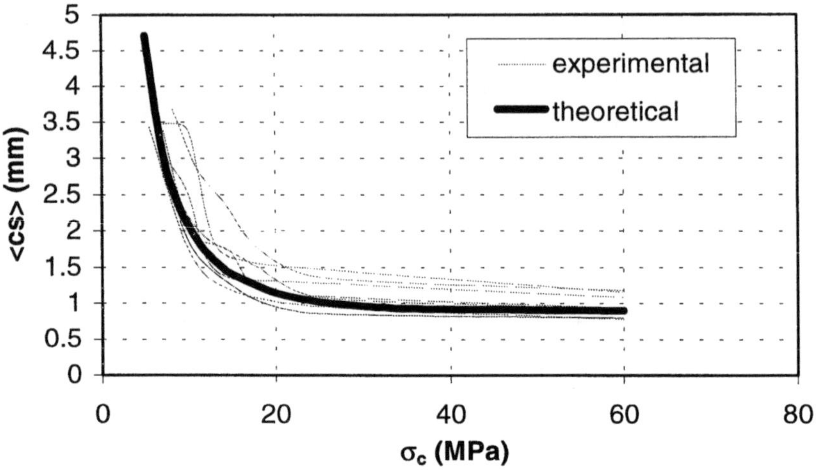

Figure 8. Average crack distance as a function of the applied load of the eight specimens of the plate UD2.

Suppose the initial value of the Weibull modulus and reference stress, obtained on pure IPC matrix, are inserted into equation 6 for the prediction of the average crack spacing. In this case, the theoretical value of the average crack spacing versus the experimental curves is shown in figure 9.
As can be noticed a higher value of m (9.3) concentrates the multiple cracking process too much at one stress level. In case the reference cracking stress is 10.5MPa (initial value, obtained on pure IPC specimens, see table 3) we only observe a shift to the left.

5. CONCLUSIONS

A stochastic cracking theory is used in this paper to describe the stress-strain behaviour of E-glass fibre reinforced IPC composite specimens under monotonic tensile loading. In case the stochastic cracking theory is used, the reference cracking stress (σ_R), the Weibull modulus (m) and the frictional matrix-fibre interface shear stress (τ_0) are difficult to obtain *a priori*.

These parameters can be obtained from "best-fitting" of a theoretical stress-strain curve with an experimental curve under monotonic tensile loading, performed on a specimen of the laminate. This method is a fast and easy method and needs no specialised equipment.

The Weibull parameters have been determined on pure IPC matrix and on IPC composite. It was noticed that the introduction of the fibres has major influence

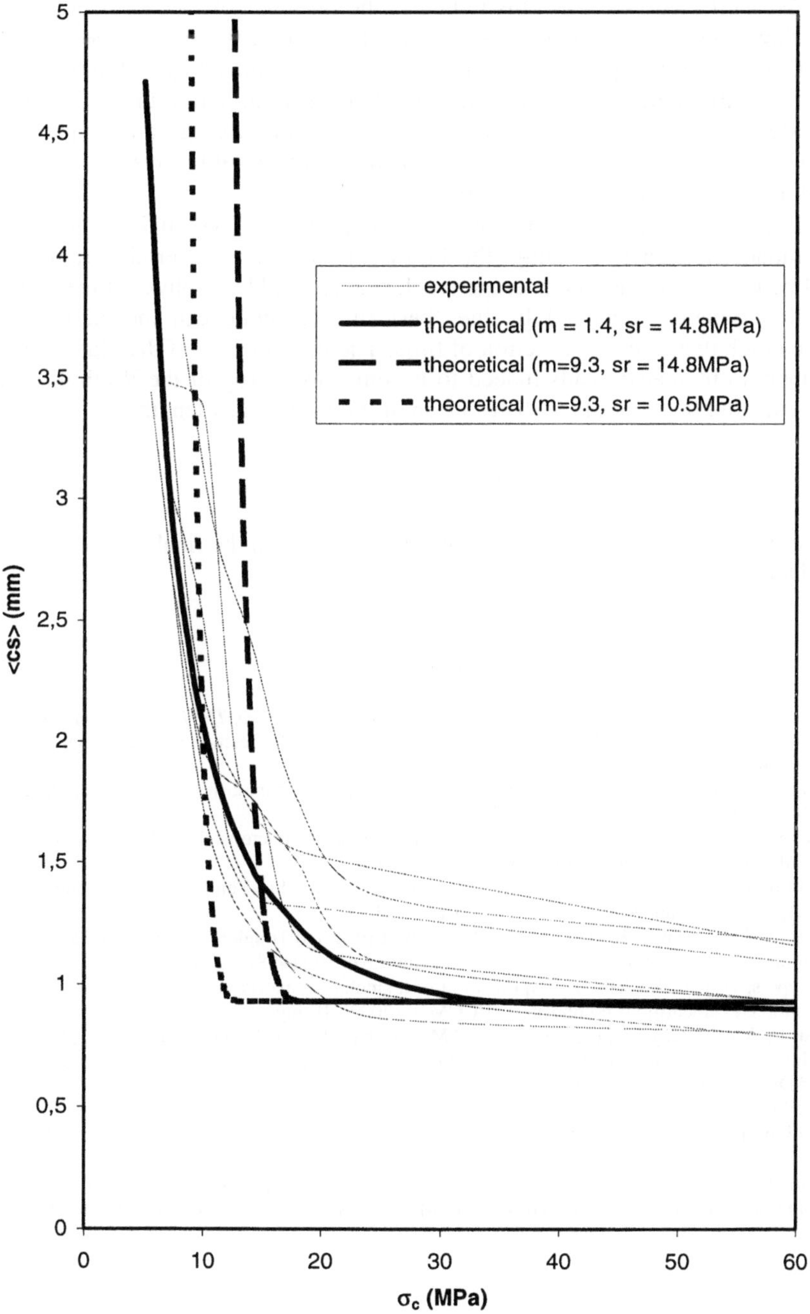

Figure 14. Influence of m on theoretical curve average crack spacing versus composite stress.

on the Weibull modulus (m). This indicates the presence of the fibres broadens the matrix flaw length distribution considerably. This statement needed to be verified.

One goal of this paper is to verify whether the value of σ_R and m, obtained by "best fit" of the experimental and theoretical stress-strain curve under loading, can be inserted into a Weibull model for prediction of the average crack spacing as a function of the external load. The theoretical predicted value of the average crack spacing is compared with observations on specimens.

If the matrix material and interface properties (m, σ_R and τ_0) are obtained under monotonic tensile loading, the knowledge of these material parameters can indeed be used to predict the average crack spacing and have physical meaning. The Weibull parameters obtained by the 'best fit' technique can indeed predict the average crack distance as a function of time. The introduction of the glass fibres into the cementitious matrix leads indeed to a tremendous drop of the Weibull modulus m, representing how broad the distribution of the matrix flaw lengths is.

ACKNOWLEDGEMENT:

Most of the material presented in this paper is taken from the final thesis work of Bram Dooms.

6. REFERENCES

1. Aveston, J., Cooper, G.A., Kelly, A., "*Single and multiple fracture, The Properties of Fibre Composites, Proc. Conf. National Physical Laboratories*", IPC Science & Technology Press Ltd. London (1971), pp.15-24
2. Aveston J., Kelly, A., "Theory of multiple fracture of fibrous composites", J. Mat. Sci., Vol. 8 (1973), pp.411-461
3. Bauweraerts, P., 1998, "Aspects of the Micromechanical Characterisation of Fibre Reinforced Brittle Matrix Composites", Phd. Thesis VUB, (1998).
4. Curtin W.A.,, Stochastic Damage Evolution and Failure in Fibre-Reinforced Composites, Advances in Applied Mechanics; Vol. 36 (1999), pp.163-253
5. Curtin, W.A., Ahn, B.K., Takeda, N., "Modeling Brittle and Tough Stress-strain Behaviour in Unidirectional Ceramic Composites", Acta mater., No. 10 (1998), pp.3409-3420
6. Cuypers, H. and Wastiels, J., "Application of a stochastic matrix cracking theory on E-glass fibre reinforced cementitious composites", ECCM10 (2002), Brugge
7. Gu, J., Wu, X., Cuypers, H., Wastiels, J., "Modeling of the tensile behaviour of an E-glass fibre reinforced phosphate cement", Computer Methods in Composite Materials VI, proceedings CADCOMP 98 (1998), pp.589-598
8. Weibull, W., "A statistical distribution function of wide applicability", ASME J. (1952), pp.293-297
9. Widom, B., Random sequential addition of hard spheres to a volume, J. Chem. Phys., 44, 1966, pp.3888-3894
10. Purnell, P., Buchanan, A.J., et al., Determination of bond strength in glass fibre reinforced cement using petrography and image analysis, J. of Materials Science, 35 (2000), pp.4653-4659

DELAMINATION DETECTION IN LAMINATED COMPOSITES USING LAMB WAVES

C. SOUTIS and S. H. DÍAZ VALDÉS
Department of Aeronautics
Imperial College of Science, Technology, and Medicine
Prince Consort Road
London SW7 2BY, United Kingdom

Abstract. This investigation examines the application of low frequency Lamb waves for the detection of delaminations in thick composite laminates. Surface mounted piezoelectric devices were excited with a tone burst of few cycles generating a stress wave that propagates along the structure. Experiments were carried out on composite beam specimens where wave propagation distances over 2 m were achieved and artificially induced delaminations as small as 1 cm^2 were successfully identified. The resonance spectrum method, which is based on the study of spectra obtained by forced mechanical resonance of samples using sine-sweep excitation, was used for measuring the A_0 Lamb mode phase velocity. Finite element analyses of wave generation and propagation in wide laminated plates are also presented. The feasibility of employing piezoelectric devices for the development of smart structures, where a small and lightweight transducer system design is required, has been demonstrated.

1. INTRODUCTION

The ability to evaluate the integrity of a structure without removing its individual components has become an important technology challenge. Several nondestructive evaluation (NDE) methods exist and are used in composite structures, where visual inspection, radiography, ultrasonics, shearography, and thermography are among the most commonly used. Despite their wide use and improvement in the last decades, the majority of NDE methods are not suitable for implementation into a smart structure. Applications that require a probe to obtain data and scanning of a large area are disregarded as the basis for the development of self-diagnostic systems since they need the direct intervention of humans to perform the inspection. In principle, an in-service health monitoring system would imitate a biological system, where attached or built-in sensors interrogate continuously the structural integrity throughout the component's life. Therefore, techniques that can operate from fixed locations in the structure while inspecting large areas are prime candidates for the development of a structural integrity monitoring system (SIMS). Furthermore, by fixing the transducer, many variables affecting the reliability and repeatability of measurements are removed allowing the precise assessment of minute changes in structural behaviour, which permit the early detection of damage occurrence.

E.E. Gdoutos and Z.P. Marioli-Riga (eds), Recent Advances in Composite Materials, 109-126

An attractive technique for the development of a SIMS is the use of Lamb waves. Their application has long been acknowledged as a potential solution for large-area nondestructive inspection, as they are able to travel relatively long distances allowing the material between transmitter and receiver to be interrogated [1]. Hence, a line-scan is achieved with each pulse rather than the comparatively slower point-scanning performance of conventional ultrasonic techniques. Fundamentally, this method involves the analysis of the transmitted and/or reflected wave after interacting with the test part at boundaries or discontinuities. The presence of damage is identified from changes in the response signal of subsequent tests when compared to the reference response of the undamaged configuration taken earlier in the structure's life.

Lamb waves can be excited and detected by a variety of methods, such as the use of interdigital transducers (IDT's) [2,3], fine point contact transducers [4], air-coupled ultrasonic transducers [5], laser-generation methods [6], and the widely employed angled Perspex wedge [7]. However, among these methods, only the use of IDT's appears suitable for implementation in smart-structures applications, where a small and lightweight, permanently attached transducer system design is required. Still, IDT's present some limitations for the Lamb wave inspection of thick sections that are commonly employed in practical structures. When used in ultrasonic applications, piezoelectric materials are normally operated at its thickness-mode (d_{33}) resonant frequency, which is determined by the thickness of the element and the longitudinal wave velocity in the material. The thickness of piezoelectric elements of practical use varies from few microns to few millimeters; thus the dynamic range of piezoelectric transducers extends from low MHz (0.5) for thick elements to few hundred MHz for very thin films. The lower operational frequency of a transducer therefore imposes an upper limit on the thickness of the plate that can be inspected at values under the cut off frequency-thickness (*fh)* product of high order Lamb modes. For instance, frequencies less than 250 and 100 kHz respectively would be necessary in order to have only the fundamental Lamb modes (A_o and S_o) propagating in typical composite aerospace laminates, which have thickness ranging from 6 to 15 mm [8]. Likewise, if a 12 mm thick steel plate is to be tested, this implies working at frequencies below 136 kHz [9].

This work investigates the generation of Lamb waves for the NDE of composite laminates using surface mounted piezoelectric elements of narrow strip shape, operated in the longitudinal mode (d_{31}). When voltage is applied to a bonded piezoelectric patch, it expands and contracts parallel to the surface inducing a bending moment in the structure. If the voltage applied is a sinusoid of few cycles, it then generates a transient flexural wave whose transmission, propagation, and subsequent reflections at the specimen's boundaries can be analyzed and used to identify the size and location of damage. The advantage of using the longitudinal or radial modes (related to width and length, or diameter) of the piezoelectric element rather than the thickness mode is that the former can be excited at much lower frequencies, which allows the inspection of thicker laminates while keeping the *fh* product low, thus generating only fundamental Lamb modes. This condition is important in the use of Lamb waves for NDE applications since the excitation of a single Lamb mode favors signal interpretation. The study also examines the modal response of narrow beam specimens for wave velocity measurements using continuous wave excitation [9, 10].

2. EXPERIMENTAL PROCEDURE AND MEASUREMENTS

2.1 *Materials and Instrumentation*

Tests were performed on narrow beam specimens made of Aluminium and composite material. The composite beam was obtained from a twenty-four ply $[\pm 45^{\circ}/0^{\circ}/90^{\circ}]_{3s}$ carbon/epoxy laminate of size 660 mm x 570 mm. The laminate was fabricated using T300-924C prepreg tapes. Individual test specimens 629 x 25 mm and 2.7 mm thick were cut from the laminate using a diamond-wheel saw. The elastic properties of the unidirectional ply were $E_{11}=162GPa$, $E_{22}=11GPa$, $\nu_{12}=0.34$, $G_{12}=5.7GPa$, and density $\rho=1536\ Kg/m^3$ [11]. The dimensions of the Aluminium specimen were 814 mm x 16 mm and 3.3 mm thick. The beams were instrumented with two piezoelectric patches, 20 mm x 5 mm, made of commercial brass-backed piezoceramic resonators; used as actuator and sensor respectively and bonded near the beam's end, as shown in Figure 1. National Instruments' LabVIEW® signal processing and an analog to digital card (PCI-MIO-16E-1) were used in conjunction with a personal computer to implement the data transmission/acquisition under an automated framework and to perform the sensor response analyses.

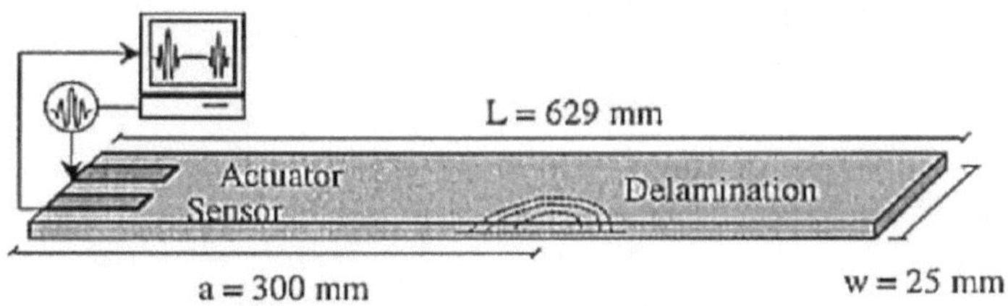

Figure 1. Experimental setup of the composite beam specimen.

2.2 *Phase Velocity Measurement*

The wave velocity is the fundamental characteristic of a Lamb wave, since wave propagation may be analyzed by the variation of its velocity as a function of the frequency-thickness *(fh)* product for each Lamb mode, as shown in Figure 2. These relations (dispersion curves) are found by numerical solution of the Rayleigh-Lamb relation for wave propagation in isotropic plates, described comprehensively by Viktorov [12]. They can also be determined experimentally using the amplitude spectrum method [13] and the phase spectrum method [14].

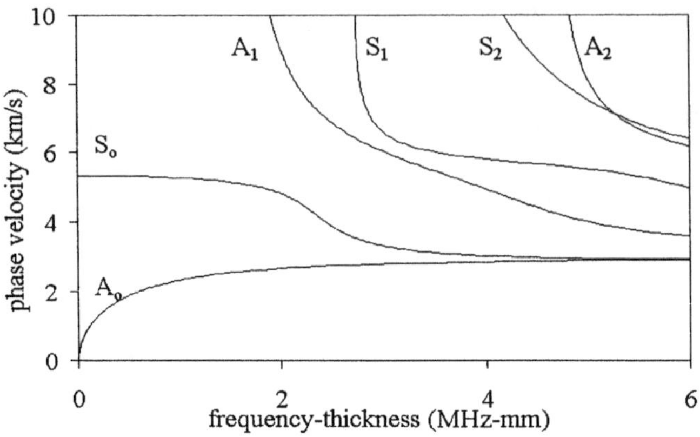

Figure 2. Lamb wave phase velocity dispersion curves for Aluminium.

In this work, the dispersion curve of the A_o Lamb mode in the low *fh* product range was determined experimentally from the mechanical resonant response of the beam specimens [15]. In a narrow beam of length *L*, the particle velocity at resonance consists of a series of standing waves whose frequency correspond to the condition that an integral number of half-wavelengths fits in the sample. Thus resonance exists when $2L=\lambda n$, where n is the harmonic integer and λ is the wavelength. Since the phase velocity is given by

$$c_p = \lambda f \tag{1}$$

then the nth mechanical resonance frequency can be expressed as

$$f_n = \frac{nc_p}{2L} \quad \text{or} \quad \omega_n = \frac{2\pi nc_p}{2L} \tag{2}$$

The resonance spectrum of the Aluminum specimen was obtained using forced mechanical vibration. The actuator was excited with a 10 V sine sweep signal [16], varying from 0.1 to 50 kHz in 0.2 s. The structural response captured with the sensor was sampled at a rate of 0.8 MHz and was then Fourier transformed to obtain the resonant frequencies shown in Figure 3.

The well-defined peaks shown in Figure 3 correspond to the flexural modes of the beam, whereas the smaller peaks appearing above 20 kHz correspond to axial modes. Below 1 kHz, the resonance maxima were only poorly excited so that the modal number of each peak could not be indexed reliably. Therefore, the appropriate value of n was estimated by comparison with theoretical values of the natural frequencies of a uniform beam in transverse flexural vibration, using the expression [17],

$$f_i = \frac{\lambda_i^2}{2\pi L^2}\sqrt{\frac{EI}{m}} \quad ; i = 1,2,... \tag{3}$$

where m is the mass per unit length of the beam, E is the modulus of elasticity, I is the second moment of area, and λ_i is the solution of the characteristic equation for the imposed boundary conditions (free-free).

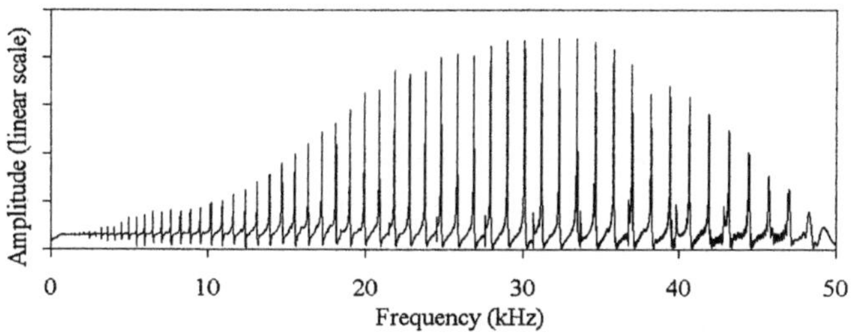

Figure 3. Resonance spectrum of the response of the Aluminium beam to sine sweep excitation.

Figure 4 shows the dispersion curve for the A_o mode in Aluminium obtained using the resonance frequency values from Figure 3, substituted in Eq. (2). Also, the beam theory results derived using Eqs. (2) and (3) are presented in Figure 4 along with the Lamb theory curve for Aluminum, showing excellent correlation with the experimental results.

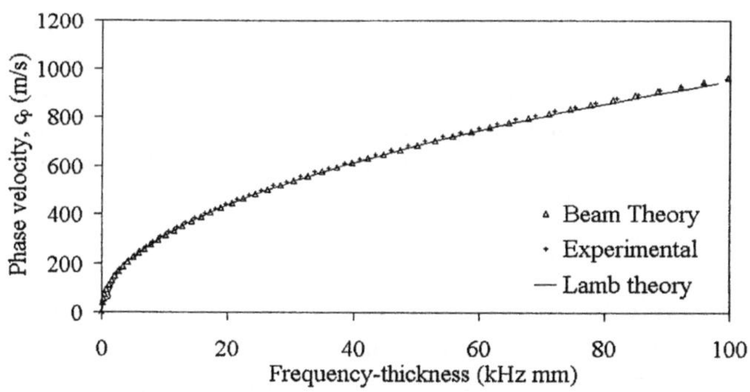

Figure 4. Phase velocity in a 3.3-mm thick Aluminium beam measured using the mechanical resonance method. Beam and Lamb wave theory estimates are also presented.

The phase velocity of the A_o Lamb mode in the composite specimen was also measured using the resonance spectrum method. The sine signal used to excite the actuator was swept from 0.1 to 25 kHz in 0.2 s, and the structural response was sampled at a rate of 0.5 MHz. Figure 5 shows the experimental curve for the laminated beam along with the Lamb theory curve calculated using the average elastic properties of the orthotropic plate [18].

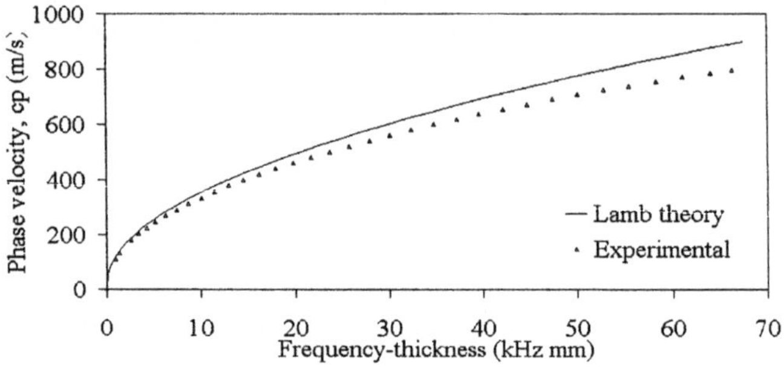

Figure 5. Phase velocity in a 2.7-mm thick $[\pm 45^o/0^o/90^o]_{3s}$ composite beam measured using the mechanical resonance method. Lamb theory estimates are also presented.

The experimental data shown in figure 5 have the characteristic of the dispersion curve of the A_o Lamb mode, although the agreement between Lamb theory and experimental results is not quite as good as for the isotropic case, especially as the *fh* product increases (decreasing wavelength). The difference can be attributed to the fact that transverse shear deformation effects, which are neglected in the formulation of the classical plate theory, are significant in the case of laminated plates due to the relatively low transverse shear modulus [19].

It can be seen from Figure 4 that Lamb wave and Beam theory can be used to successfully predict the dispersion curves of an isotropic material. However, for composite laminates there are no obvious analytical solutions and numerical techniques are required to model such systems. In this study, this has been achieved experimentally by the resonance spectrum method, which is a reliable procedure for measuring low frequency, long wavelength flexural wave phase velocity.

2.3 Delamination Detection

Due to the relatively low interlaminar strength of composite laminates, damage in the form of delaminations can be easily introduced from low-velocity impacts during service, with the subsequent degradation of the mechanical properties of the laminate that can lead to the premature failure of structural components. In practice, composite laminates are designed to tolerate certain degree of damage and it is often only necessary to find relatively large defects such as 10-20 mm diameter delaminations [20]. For instance, the typical size of a critical defect in a composite structure for a Harrier aircraft is approximately 20-25 mm [8].

In order to study the ability of Lamb waves for delamination detection at low *fh* products, the composite beam was examined at different damage scenarios. The beam specimen was excited with short sinusoidal pulses rather than with continuous wave excitation, as it was previously done for phase velocity measurements. The undamaged beam was tested and its response history was captured and kept as baseline. Then a sharp and thin scalpel blade was inserted into the beam's midplane, 300 mm from the left end, as shown in Figure 1. The blade was forced into the material producing initially a small delamination, whose dimension was increased each time the blade was forced into the midplane. In this manner the delamination area was gradually extended from a small incision located at one edge of the beam until it almost reached a full width delamination. This type of artificially induced delamination is thought to better represent damage patterns observed in fatigue loading than beams with a full width delamination of fixed length, as is commonly found in the literature. The damage area, A_d, was measured at every stage by conventional ultrasonic c-scan, as shown in Figure 6.

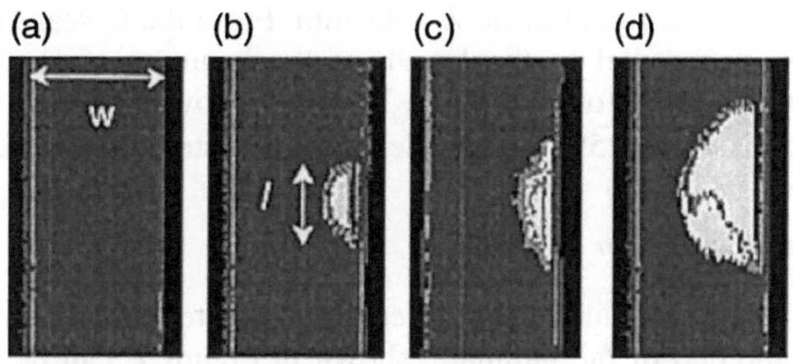

Figure 6. Ultrasonic c-scans of the composite beam: (a) no delamination, (b) $A_d = 22$ mm^2, (c) $A_d = 47$ mm^2, (d) $A_d = 220$ mm^2.

Figure 7(a) shows the response of the undamaged composite beam when the actuator was excited with a 15 kHz sinusoidal pulse of 5.5 cycles modulated by a Hanning window. The response history shows the input pulse followed by another large wavelet, which is the first reflection from the opposite end of the beam. The time delay between the transmitted and reflected signals corresponds to the propagation distance of the wave, which is twice the length of the beam. The second and third reflections were also identified using longer acquisition times, suggesting a propagation range well over 2 m. However, the response signals shown in Figure 7 were limited to capture the first reflection. It can also be observed that the shape of the wave changes as it propagates along the beam due to the dispersive nature of the A_o mode at this *fh* product.

The same test was performed after damage had been induced, thus monitoring the specimen response at different stages of delamination growth. In comparison to Figure 7(a), Figures 7(b) to 7(d) show an extra reflection between the input pulse and the first reflection from the end of the laminate. It can also be observed that the amplitude of the first reflection is affected (reduced) due to the presence of damage. These effects can be better appreciated in Figure 8, which represent the arithmetic difference between Figure 7(a) and each one of Figures 7(a) to 7(d), respectively.

This basic operation allows each time history to be compared to the undamaged configuration, eliminating the large amplitude sections of the response signals and making possible the identification of small changes in the propagating wave, such as variations in the amplitude and phase shifts. This procedure can be easily implemented in a health monitoring system, where damage occurrence would be identified by the appearance of ripples or wavelets in an otherwise straight-line signal representative of the undamaged structure.

The sensitivity of the propagating wave to even the smallest delamination can be explained in terms of the relative dimension between wavelength and defect size. In general, the sensitivity of a given Lamb mode to defects will increase as its wavelength decreases [7]. The wavelength of the excitation signal can be calculated from Eq. (1). The pulse centre frequency used was 15 kHz, and plate thickness of 2.7 mm, which give a frequency-thickness product of 40.5 kHz mm. At this *fh* value the phase velocity of the A_o mode obtained from Figure 5 is close to 640 m/s, from which the wavelength is found to be $\lambda \approx 42$ mm. From the C-scans shown in Figure 6, the *l* dimension (parallel to the length of the beam specimen) of the smallest delamination tested was found to be 10 mm, showing that even though the delamination size is only ≈25% the wavelength, the system is capable of detecting it.

2.4 Delamination Location

The position on the time scale of the reflection generated at the damage site and the reflection from the end of the laminate, shown in Figure 7, can be used to estimate the location of the defect along the beam span. The time difference (Δt_L) between the maximum peak of the input pulse and the maximum peak of the reflected signal from the end of the beam is about 1060 μs, which corresponds to twice the length of the specimen (*2L=1258 mm*). Similarly, the time difference (Δt_{dam}) between the maximum peak of the input pulse and the maximum peak of the reflected signal from the damage site is about 490 μs, which corresponds to a round trip of the wave between the receiver and the delamination (*2a*). Substituting these values in Eq. (4) the location of damage (*a*) is found to be at 290 mm, which is a fair estimate of the actual location (*a* = 300 mm) of the artificially induced delamination shown in Figure 1. This measurement is only approximate since the shape of the wave-packet does not remain the same during its propagation along the beam due to the dispersive nature of the A_o Lamb mode at this *fh* product.

$$a = \frac{L \ \Delta t_{dam}}{\Delta t_L} \tag{4}$$

These experiments demonstrate the potential use of Lamb waves at very low *fh* products for NDE applications, which along with a simple and effective signal processing can be used for the development of an in-service, health monitoring technique capable of detecting delaminations in composite components. However, the specimens used were essentially narrow beam elements where lateral wave spreading effects are minimized. In the following section the wave propagation on wide plates is studied using the FE77 finite element software [21]. Different actuation configurations were examined including the use of a linear array of actuators for the damage evaluation of large surfaces.

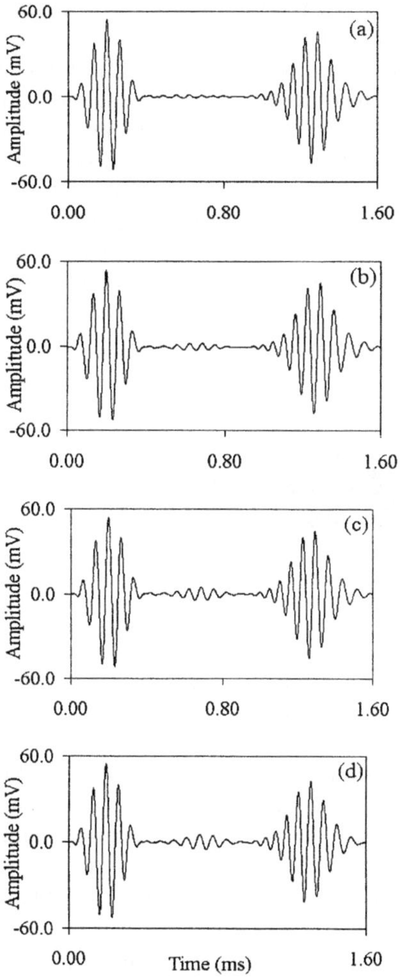

Figure 7. Measured response of the composite beam at different stages of damage: (a) no defects, (b) A_d = 22 mm^2, (c) A_d = 47 mm^2, (d) A_d = 220 mm^2.

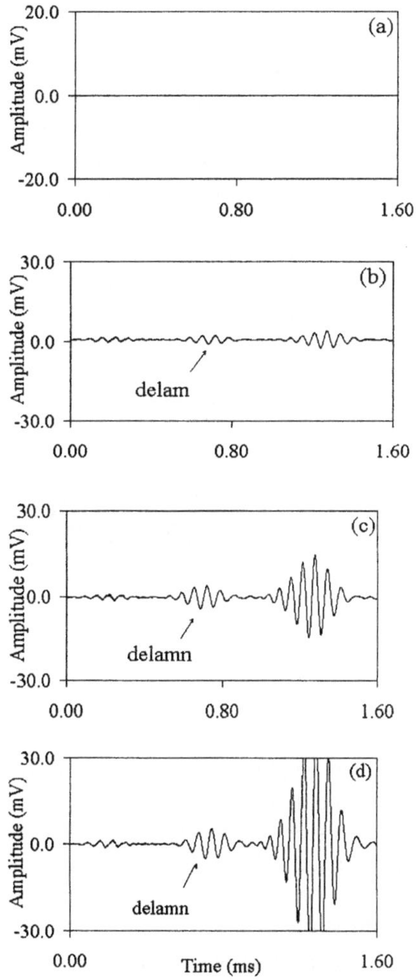

Figure 8. The arithmetic difference between the response of the undamaged specimen and the response at different stages of damage: (a) no defects, (b) A_d=22 mm^2, and (c) A_d=47 mm^2, (d) A_d= 220mm^2.

3. FINITE ELEMENT ANALYSIS

3.1 *Wave Propagation in Plates*

The finite element analysis was carried out in parallel with the experimental approach to study Lamb wave generation and propagation in quasi-isotropic beams and plates. Preliminary models were used to investigate qualitatively the interaction of the elastic wave at boundaries in order to predict the response of a surface bonded sensor. A mesh in the *xy*-plane was implemented to model the laminate using quadrilateral shell elements, where the nodes are defined on the mid-thickness of the shell and each node has both translational and rotational degrees of freedom. The *x*-coordinate was aligned along the length direction and the *z*-coordinate normal to the surface of the plate. The plate thickness is given as a geometrical parameter when the mesh is generated. The excitation of the plate using piezoelectric patches was modeled with uniformly distributed moments of opposite sign applied along two short parallel nodal lines (actuator width), separated by a distance equivalent to the length of the piezoelectric actuator, as indicated with bold lines in Figure 9.

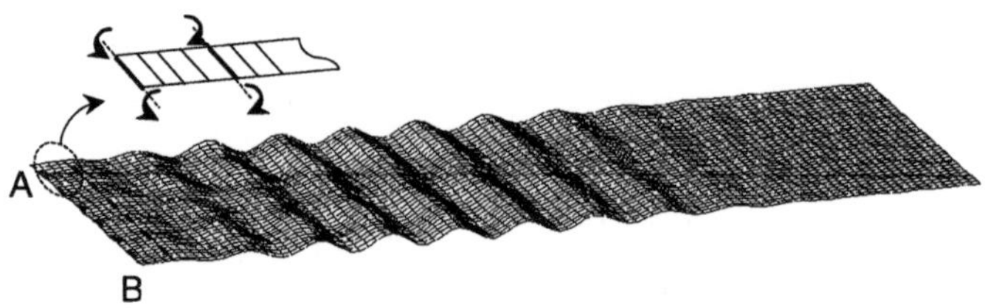

Figure 9. The predicted response of a 2 mm thick laminate to a sinusoidal load applied at the locations indicated by bold lines along the AB boundary.

The excitation signal employed in all of the simulations was a 5.5 cycle, 20 kHz sinusoidal wave modulated by a Hanning window. It was defined in the code as a time history variation of the amplitude of the applied load. Then an explicit central difference scheme was employed in the *MARCH* module of the *FE77*, which carried out a step-by-step time marching integration to solve the wave propagation simulation.

Figure 9 shows a snapshot of the predicted response of the specimen at a certain time step after being excited by the described load case. It shows the propagation of the A_o mode across the surface of a 500 mm x 200 mm plate excited with a linear array of actuators distributed along its left edge (AB boundary). This arrangement produces a fairly uniform wave front across the width of the plate, which is reflected by the right edge with minor lateral wave spreading. Such response favors the interpretation of the signal produced by a bonded sensor, since the input pulse and subsequent reflections are similar to those observed in the narrow beam and can be easily identified in the signal time history, as figure 10 shows. It presents two well-defined sinusoids and a fairly flat signal between them, similar to that observed for the narrow beam element, Figure 7(a).

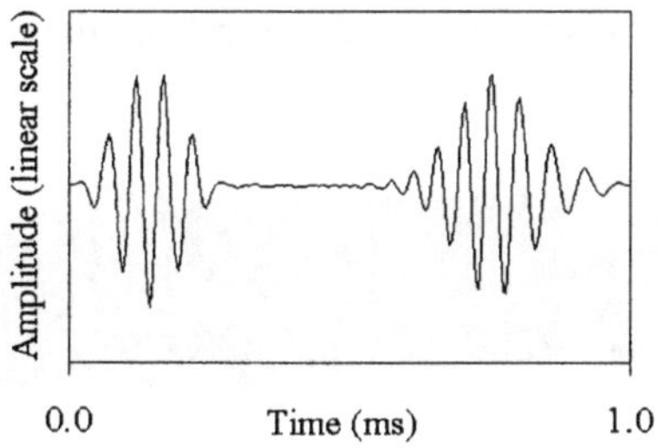

Figure 10. The predicted displacement history of a node located at the centre of the AB boundary. The plate was excited with a sinusoidal load applied the AB edge.

3.2 Damage Detection

The advantages of using a linear array of actuators for the inspection of large areas can be greatly enhanced if the actuators comprising the linear array of transducers were also used as sensors in a pitch-catch mode, thus exploiting the dual capabilities of piezoelectric materials as receiver and transmitter. This approach has been examined using a uniformly distributed sinusoidal load applied along the left edge of the plate model, the AB boundary on Figure 11, while monitoring the in-plane nodal displacement along that same edge. Both, undamaged and damaged configurations were simulated and their response compared to identify the effects of damage on the model behaviour. Damage was simulated by degrading the material properties of a selected number of elements within the mesh. This reduction of the mechanical properties varied from zero-stiffness to 50% of the elastic properties of the undamaged material, thus representing the cases of an open-hole and damage due to low-velocity impact, respectively. Also two different sizes of damage were studied, a square area of 10 mm x 10 mm and a rectangle of 20 mm x 10 mm, located at the center of the plate, as illustrated in Figure 11.

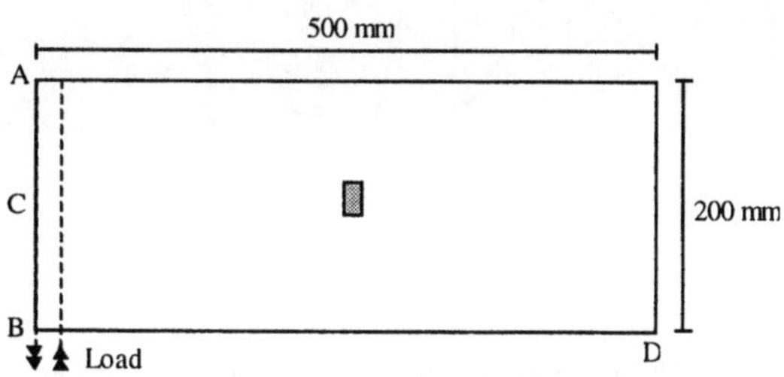

Figure 11. Sketch of the finite element model used in the calculations, showing a central area with reduced material properties.

Figure 12 shows the response of the undamaged plate presented in a 3D plot of amplitude of in-plane displacement, time, and plate width. Similar to the experimental results, the first part of the response shows the input pulse followed by another large wavelet, which is the first reflection from the opposite end of the plate.

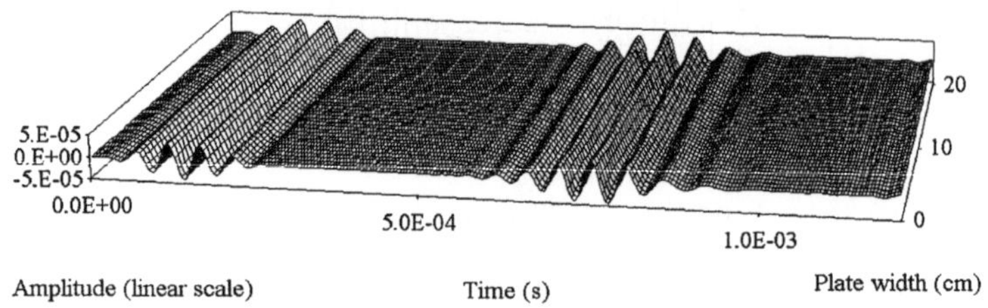

Figure 12. The predicted in-plane displacement history of the nodes located along the AB boundary of the undamaged mesh.

Figure 13(a) presents the response history of the plate containing damage in the form of a square open-hole. This displacement history is at first glance similar to that of the undamaged model, however, evident differences between the two time histories are revealed in Figure 13(b), which represents the arithmetic difference between Figures 12 and 13(a).

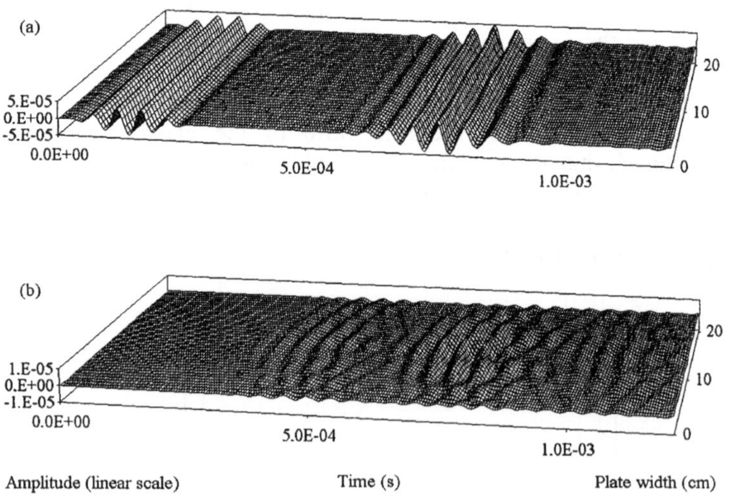

Figure 13. The predicted in-plane displacement history of the nodes located along the AB boundary of the damaged mesh. a) Time history for the 10 mm x 10 mm open hole case, b) arithmetic differece between undamaged and damaged response histories.

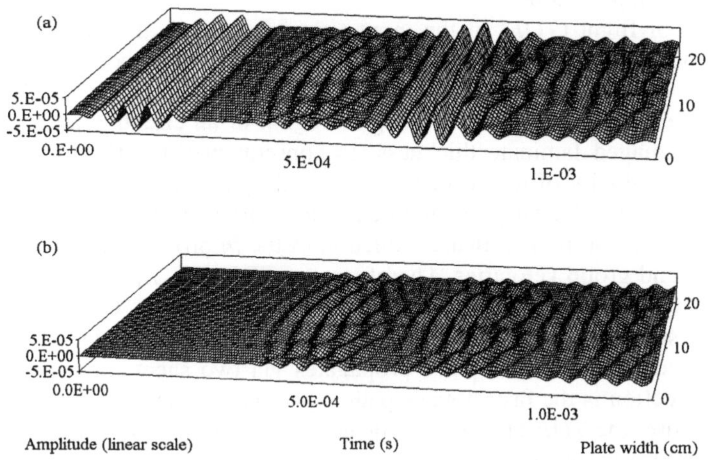

Figure 14. The predicted in-plane displacement history of the nodes located along the AB boundary of the damaged mesh. a) Time history for the 20 mm x 10 mm open hole case b) arithmetic difference between undamaged and damaged response histories.

Likewise, Figure 14(a) shows the response of the plate with the rectangular cut-out, where the effects of damage are clearly shown; Figure 14(b) presents the difference between this damage configuration and the undamaged plate. Qualitatively similar results to those shown in Figures 13 and 14 were obtained in the case where the elastic properties in the damaged area were reduced by 50%, although the amplitude of the wave reflection generated at the damage site is lower than in the open-hole case.

The presence of structural discontinuities can be easily inferred from Figures 13(b) and 14(b). In addition, the location of damage can be estimated from visual inspection of these figures or by correlating the position on the time scale of the wave reflection generated at the damage site with the laminate length, as it was done in the experimental section for the narrow beam. Furthermore, the severity of damage could also be estimated, since the amplitude of the waves appearing in Figures 13(b) and 14(b) is in direct relation with the damage size; further study is needed to establish an explicit relationship. These results strongly suggest that damage could be found in large-area specimens using a permanently bonded linear array of piezoelectric sensor/actuator elements. This concept is currently investigated experimentally.

3.3 Damage Characterization

Defect detection has been demonstrated using experimental measurements and finite element estimates. However, the determination of delamination size and position through the thickness of the laminate is still a challenge in the use of Lamb waves for NDE purposes. This subsection presents a methodology to determine these two parameters from changes in the time of arrival of a tone burst when it travels through a region with delaminations.

The time of flight (*TOF*) of the input pulse, defined as the time it takes to the input pulse to complete a round trip along the length of the specimen, is altered in the presence of damage. In general, the wave velocity changes in the delamination area, since the wave has to travel through a region of an effective thickness smaller than the undamaged laminate thickness. In consequence, the *fh* product decreases and the phase velocity of the wave varies according to the dispersion curves of the material (Figure 2). The shape of the dispersion curve of the A_o Lamb wave in the low *fh* product region is such that a reduction of the *fh* product produces a reduction of the phase and group velocities. Therefore, the *TOF* of an A_o Lamb wave traveling in a specimen with delaminations is greater than that of a wave traveling in the same undamaged specimen. This difference, *ΔTOF,* can be estimated assuming that through the delamination, the wave propagates via two sheets of material and that the thickest section is the fastest wave path, which is the one used for calculating the *TOF*. Therefore, the *TOF* of a wave propagating along an undamaged and damaged specimens are given, respectively, as

$$TOF_u = \frac{2L}{c_g} \quad \text{and} \quad TOF_d = \frac{2(L-l)}{c_g} + \frac{2l}{c'_g} \tag{5}$$

from which the *ΔTOF* is obtained, i.e.,

$$\Delta TOF = 2l\left(\frac{1}{c'_g} - \frac{1}{c_g}\right) \tag{6}$$

where L is the length of the specimen, l is delamination size, and c_g and c'_g are the group velocities in the undamaged and damaged regions, respectively. The inspection of Eq. (6) shows that *ΔTOF* is a function of delamination size (l) and its position through the thickness (d), since c_g is a function of d. Hence, for a given *ΔTOF* there are several combinations of l and d that satisfy Eq. (6).

In order to find the characteristics (l and d) of a given delamination, it is necessary to perform at least two tests of the *ΔTOF* at two different frequencies (different *fh* products), so that a unique combination of l and d would satisfy both *ΔTOF* values measured. To illustrate the method, consider a tone burst excited at 10 kHz and 20 kHz, traveling along a 1000 mm long (L) and 10 mm thick (h) aluminium plate with a 20 mm long (l) and 2 mm deep (d) defect. Using dispersion curves for aluminium, the group velocity of a 20 kHz tone burst traveling through the undamaged (h = 10 mm) and damaged regions (h-d = 8 mm) are found to be 2284 m/s and 2125 m/s, respectively. Substituting these values into Eq. (6) gives a

ΔTOF of 1.310 μs. Likewise, a 10 kHz tone burst travels at 1775 m/s and 1618 m/s through the undamaged and delaminated regions, respectively. Substituting these values into Eq. (6) gives a *ΔTOF* of 2.186 μs.

Figure 15(a) and 15(b) show *ΔTOF* as a function of damage size (*l*) and lines of constant depth (*d*) for a pulse generated at 20 kHz and 10 kHz, respectively. The possible combinations of *l* and *d* for each *ΔTOF* previously calculated are those values where a horizontal line at 1.310 μs and 2.186 μs on Figure 15(a) and 15(b), respectively, intersects the lines of constant depth. These values are listed in Table 1 along with results for similar tests performed at 40 kHz and 30 kHz. It can be observed from this table that the defect size that remains almost constant for all frequencies is that of l = 20 mm in the d= 2 mm depth column (shaded), demonstrating the use of this methodology.

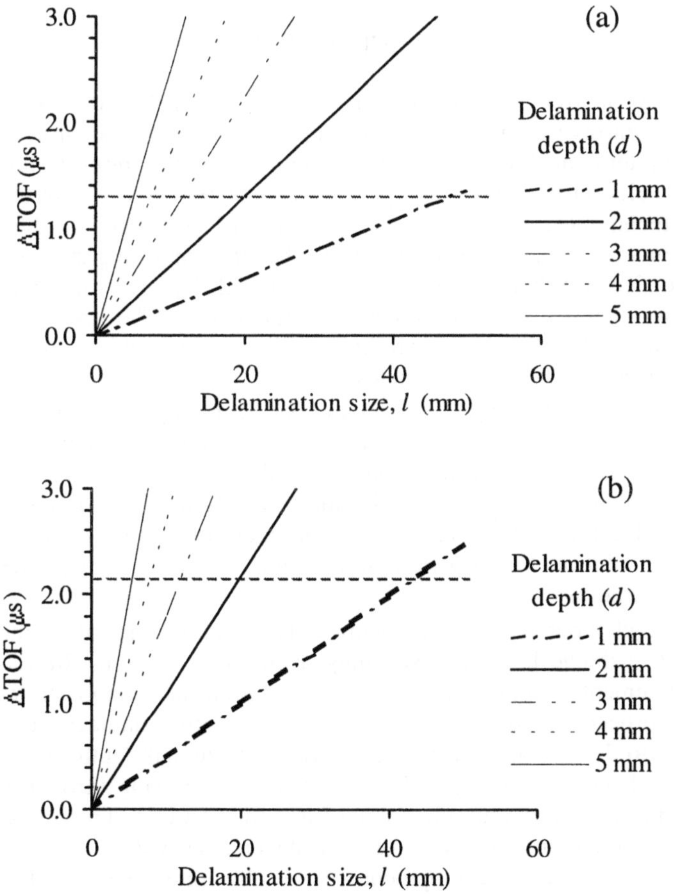

Figure 15. ΔTOF plots calculated for (a) 20 kHz and (b) 10 kHz pulse propagating in a1000 mm long, 10 mm thick aluminum plate.

Table 1. ***Values of defect size and depth for given ΔTOF calculated at different frequencies.***

Defect (mm)						ΔTOF (μs)	f (kHz)
	depth	1.0	2.0	3.0	4.0	5.0		
	Size	44.88	19.99	11.73	7.63	5.04	0.792	40
		44.51	19.99	11.83	7.76	5.32	1.022	30
		47.78	19.99	11.63	7.60	5.22	1.310	20
		44.05	20.06	12.05	8.02	5.59	2.186	10

4. CONCLUDING REMARKS

The generation of low frequency Lamb waves using small and thin piezo-ceramic patches has been investigated for the development of an on-line, structural integrity assessment system capable of detecting delaminations in composite laminates. The study offers solution to some of the problems encountered with current Lamb wave generation techniques, where the use of standard ultrasonic probes turn them unsuitable for the development of a built-in, health monitoring system due to their considerable size and shape. The inspection of CFRP laminates has been performed producing promising results at excitation frequencies in the low ultrasonic range (<100 kHz) where IDT's have certain limitations to operate that restrict their use for the damage inspection of thick laminates.

A simple and effective signal-processing has been employed for the detection of small changes in the structural response, whose minimal computational demand makes it appropriate for real-time continuous damage monitoring. Also, the good sensitivity to delaminations of relatively small size ($1cm^2$), considerable propagation distances (over 2 m), and the very low requirement of electronic hardware, offer a cost-effective solution for structural monitoring in terms of implementation and subsequent operation.

The finite element analysis has shown the potential use of a linear array of transducers to achieve large area scanning from fixed locations. Measuring the in-plane displacement response to a given excitation has demonstrated to be an appropriate procedure to detect, locate, and quantify damage. In addition, a methodology to determine damage characteristics (size and depth) was investigated, which is based on the velocity reduction of a wave passing through regions with defects that produce a measurable delay of its time-of-flight. These analyses can be easily implemented in an array of transducers, providing it with additional capabilities towards the development of a fully integrated, continuous health monitoring system for large composite structures.

Future work will involve the use of arrays of sensors, whose optimum number and spatial distribution will be determined from experimental and finite element analyses, so that a proper inspection can be realized while economizing on hardware and processing complexity. Initially the piezoelectric elements will be individually bonded to the specimen, however the concept has been conceived as to consist of a multi-element array of ultrasonic transducers bonded to a thin flexible strip that can be permanently bonded to new and existing structures. The possibility of embedding these devices into the laminate during the fabrication process will also be explored.

5. ACKNOWLEDGEMENTS

This research was financially supported in part by the overseas research studentship (ORS) awards scheme, U.K., the National Council of Science and Technology (Conacyt), México, and the British Defence Evaluation and Research Agency (DERA), Farnborough. The authors wish to thank Mr. D. Hitchings from Imperial College and Prof. P.T. Curtis from DERA for many helpful discussions.

6. REFERENCES

1. Percival, W.J. and Birt, E.A.: *A study of Lamb wave propagation in carbon-fibre composites*, Insight, 39 (10), pp. 728-735, 1997.
2. Monkhouse, R.S.C., Wilcox, P.D. and Cawley, P.: *Flexible interdigital PVDF lamb wave transducers for the development of smart structures*, Ultrasonics, 35(7), pp. 489-498, 1997.
3. Wilcox, P.D., Cawley, P., and Lowe, M.J.S.: *Acoustic fields from PVDF interdigital transducers*, IEE Proc.-Sci. Meas. Technol., 145(5), pp. 250-259, 1998.
4. Degertekin, F.L. and Khuri-Yakub, B.T.: *Hertzian contact transducers for nondestructive evaluation*, J. Acoust. Soc. Amer., 99(1), pp.299-308, 1996.
5. Farlow, R. and Hayward, G.: *Real-time ultrasonic techniques suitable for implementing non-contact NDT systems employing piezoceramic composite transducers*, Insight, 36(12), pp. 926-935, 1994.
6. Pierce, S.G., Culshaw, B., Philp, W.R., Lecuyer, F., Farlow, R.: *Broadband Lamb wave measurements in aluminium and carbon/glass fibre reinforced composite materials using non-contacting laser generation and detection*, Ultrasonics, 35, pp. 105-114, 1997.
7. Alleyne, D. and Cawley, P.: *Optimization of Lamb wave inspection techniques*, NDT & E International, 25(1), pp. 11-22, 1992.
8. Birt, A.: *Damage detection on carbon-fibre composites using ultrasonic Lamb waves*, Insight, 40(5), pp. 335-339, 1998.
9. Díaz Valdés, S.H. and Soutis, C.: *Application of the rapid frequency sweep technique for delamination detection in composite laminates*, Advanced Composites Letters, 8(1), pp. 19-23, 1999.
10. Díaz Valdés, S.H. and Soutis, C.: *Delamination detection in composite laminates from variations of their modal characteristics*, Journal of Sound and Vibration, 228(1), pp. 1-9, 1999.
11. Mackinley, C.P.: *Compressive failure of CFRP laminates containing pin-loaded holes*, Ph.D. thesis, Imperial College, Department of Aeronautics, 2000.
12. Viktorov, I.A.: *Rayleigh and Lamb waves- Physical Theory and Applications*, Plenum press, New York, 1967.
13. Pialucha, T., Guyott, C.C.H., and Cawley, P.: *An amplitude spectrum method for the measurement of phase velocity*, Ultrasonics, 27, pp. 270-279, 1989.
14. Sachse, W. and Pao, Y.H.: *On determination of phase and group velocities of dispersive waves in solids*, J. Appl. Phys., 49(8), pp. 4320-4327, 1978.
15. Díaz Valdés, S.H. and Soutis, C.: *Real-time nondestructive evaluation of fibre composite laminates using low-frequency Lamb waves*, J. Acoust. Soc. Am., 11(5), pp.2026-2033, 2002.
16. White R.G. and Pinnington R.J: *Practical application of the rapid frequency sweep technique for structural frequency response measurement*. Aeronautical Journal, 86, pp.179-199, 1982.
17. Blevins, R.D.: *Formulas for Natural Frequency and Mode Shape*, Krieger Publishing Co., 1984.
18. Datoo, M.H.: *Mechanics of Fibrous Composites*, Elsevier Science Publisher Ltd., 1991.

19. Tang B., E.G. Henneke, II, and R.C. Stiffler: *Low frequency flexural wave propagation in laminated composite plates*, Proceedings of Acousto-Ultrasonics: Theory and Application, ed. J.C. Duke, Jr., 1988, pp 45-65. Plenum press, New York, N.Y.
20. Guo, N. and Cawley, P.: *Lamb wave reflection for quick nondestructive evaluation of large composite laminates*, Materials evaluation, **52** (3), pp. 404-411, 1994.
21. Hitchings, D.: *Finite Element Package FE77*, Iperial College, Department of Aeronautics, 1997.

PERFORATION INVESTIGATIONS OF ZYLON SHEETS

W. GOLDSMITH and E. PINEDA
Department of Mechanical Engineering
University of California, Berkeley, CA 94720, USA

Abstract. This investigation is concerned with the determination of the ballistic resistance of sheets of Zylon, a polymeric PBO fiber manufactured by Toyobo Company. The current application is intended to ascertain whether this material is suitable for body armor. Ten layers of this substance were stitched together, tightly emplaced in a specially constructed holder and subjected to impact by standard 9 mm NATO bullets. Resistance to perforation is represented by the bulk deformation of the sheets, as well as fiber damage. Thus, the ballistic limit of this fabric combination depended significantly on the impact position relative to the edges of the steel frame. The magnitude of this quantity for impact near the center was found to be of the order of 350 ± 10 m/s.

1. INTRODUCTION

The protection from injury or fatality of warriors since antiquity and of police, civilians, sports participants and both industrial and other types of workers subsequently has an extended history. The role of helmets in avoiding or mitigating head injury is well described by Gurdjian [1]; modern military helmets can not stop a high-speed small caliber projectile, but rather are designed to remove the threat of shrapnel. Until the advent of the crossbow and gunpowder, shields and metal armor were widely used to safeguard all parts of the body, but were not found to be effective past the fifteenth century. The shields of Ajax and Achilles were described in sufficient detail by Homer to permit a continuum-mechanical analysis using finite element methods of the response of these devices to elastic-plastic impact when subjected to large deformations [2, 3]. The results for the former device, that consisted layer of bronze and seven layers of leather, were also verified by corresponding experiments.

The open literature concerning "modern" body armor is very sparse. The earliest citation depicts a 1931photograph of a bullet-proof vest that was, however, not published until 1988 [4]. The use of such devices is widespread among police force and certain portions of the military. A few other references that have almost exclusively dealt with ballistic fabrics or fabric composites have appeared during the last decade [5-9].

On the other hand, a plethora of information has been developed over the last three centuries concerning the penetration and/or perforation characteristics of various types of targets by projectiles and missiles. The initial velocity of these

E.E. Gdoutos and Z.P. Marioli-Riga (eds), Recent Advances in Composite Materials, 127-138

strikers extend from the sub-ballistic domain (≥ 30 m/s) through the ballistic range (450 – 1200 m/s) up into the hypervelocity region (> 2750 m/s) where special projection devices are utilized. Compilations of investigations involving the normal impact of non-rotating projectiles on stationary targets have been published [10, 11]. A recent canvas of other striker/target kinematics and geometries has examined the areas of oblique impact, yaw, yaw combined with obliquity, striker rotation and moving targets [12]. The projectile geometry, attitude and velocity must be considered together with the target properties so as to determine the effectiveness of the barrier as a protective device.

The present report deals with the determination of the ballistic limit (v_{50}) of ten sheets of Zylon that were stitched together. This limit represents the impact velocity where half of the strikers just perforate the target. The fibers of the sheets are manufactured by Toyobo Co. and were stitched together. A limit velocity for this material, to be discussed later, was also found. The strikers utilized are standard 9 mm NATO projectiles. These strikers have a nominal diameter of 9 mm (0.355 in), a length of 15.9 mm (0.625 in), a mass of 124 grains, and are rounded at the tip. They are composed of lead with a coating of copper. This work represents a component of a test matrix that further examines the protective capability of twenty and thirty sheets tied together in the same manner. Although Zylon is also being tested for shielding of other structures, such as for containment of exploding airplane engine fragments, this article is concerned with the testing of this material as a candidate for body armor. The lighter weight of this material might make it a reasonable choice to replace Kevlar, the current widely used material for personnel protection.

2. EXPERIMENTAL ARRANGEMENT

The test arrangement features a square aluminum alloy target holder with outside dimensions of 356 x 356 mm (14 x 14 in) and an interior space of 235 x 235 mm (9.25 x 9.25 in), as shown in Fig. 1. The two similar components sandwiched the fabric and were attached to each other by means of 5 high-strength steel bolts on each side. In turn, this unit was clamped vertically to a triangular steel back stop that was mounted on a heavy (680 kg) steel table. This table also supported the 50 caliber (12.7 mm inside diameter) powder gun which consisted of a 12.93 mm inside diameter barrel with a thickness of 20 mm and a length of 1.6 m, a breech capable of accommodating a 50 caliber cartridge and a steel rail frame serving as support for the unit.

The projectiles were the 9 mm NATO ammunition cited, manufactured either by Winchester or Remington. Due to the difference in inside diameter of gun and strikers, the latter were embedded in a plastic sabot. A sabot stripper attached to the gun muzzle effectively removed the sabot for each shot. A blast shield was placed 254 mm in front of the muzzle to minimize the effect of firing.

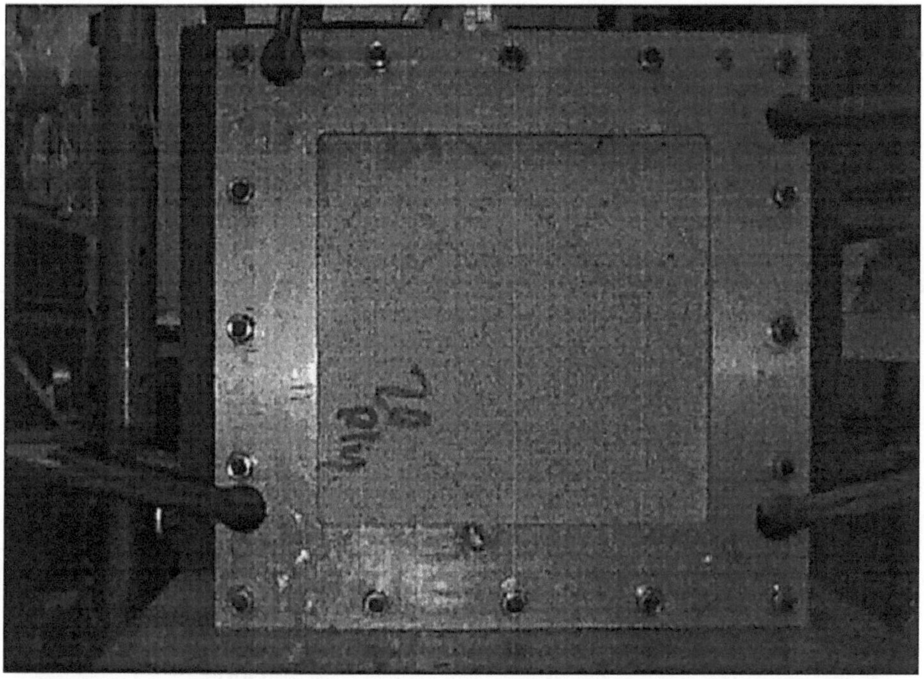

Figure 1. Target holder

The initial velocity of the projectile was measured by means of the signal from the successive interruption of two parallel helium-neon laser beams located orthogonal to the centerline of the gun and spaced 165 mm apart with the proximate beam mounted about 610 mm ahead of the breech. The beams were focused on two photodiodes that sent a voltage pulse upon beam interruption to a Hewlett-Packard 5316 time interval meter. The target holder was positioned about 1.12 m from the muzzle. The final velocity of the fastest object to appear beyond the Zylon target was found either from the completion of a circuit by projectile contact of two successive sets of 432 x 254 mm aluminum foils, attached to wooden bars, with each pair about 7 mm apart with the two set spaced at 155 mm. The signals from these make circuits were directed to another time interval meter. A back-up system consisted of two sets of silver interlocking grids painted on light cardboard spaced 153 mm apart utilized a similar make circuit and recording device as an alternate measurement. Near the end of the test sequence, a digital video recording camera operating at a speed of 10,000 fr/s was utilized to catch a perforating striker and other types of fragment, particularly when the trajectory deviated from the gun centerline. The arrangement is portrayed in Fig. 2.

A catcher box containing rags so as to not pit or deform the projectile was placed in line with the gun centerline to catch the perforating striker and other fragments. A diagram of the experimental arrangement is portrayed in Fig. 2.

3. PROCEDURE

A calibration curve was established for the mass of the as 3031 IMR powder as a function of initial velocity by shooting without the presence of a target. Firing was accomplished by placing a scaled-measured amount of powder into a shell, tamping this down with a piece of tissue and inserting the shell into the breech which was then

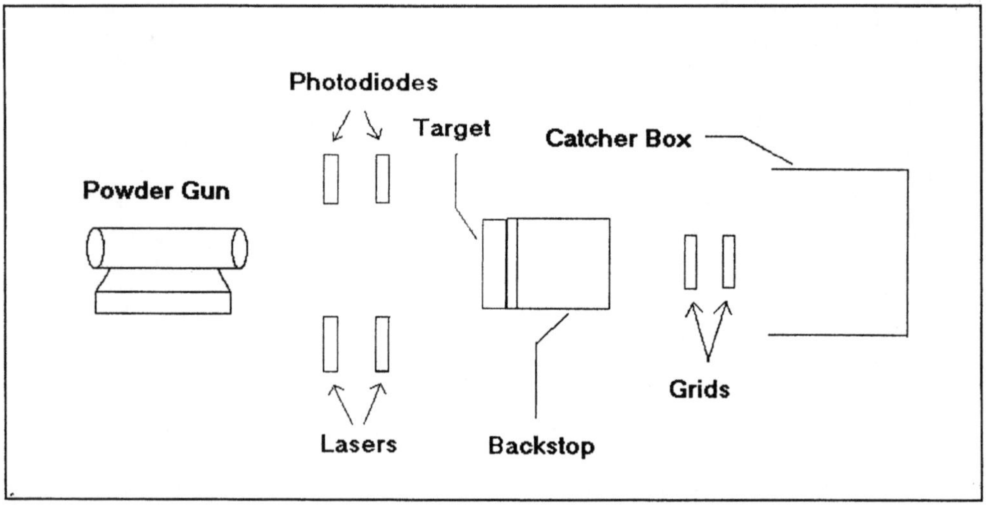

Figure 2. Experimental arrangement

closed and locked. Firing for all tests was accomplished by evacuating the projectile chamber and activating the firing pin electrically from outside. The 305 x 305 mm ten–sheet Zylon targets, which were stitched together in a chain-link manner, were secured in the holder by a tongue and groove system that reduced fabric slippage. Uniform insertion pressure was maintained by use of a 116 N-m (85 in-lb) torque wrench applied to the five bolts on each side. The desired impact position was obtained by moving the holder to a point where the light from a laser centered in the breech was concentrated. Targets were used repeatedly with contact secured so that the damage from previous shots would not overlap the domain of fiber damage of the current test, limiting central shots. Target position and the extent of the damage area were recorded.

Foils and/or grids were emplaced behind the target, and lasers, photodiodes and time-interval meters were activated, and their readings recorded. The sample was examined to ascertain whether perforation, rebound or embedment had occurred. The projectile was recovered and examined for any damage; this was often found and when actually noted, it was attributed to either contact with a metallic surface upon ricochet or the action of the target on the striker. An intact projectile was used repeatedly. The striker trajectories subsequent to perforation were estimated from the position of the holes in the foils or grids.

A static deflection test of the embedded target was performed in an Instron machine. The results for a typical test configuration at three locations: at the center, at 3.5 mm (2.5 in) and 114 mm (4.5 in) along the diagonal from the center are depicted in Fig. 3. The bolts were tightened with a torque of 9.6 N-m (85 in-lb). The resulting curves document clearly that less force is required to produce a given deflection as the loading point approaches the sheet center.

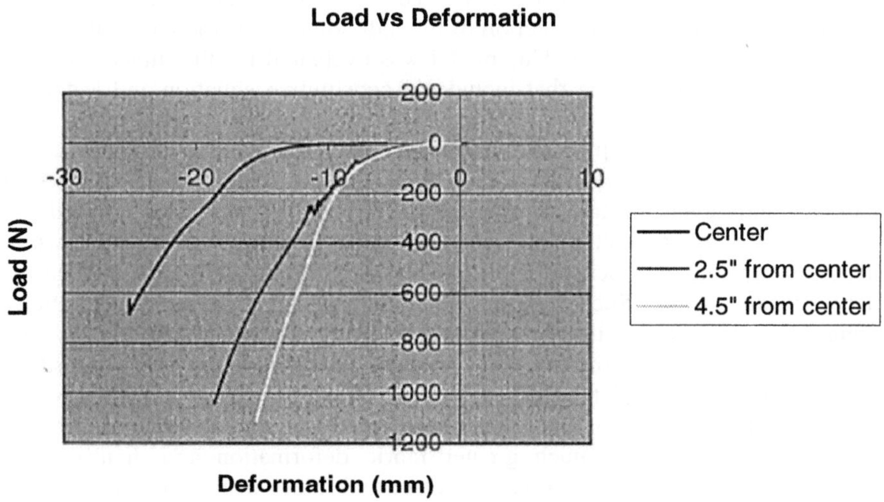

Figure 3. Quasistatic load-deformation tests

The initial velocity for the calibration and the record shots was strongly influenced by the amount of tamping experienced by the powder in the shell. Since this manual process was not precisely controlled, substantial variations of the measured v_o for identical power masses were experienced. The test program was also significantly affected by several experimentally-determined factors. The boundary conditions of the target sheets had a strong influence on fabric performance. Further, the impact location was a major factor in the determination of v_{50} and on the limit velocity, due to the bulk deformation of the target sheets which is maximum for central impact. The pull-out of the target from the holder was measured by noting the greatest deviation of a line painted on the target at its initial boundary.

Modifications of the target holder by use of a nut/bolt combination rather than a screw was executed to minimize this slippage. Higher grade bolts were employed successively to improve this condition.

4. RESULTS AND DISCUSSION

The impact transforms the initial kinetic energy into energy of bulk deformation of the target, permanent damage to the fibers that permits or accompanies penetration or perforation, deformation of the striker and translational and rotational components of the velocity in the case of perforation. A numerical analysis of the Zylon/projectile interaction has been effected in [13] which also takes account of the uncertainty in a mechanical depiction of the individual fiber orientation that strongly affects penetration resistance. This model was evaluated for the impact of which is based on an energy balance that includes a constitutive equation and accounts for slip of the fabric ensemble

The nature of the clamping mechanism for the target had a major influence on the results, as documented by differences in perforation limits for various means of clamping and levels of torque wrenching. The use of an insert as well as the employment of larger bolt torques reduced gross slippage of the sheets, with low energy absorption, and increased fiber and bundle damage with concomitant higher energy consumption. Such damage was manifested by visible discoloration and extension in a nearly circular area around the impact position and fibrous damage along a cross extending from this point orthogonally towards the holder sides.

The penetrability and the ballistic limit of the fabric targets was substantially affected by the location of the impact relative to the edge of the target holder; a central impact produced much greater fabric deformation and, hence, a larger relative energy absorption than when contact took place near a boundary. Thus, for proximate impact near the boundary, a larger fraction of the initial kinetic energy of the striker could be devoted to the penetration--perforation mechanism. This feature is illustrated in Fig. 4.

Fifty-four projectile tests were executed on the Zylon targets, some of which did not provide acceptable results due to loss of or questions concerning data. The relation between final and initial velocity, v_f vs. v_o, for all properly documented shots is depicted in Fig. 5 which is greatly affected by impact location. For example, no perforation occurred for a central shot at a speed of 566 m/s, whereas an impact near the frame corner resulted in perforation at an initial velocity of 349 m/s. Everything else being equal, the central and corner shots can be regarded as

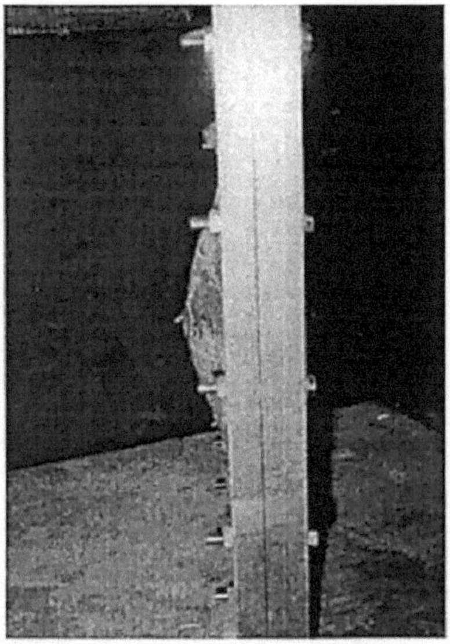

Figure 4. Target bulge after ricochet

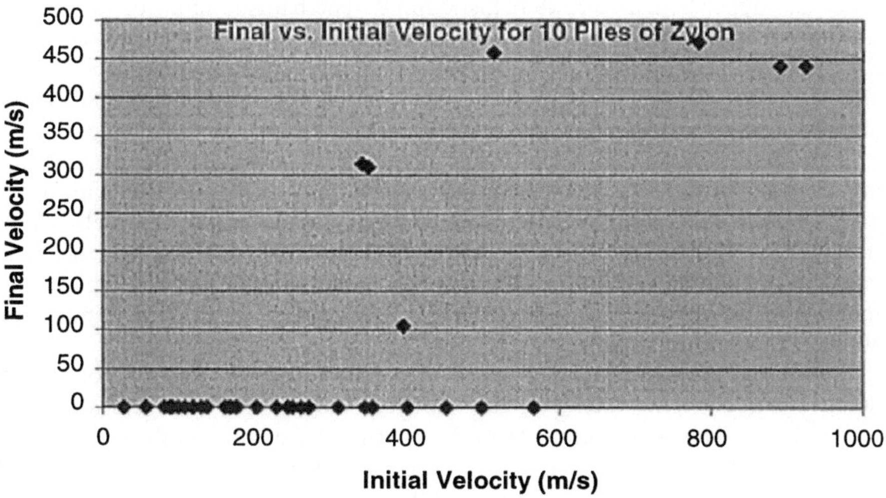

Figure 5. Initial vs. final velocity

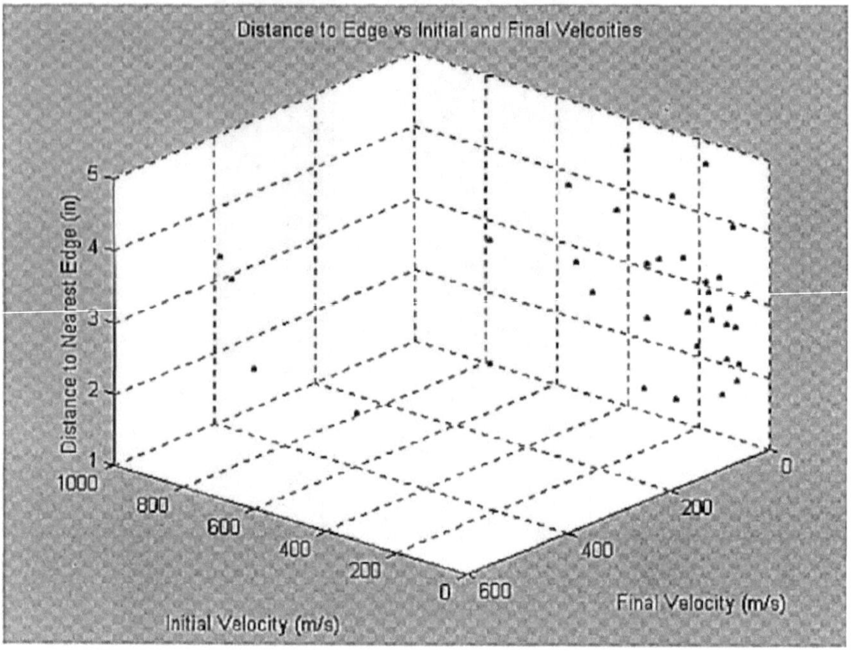

Figure 6. Initial and final velocity as a function of impact position

lower and upper bounds for the ballistic limit. The ballistic limit found from Fig. 5 for this striker-target combination point to the nearest frame edge is presented in Fig. 6. In addition to impact location and clamping tightness, penetrability is also affected by the amount of rotation of the projectiles which was noted in a significant number of cases and would absorb a small fraction of the available energy.

The three-dimensional plot of these parameters including the position of the impact point to the nearest edge for perforation shots is presented in Fig. 6. This diagram also documents that the preponderance of the tests in the velocity range from 100 – 800 m/s located 38 mm to 114 mm from the nearest edge did not produce perforation. However, the range of initial speeds producing piercing was found to be 340-890 m/s. In most instances, to pierce the 10 ply target, initial speeds above 600 m/s were required.

Recovered projectiles fired at initial velocities below 140 m/s did not noticeably deform, but all were found to have rotated and came to rest sideways before rebounding. This was inferred from the imprint of the bullet in the front surface of the target that showed a penetration profile rather than a circular hole. Some of the recovered projectiles fired above v_o = 160 m/s are depicted in Fig. 7a. In the range from 160 – 450 m/s, there *was significant distortion of the shape, increasing with initial* velocity, although perforation did not eventuate. This is seen in Fig. 7a

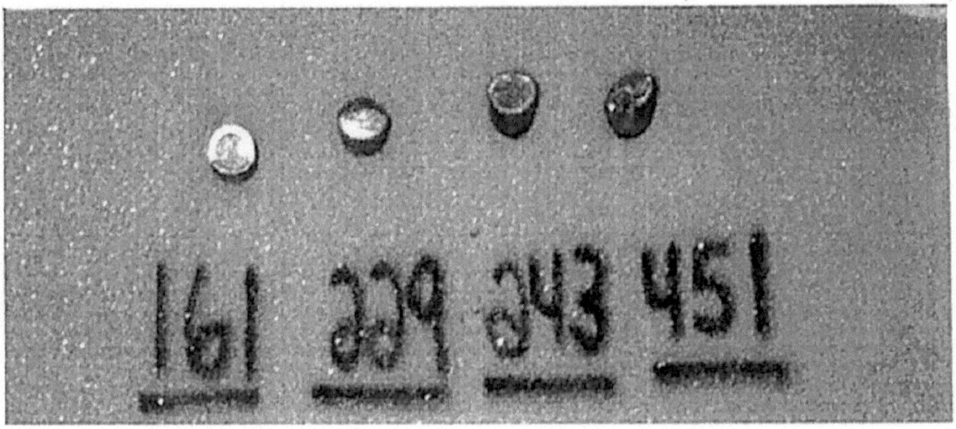

Figure 7a. Ricocheted projectiles fired above 160 m/s (strikers rotated)

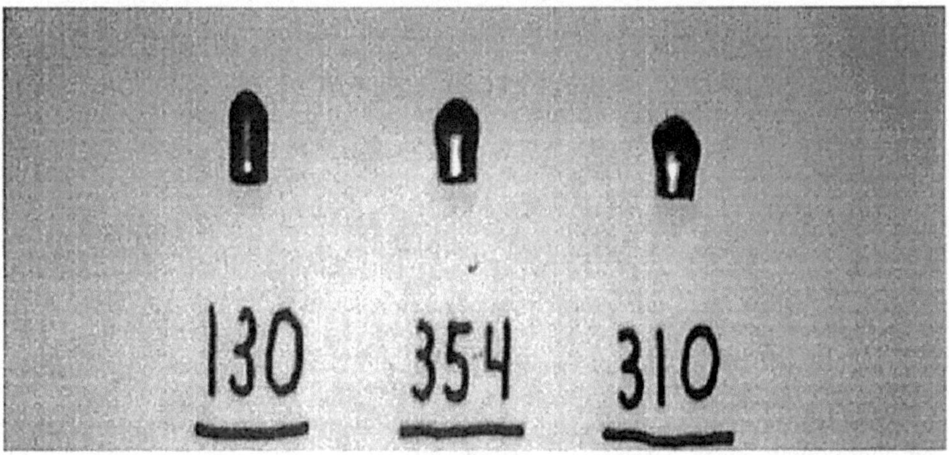

Figure 7b. Rebounding "snub-nosed" strikers

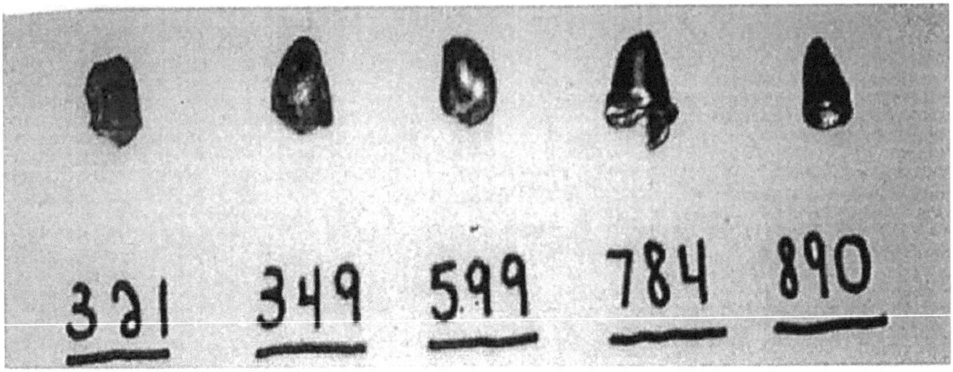

Figure 7c. Recovered perforating projectiles(Initial velocities given in m/s)

where the bullets are shown from behind, parallel to their major axes. The rotation and squishing also helps the target to absorb more energy and partly accounts for the discrepancy in perforation velocities of 320 and 349 m/s, while preventing piercing at 450 m/s. The shapes of these laterally deformed strikers may be compared with that of non-rotating projectiles, shown in Fig. 7b that exhibit a "snub-nosed" appearance expected from without rotation. Figure 7c portrays the shapes from perforating strikers fired at speeds ranging from 320 to 890 m/s that document noticeably smaller deformations with higher velocity, as might be expected a priori.

The ballistic limit v_{50} for the conditions employed in the tests was calculated from the relation

$$v_{50} = v^*(a_o + a_1X + a_2X^2 + a_3X^3 + a_4X^4) \tag{1}$$

where $X = A_fA_p/m_p$, with A_f as the fabric areal density, A_p as the striker base diameter, and m_p as the bullet mass. The values of these terms are $A_f = 1.56$ kg/m^2, $A_p = 254.5\text{x}10^{-6}$ m^2, $m_p = 0.00803$ kg and $v^* = 813.2$ m/s; the values of constants a_i are: $a_o = 0.323$, $a_1 = 6.732$, $a_2 = -26.64$, $a_3 = 68.44$ and $a_4 = -62.45$ [14].

The ballistic limit v_{50} was calculated as 486.8 m/s from Eq. (1). This represents an approximate mean between the already cited minimum perforation velocity of 340 m/s and the maximal initial non-perforation speed of 566 m/s, both of which were central shots. However, the lower piercing velocity occurred after the holder was modified to improve its clamping capability, thus diminishing the slip of the Zylon sheets and increasing the kinetic energy available for producing fiber damage that facilitates the perforation process, rather than bulk motion of the target. While further slip reduction is possible, it can never be eliminated, and represents one of the indeterminate factors, along with impact position and variability in initial striker speed that prevents an unambiguous determination of the ballistic limit. In addition, the need for multiple shots per target contributed an additional uncertainty in the potential of weakened fibers from previous tests that were not clearly visible upon target inspection.

The numerical model was evaluated for the normal impact of a 36 g steel cylinder with a diameter of 12.7 mm and an aspect ratio of three striking clamped Zylon sheets. An initial velocity of 152 m/s required 11 sheets to stop the striker, with an uncertainty of ± 2 sheets. Tests involving 6 sheets of the Zylon indicated a ballistic limit of about 111 m/s. When this is scaled to a v_o = 152 m/s by the ratio of kinetic energies, this gives a value of 11.25 sheets for stopping the projectile, in excellent agreement with the prediction [13]. If this analysis is extended to the present NATO bullets, it must incorporate both the permanent striker deformation and its rotation, although the latter may represent a minimal effect.

A 9 mm handgun has a muzzle velocity of about 388 m/s [15]. Based on the present tests, a barrier of ten plies of Zylon can not guarantee protection for the present method of suspension. However, a slight increase in the number of layers could prove effective, subject to the uncertainties already indicated. Conversely, ten plies of Kevlar 29 exhibit a value of v_o = 454 m/s [16]. Although the ballistic limit of Kevlar 29 is higher than that found for Zylon, the latter has a lower areal density of 1.56 kg/m^2 compared to 2.92 kg/m^2 for Kevlar [16]. Thus, while more layers of Zylon may be required to stop a given striker at a specified velocity, the weight of the Zylon barrier may very well be lower and hence more efficacious.

5. CONCLUSIONS

In the Zylon ballistic test program it was found that:

a) The penetrability of the sheets at a given impact velocity was enormously affected by the type of grip and the impact position relative to the edge.
b) The initial velocity of the bullet depends not only on the mass of powder, but also the nature of tamping.
c) For central impact, the lowest perforation was found to be 340 m/s when gripped tightly, while a non-perforation velocity of 566 m/s occurred with substantial slip in the holder. The lowest perforation speed was 340 m/s. The calculated value of v_{50} was found to be 487 m/s.
d) For non-perforating shots, damage was limited to the immediate area surrounding the impact point. For tests involving perforation, a cross of fibers parallel to the holder edges was visibly damaged, of variable extent.
e) A computer program has been created in a parallel investigation that predicts the number of sheets of Zylon required to prevent perforation for a given projectile mass and initial velocity.
f) While Kevlar 29 sheets as a barrier might yield higher ballistic limit, its much greater weight may make Zylon a lighter body armor.

6. ACKNOWLEDGMENT

The authors would like to express their appreciation to the Army Research Laboratories for providing the materials and a stipend and to Chris Hogue for major assistance in the conduct of the tests. The help of Mr. Dennis Orphal of International Research Associates, Pleasanton, California, is gratefully acknowledged.

7. REFERENCES

1. Gurdjian, E.S.: *Head Injuries from Antiquity to the Present,* C. C. Thomas, Springfield, IL, 1973.
2. Paipetis, S. A. and Kostopoulos, V.: Defensive weapons in Homer, Part I: The shield of Achilles. *Personal Communication.*
3. Paipetis, S. A. and Kostopoulos, V.: Defensive weapons in in Homer: Part II. The shield of Ajax. Personal Communication.
4. Patton, A.: Images of the Past– Who cares about Bullets?, Amer. History 33 (1988).
5. Anderson, K. J.: Ceramic-metal composites -- Bulletproof bulletproof strength, Mech. Engng. 112 (1990) 46-51
6. Roberts, G. K. and Bullian, M.E.: Protective ability of the United States military personal armor system, Ground Troops (PAGT): Fragmentation vest against common small arms small aprojectiles, Milit. Med. 158 (1993) 560-563
7. Bazhenov, S.: Dissipation of energy by bullet-proof aramid fabric, J. Mater. Sci. 32 (1997) 4167-4173
8. Scott, B.: Bullet-resistant vests. Dressed to the 9 millimeters, Sci. Amer. 176 (1997 132
9. Ashley, S.: Aramid-fiber reinforce Nylon, Mech. Engng. 119 (1997) 30
10. Backman, M. E. and Goldsmith, W.: The mechanics of penetration of projectiles into targets, Int. J.Engng. Sci. 16 (1978) 1-99
11. Corbett, G.G., Reid, S.R., and Johnson, W.: Impact loading of plates and shells by free-flying projectiles: A review, Int. J. Impact Engng. 18 (1996) 141-230
12. Goldsmith, W.: Non-ideal projectile impact on targets, Int. J. Impact Engng. 22 (1999) 95-395
13. Zohdi, T. and Goldsmith, W.: Micro-macro uncertainty in ballistc fabric shielding, Submitted.
14. Orphal, D. Personal Communication Koeing, B., Hoffman, S., Nirotaka, N. and Beck,S.: Signal convolution of recorded free field gunshot sounds, J. Audio Engng. 46 (1998) 7-8, 638
15. Cuniff, P. M.: The performance of Poly (Para-phenylenebenzobizoxazole) (PBO) fabric for fragmentation protective body armor, Proc. 18th Symp. Ballistics 2 (1999) 837-844

STRAIN REDISTRIBUTION IN COMPOSITE LAMINATES RESULTING FROM OFF–AXIS PLY CRACKING

D.G. KATERELOS[1], J. PARTHENIOS[1] and C. GALIOTIS[1, 2]
[1]*Institute of Chemical Engineering & High Temperature Chemical Processes, Foundation of Research & Technology-Hellas, PO Box 1414, 265 04 Patras, Greece*
[2]*University of Patras, Materials Science Department, 265* 04 Patras, Greece

Abstract. The initial mode of damage in multidirectional composites is the accumulation of matrix cracks in the off–axis plies. This kind of damage develops until reaching a so-called saturation point. Several other damage modes, such as delamination, may appear concurrently or just upon termination of the matrix cracking process due to joining and growth of these cracks. Remote Laser Raman Spectroscopy (ReRaM) has been employed to monitor the local strains in cracked cross–ply composites. The experiment involves the incorporation of an aramid (Kevlar 49®) fibre Raman strain sensor into the 0° ply and near the 0°/θ° interface of glass–fibre reinforced epoxy resin laminates. This type of laminate is transparent due to the matching of the refractive indices of glass fibres and epoxy resin. Thus, the changes in the longitudinal strain in the 0° ply, caused by transverse cracking in the θ° ply, are quantified.

In this work, the results obtained from $[0/90/0]_t$ and $[0/45/0]_t$ laminates that have been produced under identical conditions, are reported. The strain magnification in the 0° ply caused by 90° or 45° cracking are measured at different levels of loading. A relationship between the values of strain magnification and the distance between the fibre and the 0°/θ° interface has been derived. Finally, important conclusions concerning the mechanisms of strain redistribution as well as the conditions of crack growth and crack–crack interaction are drawn from this work.

1. INTRODUCTION

As is well known, damage tolerance is broadly defined as the attribute of a structure that permits it to retain its required residual strength for a period of usage after the structure has sustained specific levels of damage. In this context, the knowledge of the damage evolution during the life of a structural component that has been designed and fabricated from composite materials is of paramount importance.

The performance of laminated composite materials structures is guided by the thermo-mechanical properties of the reinforced fibres. However, the generally inferior properties of the matrix can play a dominant role in the transferring of stress and strain between the constituents of the structure. Several researchers have focused their effort on modelling the behaviour of glass fibre reinforced epoxy resin systems (GFRP) in a number of pertinent works [1–6]. The use of this particular composite system is due to the high ratio of the longitudinal to the transverse modulus of elasticity, as compared to other materials systems, and the consequent cracking pattern, which develops during quasi-static loading. By matching the

E.E. Gdoutos and Z.P. Marioli-Riga (eds), Recent Advances in Composite Materials, 139-150

refractive indices of fibre and resin, transparent composites can be produced for which crack or, in general, damage development can be easily monitored by optical means. Finally, shear–lag, thermodynamic or energy-based considerations have been applied to predict the stiffness reduction of the material, as well as, the residual strain resulting from damage onset and growth in these composites [5,7].

One of the most important questions arising when dealing with the problem of damage development is the way in which the stress and / or strain is redistributed between plies at the vicinity of a matrix crack. In previous work [1–3] a new method was proposed for the monitoring of the local strain arising in the 0° plies due to 90° ply cracking. The method uses the well-known stress (strain) sensitivity of the axial Raman vibrational modes of crystalline fibres and is a direct consequence of the change in the bond stiffness with bond extension or contraction [8, 9]. Thus aramid fibres can be embedded near the 0/90 interface of a cross–ply laminate to serve as *in-situ* stress (or strain) sensors. The application of aramid fibres as stress and strain sensors in tandem to their reinforcing role has been recently reviewed [10]; the main advantage of this technique is its ability to yield results at a spatial resolution of up to 1 μm, which is not achievable by other sensors.

In the present work the application of laser Raman spectroscopy (LRS) in measuring the local strain arising due to crack development in the off–axis plies is expanded, using two fibre–sensors one at the 0/90 interface and a second at the middle of the 0° ply. This makes possible to extend the range of measurements within 0° ply in cross–ply laminates, as well as, to angle–ply $[0/45/0]_t$ plates with induced edge notching. The presence of notches in addition to a low loading rate, causes a more stable crack propagation, which allow us to determine the strain fields arising in the 0° ply in various phases of damage evolution. Thus important conclusions are extracted for the development of strain fields and the corresponding evolution of damage.

2. EXPERIMENTAL

2.1. Materials Fabrication

The transparent cross-ply laminates containing the Kevlar fibres were manufactured using a modified frame-winding technique. Glass fibres were first wound onto an open steel frame, 300 mm–square, to form the 90° ply. To locate the Kevlar fibres at the 0/90 interface and parallel to the 0° fibres, two thin grooved metal cross-pieces were glued to the frame parallel to the 90° fibres and about 240 mm apart. The grooves allowed the accurate location of the Kevlar fibres at spacing of about 24 mm and parallel to the 0° direction. These fibres were held in place with the aid of silicone rubber, which sets to hold the fibres in position, but allows some flexibility so that the fibres do not break where they are held. The steel frame could now be positioned so that the 0° glass fibres were wound over both the 90° glass fibres and the individual Kevlar fibres. Before winding the 0° fibres, two tows of glass fibres were fixed perpendicular to the Kevlar fibres in the grip areas. These two tows, one each near the top and the bottom of the frame, ensured that the Kevlar fibres did not move out of alignment during the overwinding of the 0° fibres and also helped to ensure that the Kevlar fibres did not drift unduly into the 0° tows during winding and curing.

To make a composite panel, the wound fibres were impregnated with the epoxy resin (Shell "Epikote" 828 cured with nadic methyl anhydride and accelerator K61B in the ratio 100:60:4), which was first mixed and degassed before impregnation. Full wetting of the fibres was achieved by placing the uncured laminate inside a vacuum chamber for about 20 minutes before squeezing out any excess resin and entrapped air bubbles. The laminate was cured between thick glass plates under 100 Kg weight for 3 hours at 100° C, followed by a post-cure at 150° C for 3 hours. The overall dimensions of the panel were 360 mm square, from which coupons 230 mm long by 20 mm wide were cut, each specimen containing one or two Kevlar fibres as shown in Figure 1. The final specimens have a transverse ply thickness of 1.22 mm and 0^0 ply thickness of 0.64 mm.

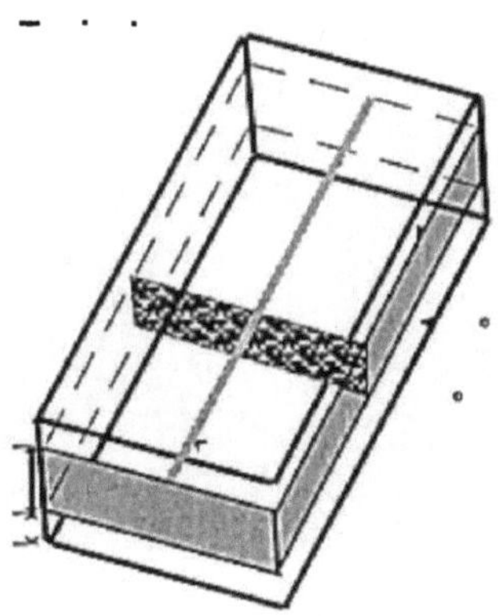

Figure 1 A schematic drawing of the test specimen showing the positions of the Kevlar® 49 fibre – sensors and a typical transverse crack

2.2. Remote Laser Raman Spectroscopy

A novel remote laser Raman microprobe, which has been built and tested in our lab, was used in this work (Figure 2). The main feature of the new probe is the use of flexible fibre for laser delivery and collection, which brings about a complete separation of the spectroscopic and the testing stages. The use of flexible fibre optic cables permits operation of the microprobe in horizontal, vertical and multi-angle positions. Such an arrangement allows the interrogation of specimens of any size and shape under a variety of different environments. In addition, the presence of a miniature CCTV imaging camera on the back of the microprobe allows it to operate simultaneously both as Raman and optical microscope. This arrangement is particularly useful for the experiments described here as it allows measurements on the composite specimens while they are being deformed on the mechanical tester.

In this work the source of excitation was the 514.5 nm line of an Ar^+ laser. The light was focused onto the specimen via a conventional objective (Olympus 50×). The 180° scattered light was collected by the same objective and transmitted through a holographic beam splitter, which attenuated the 514.5 nm line. The inelastically scattered light was guided through another optical fibre to a SPEX 1000M single monochromator and was dispersed onto a Wright Instruments air-cooled CCD (Charge Coupled Device). The dispersed Raman signal was then converted to an electrical signal and stored in a PC.

Figure 2. The MTS hydraulic mechanical tester and the attached Remote Raman System.

The 1611 cm^{-1} band of a Kevlar 49® fibre, which is primarily attributed to the aromatic ring C-C stretching vibrational mode, has been employed in this work for strain measurements. The raw Raman data for this particular band are best fitted with a Gaussian distribution and a quadratic background. The laser power on the Kevlar 49® fibre was maintained at 1.0 mW to avoid laser-induced damage on the fibre.

2.3. Mechanical Testing

The composites were tested in tension following the ASTM D3039 standard procedure. Prior to testing the ends of the tensile coupons were sand blasted and end-tabbed with standard, 2.4 mm thick, glass-reinforced-plastic tabs. Strain gauges of gauge resistance of 120.4 ± 1.0 Ω and of gauge factor of 2.03 were attached to the middle of the gauge section for each coupon. All specimens were strained using a 25 kN hydraulic MTS (Figure 2) mechanical tester at a strain rate of approximately 0.05 microstrain / min. In general, transverse cracks were formed at composite strains higher than 0.2% for the cross–ply laminates and above 0.8% for the [0/45/0] laminates. In all cases measurements were made within a specified length of the coupon (Raman "window") at applied strains required for transverse crack formation. To avoid unstable crack growth during Raman data acquisition all measurements were made by unloading the specimens to 0.2% applied strain.

3. RESULTS/ DISCUSSION

3.1. Local Strain Arising in the 0° Ply

The strain arising in the 0° ply as a function of distance along a Raman window of varying length for the case of a single embedded fibre–sensor and for increasing levels of applied strain, has been determined previously [1-3, 10]. Figure 3 presents a similar diagram showing the arising strain measured using two fibre–sensors. As mentioned earlier the first fibre has been located near the 0/90 interface while the position of the second is at the middle of the 0° ply. The specimen has been loaded to various levels of strain; to prevent the growth of additional cracks during the Raman observations, the specimen was relaxed to 0.2% applied strain and the measurements shown were then taken. The results show clear strain magnifications on the crack plane of approximately 4 to 8; the variability of the maximum values of strain magnification is mainly due to sensor location vis-à-vis the crack front both in the through–thickness and along the width directions. In Table 1 below, the applied strain levels and the number of the forming cracks as well as their distance from the sensors are shown.

Table 1. Applied strain levels and corresponding crack formation. (The number in brackets indicates the measured distance from the corresponding fibre sensor)

Loading Level	*No of Cracks*	*Distance from fibre 1 (μm)*	*Distance from fibre 2 (μm)*
0.54%	3	0	0
0.6%	6	0	571.25 μm (4)
0.61%	7	0	350 μm (4)
0.64%	9	0	70 μm (4)
0.65%	11	0	0
0.64%	13	0	0
0.68%	15	0	0
0.72%	17	0	0
0.74%	18	0	0

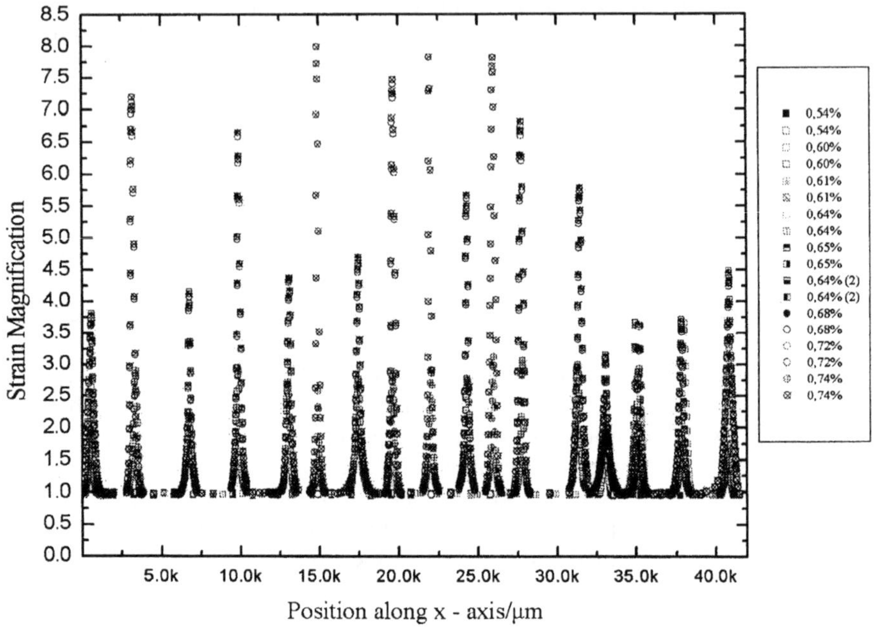

Figure 3. Strain Magnification Factor as a function of distance along the fibre sensors for increasing levels of applied strain.

3.2. Distance Dependence of the Strain Magnification Factor (SM) for the [0/45/0] Laminate

The importance of relative position between the fibre–sensor and the crack front has been observed in the cross–ply laminates became evident in the case of a [0/45/0] laminate [1]. The asymmetry in the form of the strain pattern of the cracked laminate, which was beyond a possible experimental error led to investigation of the behaviour of a single crack as it develops during increasing loading. A [0/45/0] specimen was tested under the same conditions as previously while the measurements were taken within a Raman window. The size of the Raman window is tailored to the size of the trace of the crack on the fibre – sensor. The specimen was loaded at three levels of applied strain 0.9%, 1.1% and 1.25%. The distance between the fibre-sensor and the crack front was 4981 μm, 2853 μm and 0 (the plane of the crack crossing the fibre sensor).

In Figure 4 below the behaviour of this crack is shown. As can be seen the form of strain pattern in all cases differs both in magnitude and in shape. The SM varies between 1.48 and 2.13.

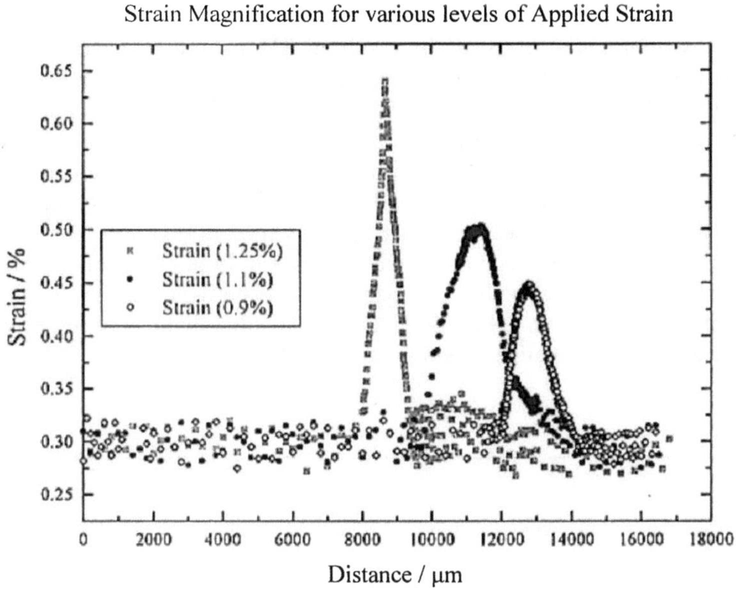

Figure 4.Single Crack behaviour in a [0/45/0] specimen

The observation mentioned above can be explained by the term of a "projection" of the crack front onto the fibre–sensor. Figure 5 below shows a schematic representation of this assumption.

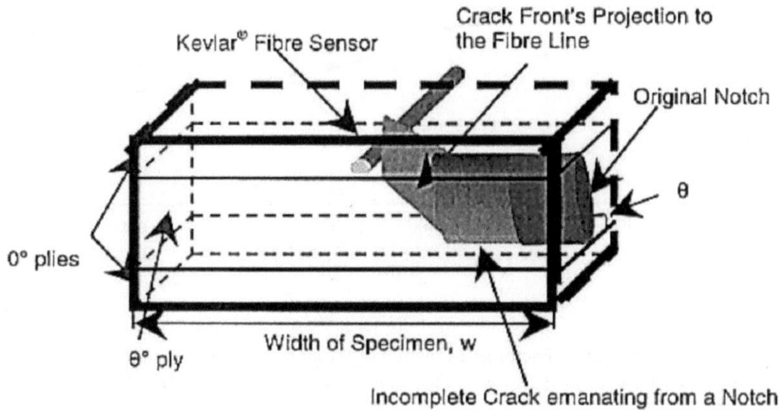

Figure 5. The "projection issue". A schematic drawing (not to scale) showing a characteristic volume element containing a crack emanating from a notch and its projection to the fibre – sensor.

3.3. Distance Dependence of the Strain Magnification Factor (SM) for the Cross-Ply Laminate

The strain distributions as well as the resulting SM patterns were derived for the specimens that contained two fibre sensors by employing the methodology employed previously [1-3]. The distance between the fibre sensors and the crack front were measured with an optical microscope. The results were also verified by cutting the specimen at the crack positions and by measuring again the distance between the edge of the crack and the fibre sensor. The results of the SM as a function of the through–thickness distance are presented in Figure 6. The data points have been fitted with an exponential decay function in broad agreement with earlier analytical treatments [11]. Finally, the magnitude of the factor at the vicinity of the crack, which is about 8 is in good agreement to the value that has been derived both analytically and by strain gauge experimentation [11].

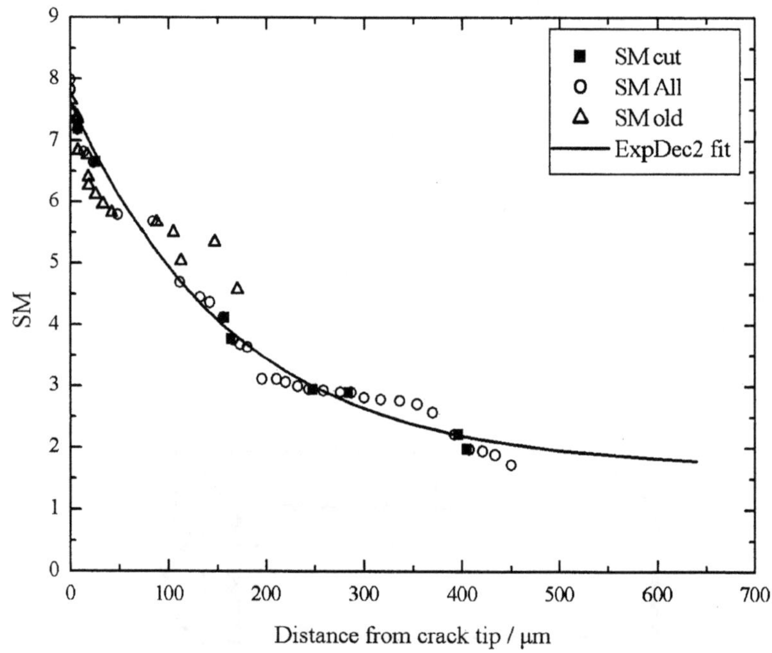

Figure 6. Strain Magnification Factor as a function of through – thickness distance between the crack front and the aramid fibre–sensors. Results measured in two specimens an verified by optical microscopy

The experiments showed a dependence of SM upon both the through–thickness and the width dimensions. An attempt has been made to correlate them into a 3-D diagram that is shown in Figure 7. The experimental results are fitted with a negative exponential smoothing surface. The selection of this fitting function was based on the trend observed in the 2-D diagrams. The *y* and *z*–directions are normalised to the width of the specimen and the thickness of the ply, respectively. The symmetric shape of the curve obtained by the normalization is changed to the characteristic lobe if the absolute values are used.

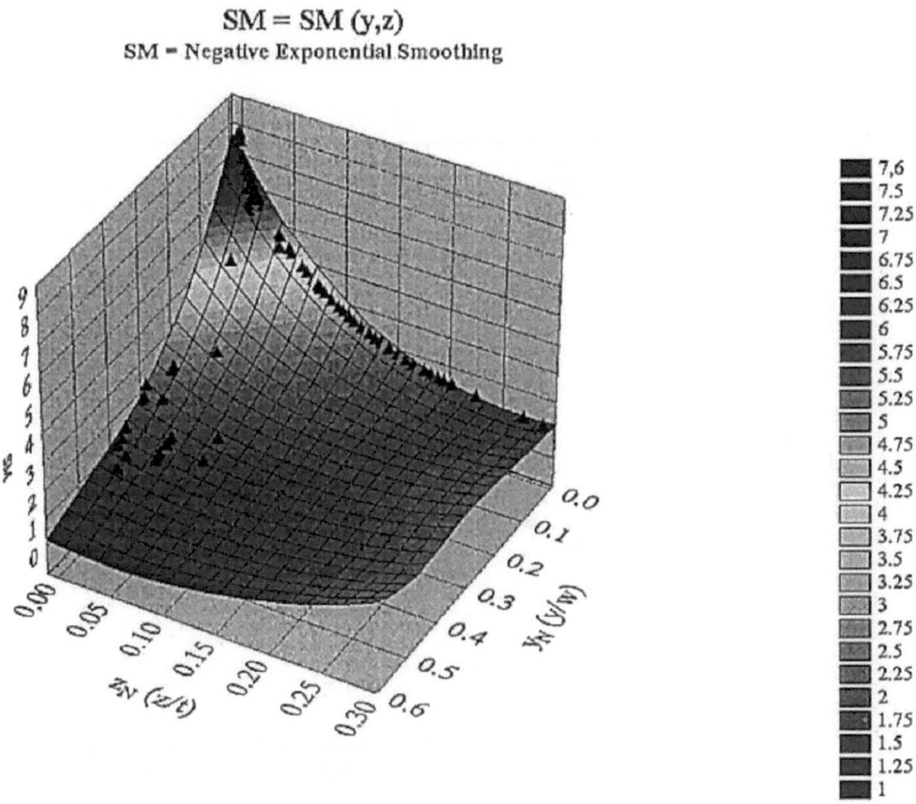

Figure 7. 3 – Dimensional representation of the distance dependence of the Strain Magnification Factor.

3.4. Saturation Point

With increasing strain, more transverse cracks grew until a reasonably high crack density was achieved. This point has been labelled as "saturation point" and is a step just before the formation of other modes of damage such as delamination. An explanation of this behaviour has been attempted by the use of an energy–based term, namely the dissipated energy during the loading and unloading paths. In Figure 8, the accumulation of the dissipated energy with increasing loading levels is shown. The amount of energy dissipated during each loading and unloading cycle follows an increasing exponential law and leads to a constant, which characterises the so–called “saturation point”. The dissipation of energy is mainly due to formation of cracks or the growth of existing ones. Consequently, as the accumulation of this energy decreases fewer cracks develop. This process leads to a point where no crack formation occurs. From that point onwards, it is reasonable to assume that the excess energy will be directed to the creation of other modes of damage such as delamination until the final fracture of the specimen (0° ply failure).

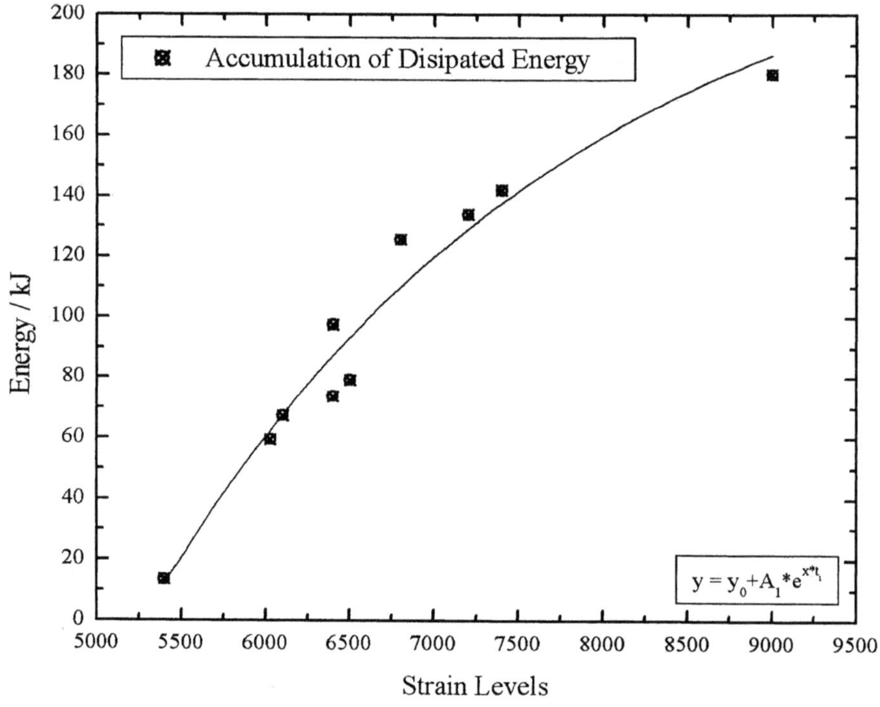

Figure 8. The accumulation of dissipated energy during each loading – unloading cycle

4. CONCLUSIONS

The strain redistribution in the 0° ply resulting from crack development in the off–axis ply has been examined for both cross–ply and angle–ply laminates. Similar to procedures established earlier, the pattern of SM was monitored with increasing applied strain levels. The variation of both magnitude and shape of this pattern let to the examination of its dependence with the distance in both *y* and *z*–directions. A 3–d representation of the dependence showed that a negative exponential function fits the experimental results as postulated by existing analytical treatments. The non-symmetric shape of the strain magnification pattern, in the case of angle–ply laminate has been attributed to geometry of the advancing crack vis-à-vis the 0^0 direction. In future work, the effect of a non–fully developed crack through the thickness of the off- axis ply and the role that it plays to stress and strain redistribution will be examined.

5. ACKNOWLEDGEMENT

The authors wish to acknowledge Dr. S. L. Ogin and Mr. R.D. Whattingham of the University of Surrey, School of Engineering, Department of Materials Science and Engineering, for the invaluable contribution in preparation of the specimens, as well as, for their significant remarks during this work. Mr. C. Koimtzoglou of ICEHT-FORTH is thanked for assisting with the MTS measurements.

6. REFERENCES

1. Katerelos D.G., Galiotis C., Ogin S.L., Whattingham R.D.: Determination of Stiffness Reduction and Residual Strain Caused by Transverse Cracking using an Embedded Aramid Fibre Strain Sensor, In *ECCM 9. Composites from Fundamentals to Exploitation,* Proc. of the conference in CD - ROM, 2000
2. Arjyal B.P., Galiotis C., Ogin S.L., Whattingham R.D.: Monitoring Local Strains in Cracked Cross-Ply Composites using an Embedded Aramid Fibre Strain Sensor, *J of Materials Science* **33** (1998), 2745 – 2750
3. Arjyal B.P., Galiotis C., Ogin S.L. and Whattingham R.D., Residual strain and Young's modulus determination in cross-ply composites using an embedded aramid fibre strain sensor, *Composites part A* **29A** (1998), 1363 – 1369
4. Garrett K.W. and Bailey J.E.: Multiple transverse fracture in 90 cross-ply laminates of a glass fibre-reinforced polyester, *J of Materials Science* **12** (1977), 157–168.
5. Highsmith A.L. and Reifsnider K.L.: Stiffness reduction mechanisms in composites. In *Damage in Composite Materials*, AST STP 775, ed. K. L. Reifsnider. American Society for Testing and Materials, Philadelphia, (1982), 103–117.
6. Parvizi A., Garrett K.W. and Bailey J. E.: Constrained cracking in glass fibre-reinforced epoxy cross-ply laminates, *J of Materials Science* **13** (1978), 195-201
7. Talreja R.: Stiffness Properties of Composite Laminates with Matrix Cracking and Interior Delamination, *Engineering Fracture Mechanics* **25** (1986) 751 – 762
8. Schadler L. and Galiotis C.: A Review of the Fundamentals and Applications of LRS Microprobe Strain Measurements, *International Materials Reviews* **40** (1995), 116-134
9. Galiotis C.: Interfacial Studies on Model Composites Using Laser Raman Spectroscopy, *Composites Science and Technology* **42** (1991), 125-150.
10. Parthenios J., Katerelos D.G., Psarras G.C. and Galiotis C.: Aramid fibers; a multifunctional sensor for monitoring stress/strain fields and damage development in composite materials, *Engineering Fracture Mechanics* **69** (2002), 1067 – 1087
11. Tan S.C.: *Stress Concentrations in Laminated Composites,* Technomic PublishingCo. Inc., USA, 1994

12. Arjyal B.P., Katerelos D.G., Filiou C. and Galiotis C.: Measurement and Modelling of Stress Concentration around a Circular Notch, *Experimental Mechanics* **40** (2000), 248 – 255

3. *Mechanical Properties*

MECHANICAL PROPERTIES OF NANOCOMPOSITE MULTILAYERS

Z.Q. QI, X. NIE and E.I. MELETIS
Department of Mechanical Engineering
Louisiana State University
Baton Rouge, Louisiana 70803, USA

Abstract. This work s concerned with the mechanical properties of nanocomposite metal/ceramic multilayers. Three thin film nanocomposite systems, namely Al/Al_2O_3, Ti/TiN and Cr/C, have been investigated. The nanocomposites were synthesized on Si (100) wafers by using an electron-beam physical vapor deposition method. The microstructure of the films was characterized by transmission electron microscopy. The deposited metal layers Al, Ti and Cr possessed fcc, hcp and bcc structures, respectively, where the ceramic TiN layers had fcc structure and the Al_2O_3 and C layers were amorphous. The mechanical properties of the nanocomposite thin films were studied by conducting nanoindentation and microhardness experiments. The nanosize of the components was found to affect significantly the mechanical properties that were dominated by the thickness of the metal layer. The hardness response in all three systems can be described as a function of the metal layer thickness by a Hall-Petch relationship. A hardness increase was observed up to a 40 nm, 5 nm, and 20 nm layer thickness for the Al, Ti, and Cr layers, respectively. The effects by the ceramic phase were found to be significant only in films with metal layer thickness at or just before the critical thickness.

1. INTRODUCTION

Nanocomposite, multilayered thin films may offer novel physical, mechanical and electronic properties and thus, attract considerable interest for a number of potential applications. There is initial evidence suggesting that multilayered films consisting of a ductile (metal layer) and a hard component (ceramic layer) may offer an attractive combination of mechanical properties [1-2]. Multilayers such as Ta/TaN [3], Ti/TiC [4], Cr/carbon [5], Ti/TiN [6-12] and Al/Al_2O_3 [13-16] have been studied mainly in terms of mechanical and tribological properties. However, the effect of the layer thickness at the nanoscale, on the mechanical behavior is still not understood and remains to be elucidated.

2. THEORATICAL CONSIDERATIONS

2.1. Hall-Petch Equation

The Hall-Petch (H-P) equation [17, 18], which is typically used to predict the flow strength in polycrystalline materials as a function of grain size, has been used in the past for strength predictions in multilayers [8, 16, 19]:

E.E. Gdoutos and Z.P. Marioli-Riga (eds), Recent Advances in Composite Materials, 153-166

$$\sigma = \sigma_o + kd^{-0.5} \quad (1)$$

where σ is the flow strength of the material, σ_o is the friction stress, k is a material-dependent constant, and d is the grain diameter or the metal layer thickness in a metal/ceramic multilayered system. However, experimental observations show that when the layer thickness decreases beyond a certain point, no further increase or even a decrease in hardness is observed contradictory to the H-P equation [6, 11]. This phenomenon has been termed inverse H-P relation and has been attributed to nanoporosity and interfacial effects from the large fraction of atoms residing at the grain boundaries [20-22]. Koehler [23] proposed a theory for predicting mechanical response in thin films with a laminated structure.

2.2. *Koehlers's Predictions*

Koehler proposed that a laminated structure which is composed of two metal layers, A and B, where metal A has a high dislocation-line energy while metal B has a low dislocation-line energy, should exhibit a resistance to plastic deformation and brittle fracture well in excess of that of the individual metals. He predicted that Frank-Read dislocations would require a large externally applied force to cause them to move from the material of lower elastic constant into the material with the higher elastic constant. When the layer thickness is thin enough, the Frank-Read source (Figure 1) may not operate at all, in either layer, leading to a situation in which a single dislocation is the mechanism by which plastic deformation occurs. This suggests that the thin nanocomposite multilayers should exhibit a resistance to plastic deformation and brittle fracture well beyond levels exhibited by monolithic homogeneous metals. For a thickness less than l, the Frank-Read dislocation source could not operate. The critical thickness l is given by:

$$l < 32\pi \boldsymbol{b}_B(\mu_A + \mu_B)/(\mu_A - \mu_B) \quad (2)$$

where μ_A and μ_B are the moduli of rigidity of metals A and B respectively; $\boldsymbol{b}_B$ is the Burger's vector of metal B. These theoretical predictions have been experimentally confirmed by Lehoczky [24, 25] in Al-Cu multilayered thin film system. Koehler's predictions can be used to explain the effect of alternate thin layers of metal/ceramic multilayers [13] with differing elastic constants on the ability of a Frank-Read source to operate.

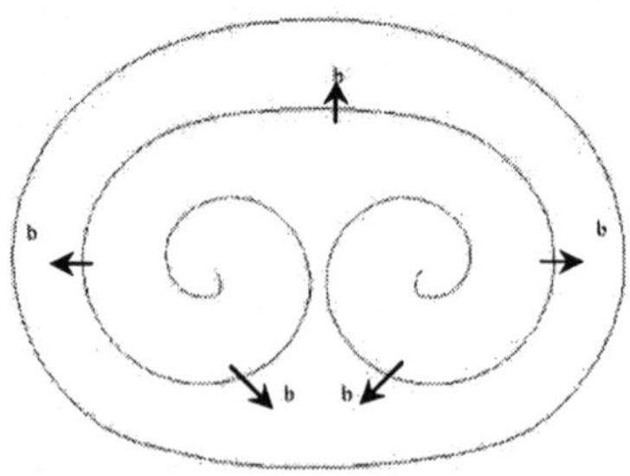

Figure 1. Frank-Read dislocation source.

3. EXPERIMENTAL

3.1. Deposition of multilayers

A dual electron beam physical vapor deposition (EBPVD) system (Edward Auto 306) was utilized to deposit Al/Al_2O_3, Ti/TiN and Cr/C nanocomposite multilayers. In the Al/Al_2O_3 and Ti/TiN systems, three groups of multilayered thin films were deposited by EBPVD with the thickness of the metallic and ceramic layer changing from 5 nm to 100 nm. In the first group, the metallic and ceramic layers were of equal thickness, ranging from 5 nm to 100 nm; in the second and third group, the layer thickness of one component (metal and ceramic, respectively) was constant (40 nm), while that of the other varied from 5 nm to 40 nm. Pure Al, Al_2O_3, Ti and TiN films were also deposited for comparison. In the Cr/C multiplayer system, two groups of thin films were deposited with the thickness of Cr and C layer changing from 10 nm to 40 nm. In the first group, the Cr and C layers were of equal thickness, ranging from 20 nm to 40 nm; in the second group, the layer thickness of C was constant at 40 nm while that of Cr was varied from 10 nm to 40 nm. The total thickness of each film was about 1 μm. Therefore, depending on the thickness of the individual layers, the films were made up of a minimum of 5 layer pairs (e.g., in Al (100 nm)/Al_2O_3 (100 nm)) to 100 layer pairs (e.g., in Ti (5 nm)/TiN (5 nm)).

Al (99.99%) and Al_2O_3 (99.9%) vapor source materials were placed in two graphite crucible liners separately. Si (100) wafers were used as the substrate. The substrate holder was 6 cm away from the source crucibles, and the substrate holder was actuated by a solenoid motor and was able to rotate 90° and stop at two points above the depositing source materials. Two thickness monitors were used to control the deposition rate and the layer thickness from the different evaporation sources. The vacuum system was pumped to a pressure of 4×10^{-6} Torr prior to deposition. During processing, the electron gun voltage and beam current were adjusted to 4.8 kV and 30 mA for both Al and Al_2O_3; the vacuum pressure maintained at 6.0 - 7.0×10^{-6} Torr for Al and 6.5 - 9.0×10^{-6} Torr for Al_2O_3. The deposition rate of Al and Al_2O_3 was about 0.86 and 15 nm/min, respectively. Ti/TiN and Cr /C nanocomposite multilayers were also deposited using this method. Details of deposition conditions are given in Table 1. For each case, the metallic layer (Al, Ti or Cr) was the first layer on the Si wafer and the ceramic layer (Al_2O_3, TiN or C) was the last layer.

Table 1. Processing parameters of the nanocomposite multilayers.

	Al/Al_2O_3 nanocomposite multilayers		Ti/TiN nanocomposite multilayers		Cr/C nanocomposite multilayers	
	Al	Al_2O_3	Ti	TiN	Cr	C
Source Material	Al (99.99%)	Al_2O_3 (99.9%)	Ti (99.995%)	TiN (99.5%)	Cr (99.95%)	C (99.5%)
Gun Voltage (kV)	4.80	4.80	4.80	4.80	4.80	4.80
Beam Current (mA)	30	30	35	35	15	95
Pressure (10^{-6} Torr)	6.0~7.0	6.5~9.0	1.0~4.0	6.5~8.0	3.0	0.1
Deposition Rate (nm/min)	0.86	15	30	4	16	9

3.2. Microstructure of nanocomposites

The microstructure of the nanocomposite multilayers was examined by transmission electron microscopy (TEM). Cross-sectional TEM (XTEM) samples were prepared by mechanical thinning, polishing and dimple grinding, followed by Ar^+ ion milling under 4.2 keV beam energy and 18 μA gun current at an incident beam angle of 6° using a Gatan precision ion polishing system. TEM images were obtained using a JEOL TEM 2010 electron microscope operated at 200 keV.

3.3. Mechanical properties

Microhardness measurements were performed using a Knoop microhardness tester at a load of 10 g. The indentation depth was around 500 nm. At least seven measurements of the microhardness were performed on each specimen.

Nanohardness and reduced elastic modulus were obtained using a nanoindenter (Hysitron Triboscope™) at a load of 2 mN. Nanoindentations were made by a triangular Berkovich-type diamond pyramid indenter with an angle of 65.3° between the tip axis and the faces of the triangular pyramid, and with the same area (A) to the square of depth (h) ratio as the traditional Viker's pyramid ($k=A/h^2=24.5$).The nanohardness of the thin films was determined using the relationship $H=P/A$, where P was the maximum load applied during the indentation, A was the contact area between the indenter and specimen, $A=k\,h_c^2$, h_c being the contact depth determined by the method of Oliver and Pharr [26], and $h_c=h_{max}-h_s$, where h_{max} is the total measured displacement and h_s is the elastic surface displacement.

The reduced elastic modulus, E_r can be given by:

$$1/E_r = (1-v_i^2)/E_i + (1-v_s^2)/E_s \tag{3}$$

where E_i, E_s and v_i, v_s are the elastic modulus and Poisson's ratio of the indenter and the specimen, respectively. The unloading portion of the applied force-displacement curve was used to determine the reduced elastic modulus of the thin films. An average of seven indentations were taken to determine hardness and reduced elastic modulus. All of the contact depths were around 100 nm, which is below 10% of the total thickness of the multilayers.

4. RESULTS AND DISCUSSION

4.1. TEM observations

Figure 2 is a bright-field TEM image, showing a cross-sectional view of a typical multilayer structure of Cr (20 nm)/C (20 nm). The periodicity and continuity of the Cr and C layers is clearly shown in the micrograph. The bright and dark areas show C (20 nm) and Cr (20 nm) layers, respectively. It was found that Cr has a columnar structure and C is amorphous (*a*-C). The microstructure of the Al/Al_2O_3 and Ti/TiN nanocomposite multilayers was also examined by TEM. The Al, Ti and Cr layers were found to have fcc, hcp and bcc crystalline structures, respectively. The TiN layer had fcc nanocrystalline structure while Al_2O_3 and C layers were amorphous.

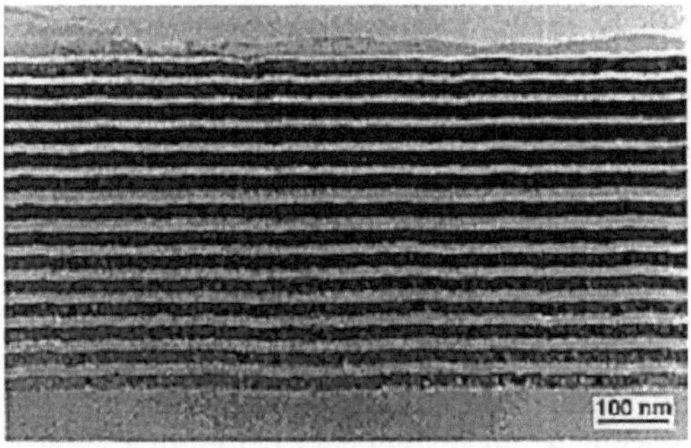

Figure 2. Bright-field XTEM image of Cr (20 nm)/a-C (20 nm) multilayered nanocomposite.

4.2 Mechanical properties of Al/Al_2O_3 nanocomposite multilayers

The results for Al/Al_2O_3 nanocomposite system are listed in Table 2. The predictions of elastic modulus and hardness by the rule of mixture (ROM) are also given in Table 2. The results show that the predictions by the ROM are in relative good agreement for the elastic modulus but they underestimate the hardness of the nanocomposite.

Table 2. Reduced elastic modulus, nanohardness and microhardness of Al/Al_2O_3 films.

Al/Al_2O_3	λ_m/λ_c (nm)	E_r (GPa)	E (ROM) (GPa)	HN (GPa)	H (ROM) (GPa)	HK (GPa)
Group I	100/100	115.2	117.2	5.5	6.5	9.49
	50/50	115.0	117.2	7.67	6.5	10.39
	40/40	120.2	117.2	8.29	6.5	10.38
	30/30	116	117.2	7.98	6.5	10.24
	20/20	114.4	117.2	7.99	6.5	10.50
Group II	40/30	108.0	114.5	6.31	6.19	failed
	40/20	106.6	110.8	6.77	5.65	7.56
	40/10	91.6	106	5.90	4.91	7.25
Group III	30/40	122.8	120	7.79	7.11	11.0
	20/40	127.9	124.4	8.47	7.50	10.83
	10/40	127.4	131	8.32	8.23	11.87
Al Monolithic Film	783.6	99.9	-	3.81	-	5.89
Al_2O_3 Monolithic Film	786	141.6	-	9.34	-	12.44

λ_m and λ_c metallic and ceramic layer thickness, respectively; E (ROM) and H (ROM) the elastic modulus and hardness as predicted by the Rule of Mixture (ROM); HN = nanohardness, HK = Knoop hardness.

The layer thickness effect on hardness and reduced elastic modulus (E_r) for the Al/Al_2O_3 multilayer system is shown in Figure 3. For Group I Al/Al_2O_3 multilayers (Λ=40, 60, 80, 100, 200 nm), the hardness and E_r exhibit a maximum value at 40 nm. According to Eq. (2), the theoretical critical thickness for Al is 41 nm [13], after which the hardness is not expected to increase by further decreasing the layer thickness. Thus, the present results are in good agreement with the theoretical predictions. It should be noted that all microhardness values are higher than those by nanohardness since the indentation depth in the former was around 500 nm and thus it contained substrate contributions. As mentioned earlier the nanoindentation depth was less than 10% of the film thickness.

Figure 3(b) presents the results for Group II Al/Al_2O_3 multilayers with the layer thickness of Al constant at 40 nm and that of Al_2O_3 changing from 10 nm to 40 nm. The results show that as the thickness of the Al_2O_3 layer increases, both the hardness and E_r increase mainly due to increase in the volume fraction of the ceramic (f_c).

Figure 3(c) presents the results for Group III Al/Al_2O_3 multilayers with the layer thickness of Al_2O_3 constant at 40 nm and that of Al changing from 10 nm to 40 nm. The results show that hardness is not changing significantly. If 40 nm is a critical thickness, then decreasing Al layer thickness below that level will tend to decrease hardness. At the same time, however, f_c increases, which will tend to increase hardness. Thus, there are two competing effects resulting in a mixed hardness behavior. As the Al layer thickness decreases, E_r increases due to the increase in f_c.

4.3 Mechanical properties of Ti/TiN nanocomposite multilayers

Nanoindentation and microhardness values for the Ti/TiN nanocomposites are listed in Table 3. The results show that ROM significantly underestimates both the elastic modulus and hardness of this system. The layer thickness effects on hardness and E_r of this system are shown in Figure 4. In Group I Ti/TiN multilayers, the results clearly show an increase in hardness with decreasing layer thickness λ_m while f_c remains constant. All E_r values in this group remain at the same level since f_c remains constant. It is interesting to note that the hardness increase continues down to a modulation period of Λ=10 nm (λ_m=λ_c= 5 nm). Previous studies on this system reported maximum hardness values of 23 GPa and 29 GPa for modulation periods of 50 nm [6] and 20 nm [12], respectively. Hardness was found to decrease by decreasing further the modulation period. The present results are in general agreement with those by Kikuchi et al [11]. They observed a maximum hardness of 11.2 GPa, compared to 15 GPa in the present study, also for a smaller modulation period of 10 nm. Further more, they determined the dissipated energy that is consistent with the energy to propagate dislocations, as a function of modulation from the nanoindentation experiments. A reduction in dissipated energy was observed beyond a modulation of 10 nm consistent with the present results.

For Group II Ti/TiN multilayers the results, Figure 4(b), show an increase in E_r with increasing f_c (increasing thickness of TiN layer). It is interesting to note here that the hardness remains at the same levels (~10 GPa) regardless of the increase of the volume fraction of the ceramic phase or its decrease in the layer thickness. This behavior along with the results in Figure 4(a) suggests that the hardness response of this system is dominated by the layer thickness of Ti.

For Group III Ti/TiN multilayers a general downtrend in Er is observed with decreasing fc (increasing thickness of Ti layer). Further more, the results show an increase in hardness upto ~16 GPa for a Ti thickness layer of 10 nm followed by a

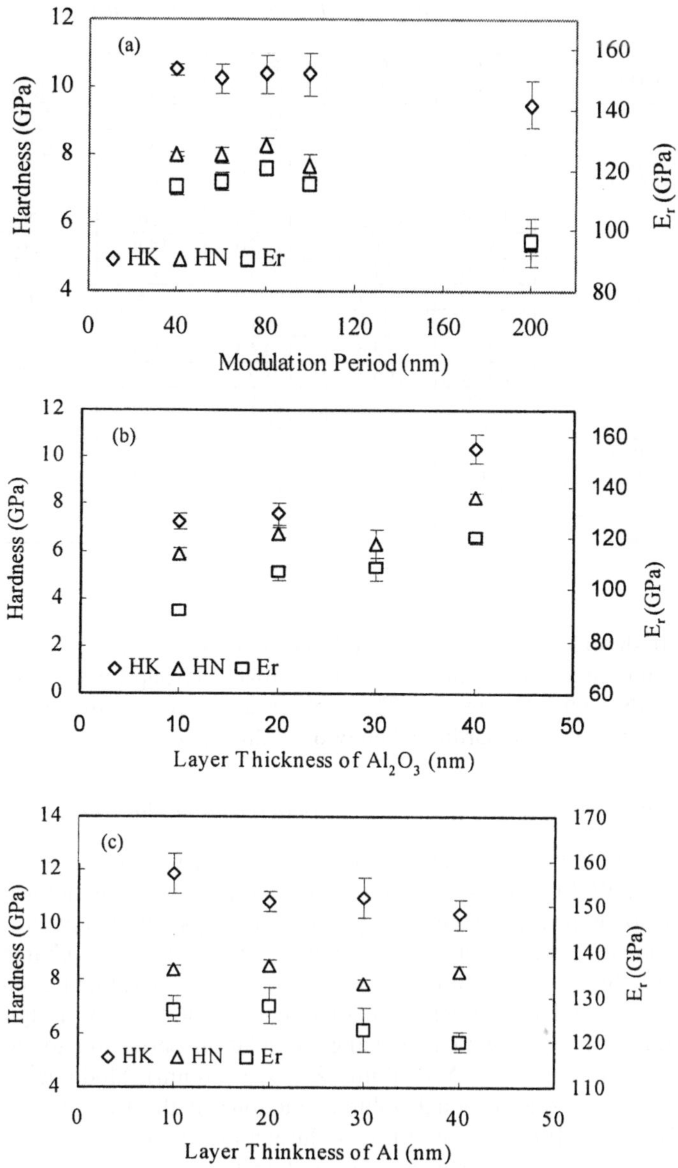

Figure 3. Reduced elastic modulus, nanohardness and microhardness versus thickness of Al/Al_2O_3 nanocomposite multilayers; (a) equal thickness of Al and Al_2O_3 layers, (b) Al layer thickness is constant at 40 nm, and (c) Al_2O_3 layer thickness is constant at 40 nm.

Table 3. Reduced elastic modulus, nanohardness and microhardness of Ti/TiN films.

Ti/TiN	λ_m / λ_c (nm)	E_r (GPa)	E (ROM) (GPa)	HN (GPa)	H (ROM) (GPa)	HK (GPa)
Group I	100/100	136.2	110.20	9.4	7.26	10.11
	40/40	138.2	110.20	9.86	7.26	13.09
	30/30	133.2	110.20	10.40	7.26	13.58
	20/20	142.7	110.20	11.35	7.26	15.56
	10/10	142.5	110.20	12.07	7.26	18.99
	5/5	151	110.20	14.9	7.26	25
Group II	40/30	134.0	106.15	10.47	7.16	16.13
	40/20	129.7	101.00	10.40	5.95	17.50
	40/10	123.6	94.65	10.20	4.91	15.79
Group III	30/40	139.8	114.63	12.46	7.80	16.65
	20/40	166.5	121.33	15.58	8.57	20.40
	10/40	153.1	131.94	16.42	9.62	23.84
	5/40	164.9	140.00	12.9	10.33	15.2
Ti Monolithic Film	1310	86.5	-	3.34	-	6.43
TiN Monolithic Film	937	151.9	-	11.19	-	14.71

decline. It should be noted that the TiN layer thickness in this group is 40 nm and the maximum hardness attained is similar to that of Group I with a Ti layer thickness of 5 nm (TiN 5 nm), Figure 4(a). These observations further suggest that the Ti layer thickness dominates the hardness behavior of this system.

4.4 Mechanical properties of Cr/a-C nanocomposite multilayers

Nanoindentation and microhardness results for Cr/*a*-C nanocomposites are listed in Table 4. The ROM was also found to underestimate elastic modulus and hardness in Cr/*a*-C nanocomposites. A peculiar point for this system, compared to the previous nanocomposites, was that the metallic component was characterized by a significantly higher elastic modulus than the non-metallic (graphite) component.

The layer thickness effects on hardness and E_r are shown in Figure 5. Hardness was found to increase with increasing modulation period up to the lowest value tested in the present study, $\Lambda = 40$ nm ($\lambda_m = \lambda_c = 20$ nm). Modulus was not found to vary significantly due to equal volume fractions of the two phases involved. The results for Group II Cr/*a*-C multilayers show a maximum value for E_r and hardness at a layer thickness of 30 nm. It is interesting to note that this maximum hardness value (11.6 GPa) at 30 nm/ 40 nm is also exhibited by the 20 nm/20 nm multilayers (11.4 GPa). The reasons for this behavior are not clear at present but may be attributed to the fact that unlike the previous systems, the non-metallic component in Cr/*a*-C has higher hardness but lower modulus.

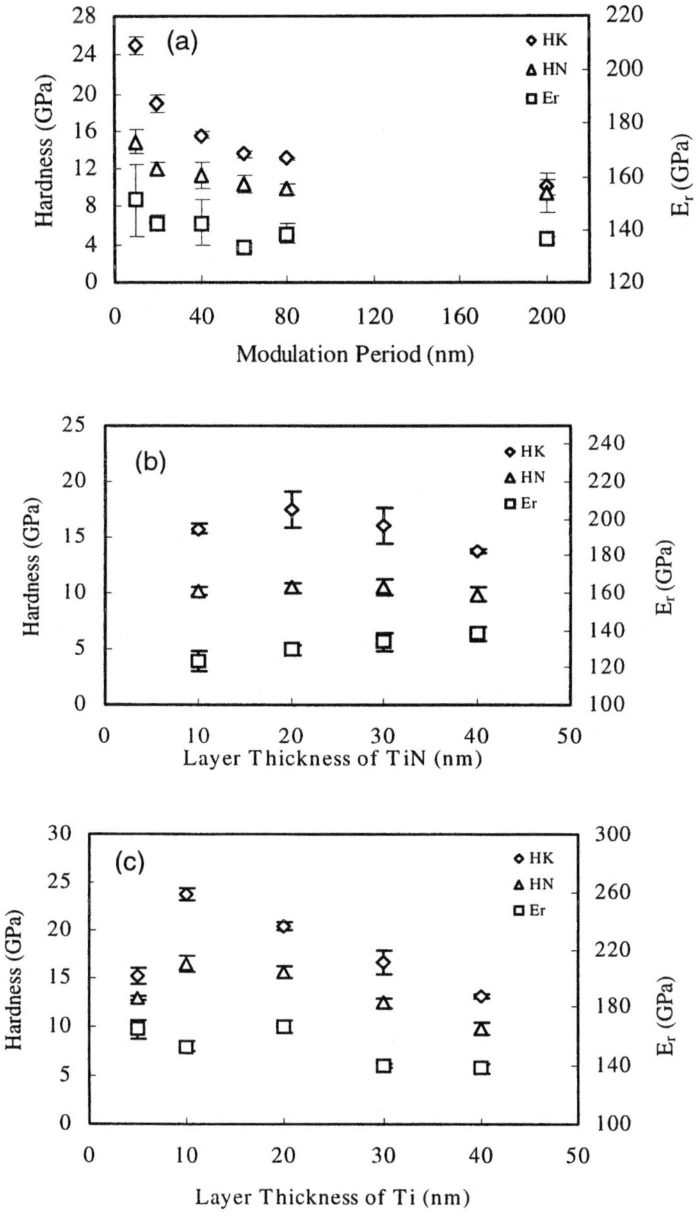

Figure 4. Reduced elastic modulus, nanohardness and microhardness versus thickness of Ti/TiN nanocomposite multilayers; (a) equal thickness of Ti and TiN layers, (b) Ti layer thickness is constant at 40 nm, and (c) TiN layer thickness is constant at 40 nm.

Table 4. Reduced elastic modulus, nanohardness and microhardness of Cr/a-C films.

Cr/*a*-C	λ_m/λ_c (nm)	E_r (GPa)	E (ROM)$^+$ (GPa)	HN (GPa)	H (ROM) (GPa)	HK (GPa)
Group I	40/40	89.9	72	9.62	6.42	13.98
	30/30	99.5	72	9.96	6.42	15.45
	20/20	89.4	72	11.4	6.42	16.23
Group II	30/40	114.1	74.1	11.6	6.70	17.79
	20/40	82.4	77.3	9.82	7.12	16.27
	10/40	71.5	81.8	8.6	7.76	13.63
monolithic Cr film	-	144	-	5.0	-	-
a-C film	-	60	-	9.0	-	-

4.5 Relationship between nanohardness and metallic layer thickness

Figure 6 presents the variation of nanohardness H *vs.* the inverse of the root square of the metallic layer thickness (i.e., H-P formalism) for all present data. The results show that the H-P relation holds up to a certain point for all three nanocomposite systems. For the Al/Al_2O_3 multilayers, the hardness response can be described by the H-P relation up to a critical thickness of λ_{Al} = 40 nm beyond which hardness does not increase further by decreasing metal layer thickness. This value is in agreement with theoretical predictions [13]. It is important to note that beyond this critical level, hardness is not increasing even if the f_c increases. An effect of f_c seems to be present only for metal layer thicknesses at or close to the critical thickness. For example, compare hardness values at λ_{Al} = 40 nm for various f_c (values marked on the graph). Further more, the 70 nm/10 nm multilayered film exhibits a lower hardness since it is close to the critical thickness but it has a very low f_c (12.5%). On the other hand, the 200 nm/20 nm film has lower f_c (10%) than the latter film but follows the hardness trend since it is further away from the critical thickness. Thus, the hardness behavior for this system can be described by

$$H\,(Al/Al_2O_3) = 1490.6\ (\mathrm{MPa}) + 1.3\ (\mathrm{MPa\ m^{1/2}}).\lambda^{-1/2}\ (\mathrm{m^{-1/2}}) \tag{4}$$

Hardness in the Ti/TiN system increased all the way down to the film with the lowest modulation (Λ= 10 nm) tested, Figure 6(b). Regarding f_c contributions, this system exhibited similar behavior with that observed for the Ti/TiN system. For example, the 100 nm/100 nm film has the same hardness with the 100 nm/10 nm film since in both the metal layer thickness is far away from the critical thickness (i.e., 5 nm, the one exhibiting the highest hardness). An increase in f_c close to the critical thickness, increases hardness as long as there is significant Ti presence in the composite (compare films 10 nm/10 nm and 10 nm/40 nm). On the other hand, when f_c reaches high values ($\geq$0.85), hardness is reduced and approaches that of TiN (compare films 5 nm/5 nm and 5 nm/40 nm). The hardness response of this system can be described by

$$H\,(Ti/TiN) = 6354.2\ (\mathrm{MPa}) + 0.6\ (\mathrm{MPa\ m^{1/2}}).\lambda^{-1/2}\ (\mathrm{m^{-1/2}}) \tag{5}$$

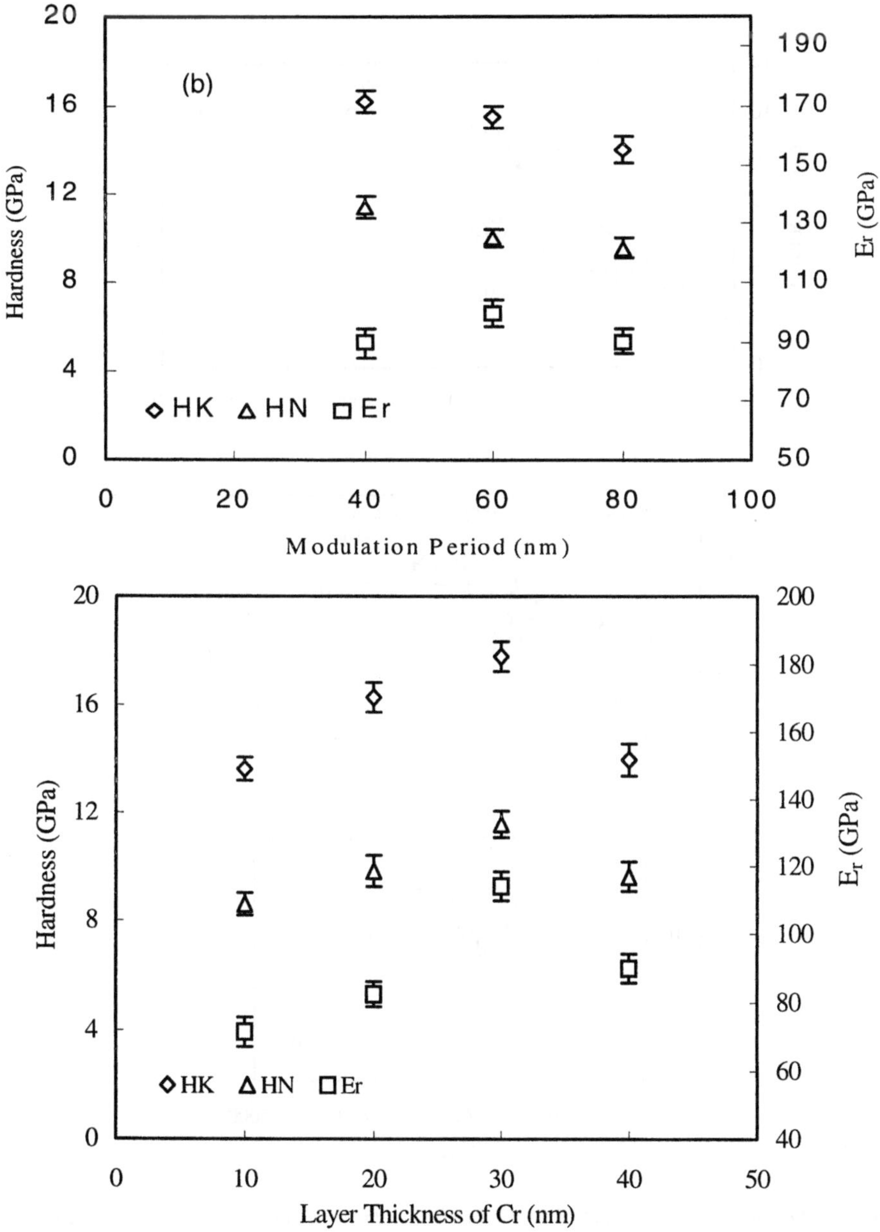

Figure 5. Reduced elastic modulus, nanohardness and microhardness versus thickness of Cr/a-C nanocomposite multilayers (a) equal thickness of Cr and a-C layers, (b) a-C layer thickness is constant 40 nm.

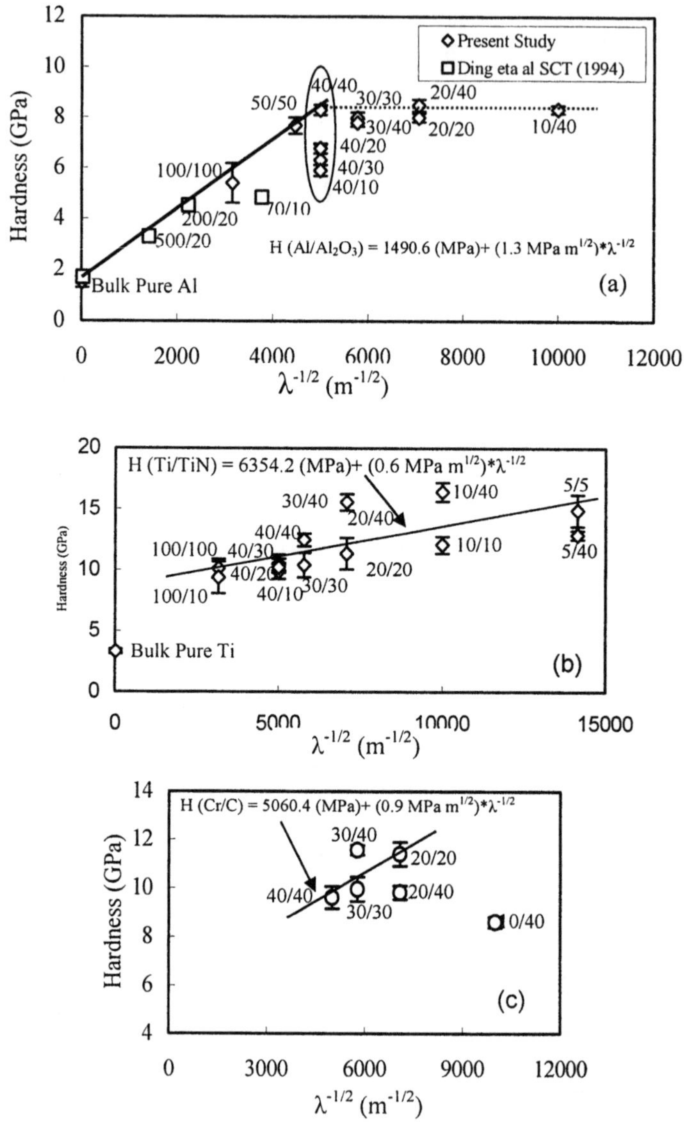

Figure 6. Hardness versus $\lambda^{-1/2}$ of metal layer for (a) Al/Al_2O_3, (b) Ti/TiN, and (c) Cr /a-C multilayered nanocomposites.

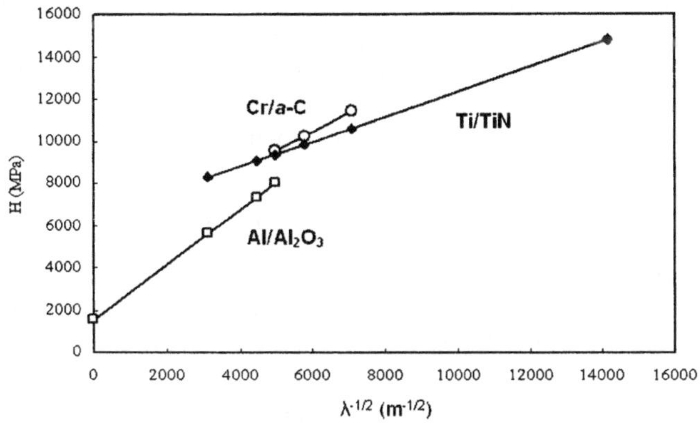

Figure 7. Hall-Petch relation describing hardness behavior of multilayered nanocomposites.

Similarly, the Cr/*a*-C system exhibited the highest hardness for the film with the lowest modulation tested (Λ= 40 nm), Figure 6(c). Even though a smaller variation of films was tested for this system, some of the previous characteristics were still present. For example, film 30 nm/40 nm exhibited higher hardness than the 30 nm/30 nm film due to higher f_c of the former. At or beyond the critical thickness (i.e., 20 nm), hardness is reduced probably due to significant reduction in f_m. For example, the hardness of the film 10 nm/40 nm approaches that of *a*-C (9 GPa). The hardness behavior of this system can be described by

$$H\ (\mathrm{Cr}/a\text{-C}) = 5060.4\ (\mathrm{MPa}) + 0.9\ (\mathrm{MPa\ m^{1/2}}).\lambda^{-1/2}\ (\mathrm{m^{-1/2}}) \tag{6}$$

Figure 7 summarizes the hardness behavior of all three multilayered systems. It is evident that the Al/Al_2O_3 system exhibits the lowest friction stress (i.e., Al has fcc structure) but the highest slope or ability to strengthen faster as a function of metal layer thickness. On the other hand, the Ti/TiN system was the one being able to attain the highest hardness among the three systems tested.

5. SUMMARY

(1) Three multilayered nanocomposite systems were synthesized and their mechanical properties were studied as a function of layer spacing. The systems involved were Al/Al_2O_3, Ti/TiN and Cr/C. The metallic Al, Cr, and Ti layers were found to possess fcc, hcp, and bcc structures, respectively. The ceramic TiN layer had fcc structure while the Al_2O_3 and C layers were amorphous.
(2) The mechanical properties of the three systems were significantly affected by the nanoscale size of their components. The rule of mixture was found to underestimate their hardness and elastic modulus.
(3) Even though in all three multilayered systems an increase in the volume fraction of the ceramic phase was found in general, to increase the elastic modulus and hardness, the dominant role in the mechanical response was played by the metal layer thickness.

(4) The hardness response in all three systems can be described as a function of the metal layer thickness by a Hall-Petch relationship. A hardness increase was observed up to a 40 nm, 5 nm, and 20 nm layer thickness for Al, Ti, and Cr layers, respectively. The effects by the ceramic phase were found to be significant only at or just before the critical metal layer thickness.

6. ACKNOWLEDGEMENT

The authors would like to thank Dr. J.C. Jiang for his help with TEM. The TEM work was performed at the LSU Materials Characterization Center. This work was supported by ARO Grant DAAG55-98-1-0279 and the Louisiana Board of Regents.

7. REFERENCES

1. F.H. Froes and C. Suryanarayana, *JOM*, June (1988) 12-17.
2. B.M. Clemens, H. Kung, and S.A. Barnett, *MRS BULLETIN*, Feb. (1999) 20-25.
3. Y. Kang, C. Lee, J. Lee, *Mater. Sci. Eng.*, B75 (2000) 17-23.
4. I. Dahan, U. Admon, N. Frage, J. Sariel, M.P. Dariel, M.P. Dariel, J.J. Moore, *Surf. Coat. Technol.*, 137 (2001) 111-115.
5. S. Yang, D.G. Teer, *Surf. Coat. Technol.*, 131 (2000) 412-416.
6. T.S. Li, H. Li, F. Pan, *Surf. Coat. Technol.*, 137 (2001) 225-229.
7. R.F. Huang, L.S. Wen, L.P. Guo, J. Gong, B.H. Yu and H. Bangert, *Surf. Coat. Technol.*, 50 (1992) 79-101.
8. L.S. Wen, R.F. Huang, L.P. Guo, J. Gong, T.Y. Wei and Y.Z. Chuang, *J. Magn. Magn. Mater.*, 126 (1993) 200-202.
9. K.J. Ma, A. Bloyce, T. Bell, *Surf. Coat. Technol.*, 76-77 (1995) 297-302.
10. M. Larsson, M. Bromark, P. Hedenqvist, S. Hogmark, *Surf. Coat. Technol.*, 76-77 (1995) 202-205.
11. N. Kikuchi, M. Kitagawa, A. Sato, E. Kusano, H. Nanto, A. Kinbara, *Surf. Coat. Technol.*, 126 (2000) 131-135.
12. E. Kusano, M. Kitagawa, H. Nanto and A. Kibara, *J. Vac. Sci. Technol.* A 16(3), May/Jun 1998, 1272-1275.
13. Y. Ding, Z. Farhat, D.O. Northwood, A.T. Alpas, *Surf. Coat. Technol.*, 68/69 (1994) 459-467.
14. A.T. Alpas, J.D. Embury, D.A. ardwick, R.W. Springer, *J. Mater. Sci.*, 25 (1990) 1603-1609.
15. Y. Ding, D.O Northwood and A.T. Alpas, *Surf. Coat. Technol.*, 62 (1993) 448-453.
16. Y. Ding, D.O. Northwood and A.T. Alpas, *Mater. Sci. Forum,* 189-190 (1995) 309-314.
17. E. O. Hall, *Proc. Phys. Soc. London*, B64 (1951) 747.
18. N. J. Petch, *J. Iron Steel Inst.*, 174 (1953) 25.
19. Z. N. Farhat, Y. Ding, A. T. Alpas and D. O. Northwood, J. Mater. Proc. Technol., 63 (1997) 859-864.
20. H.S. Kim and M.B. Bush, *Nanostruct. Mater.,* 11 (1999) 361-367.
21. N. Wang, Z. Wang, K.T. Aust and U. Erb, *Acta Metall. Mater.,* 43 (1995) 519.
22. J.E. Carsley, J. Ning, W.M. Milligan, S.A. Hackney and E.C. Aifantis, *Nanostruct. Mater.,* 5 (1995) 441.
23. J. S. Koehler, *Phys. Rev.*, B, 2 (1970) 547.
24. S.L. Lehoczky, *Phys. Rev. Lett.,* 41 (26) 25 December (1978) 1814-1818.
25. S.L. Lehoczky, *J. Appl. Phys.*, 49(11) (1978) 5479-5485.
26. W.C. Oliver and G.M. Pharr, *J. Mater. Res.*, **7** (1992) 1564.

ULTRASONIC STIFFNESS MATRIX MEASUREMENTS OF OXIDE/OXIDE COMPOSITES

A. PAIPETIS and V. KOSTOPOULOS
Applied Mechanics Laboratory
University of Patras
Patras University Campus
GR-265 00 Patras, Greece

Abstract. The stiffness matrix of Al_2O_3/ Al_2O_3 composites has been non-destructively evaluated using ultrasonics. The method is based on the propagation velocities of sound waves in the material. These were estimated experimentally using a custom pulser-receiver setup which allows control of the angle of the incident pulse on the sample, while the latter is immersed in a water bath. The recording of the time delay of the longitudinal and transverse wave components as they travel through the composite allows the determination of their respective propagation velocities within the composite, as a function of the incident angle for a given recording plane. .The derivation of the elastic constants in order to reproduce the stiffness matrix of the composite is an inverse wave propagation problem described by the Christoffel equation. All 9 elastic constants are derived assuming an orthotropic medium. The derived elastic constants are compared to experimental data acquired destructively.

1. INTRODUCTION

Ceramic matrix composites offer the best alternative to superalloys which are now very close to their physical limits. Their temperature stability combined with the design of the reinforcement for enhanced mechanical properties [1] and the tailoring of the interface to achieve the desired toughness [2] makes them ideal candidates as the next generation structural materials for hot section components (blades, nozzles, combustion liners, shrouds) [3]. Oxide-Oxide (Al_2O_3/ Al_2O_3) composites are recently chosen as candidate materials for such applications as they offer high temperature stability in oxidising environments up to 1200 °C [4]. However, the extreme thermomechanical loading these structural components are subjected to compare with their high manufacturing cost imposes the non-destructive evaluation of their mechanical properties throughout their service life in order to quantify the damage processes and reliably assess their useful life [5].

The use of a non–destructive method for evaluating composite material properties allows the determination of the material state throughout its service life. Such a method should allow the fast and reliable material characterization and account for property variation as a function of position. Moreover, in the case of anisotropic materials, the reliable property determination as a function of position is crucial, as the material anisotropy is expected to change according to the material

E.E. Gdoutos and Z.P. Marioli-Riga (eds), Recent Advances in Composite Materials, 167-180

loading history [6]. As has recently been shown [7] composite material properties may be reliably determined with the use of ultrasonics.

In the case of an infinite isotropic medium, three spherical waves emanate from a point source within it [8]. The waves consist of one longitudinal component and two shear components which coincide. The wave propagation is directly dependent on the elastic properties of the medium and is independent of the direction of propagation. This is not the case in anisotropic media, where the propagation velocities of individual wave types are intrinsically dependent on the direction of propagation; the level of anisotropy is directly influencing the propagation and the two shear wave components no longer coincide. It is then obvious that the problem of estimation of the elastic properties of the material is reduced to solving the constituent equations for any anisotropy level.

In this work the stiffness matrix of oxide –oxide orthotropic laminates is derived using ultrasonics [9] and is compared with the properties of the laminate acquired with conventional methods.

2. THEORETICAL BACKGROUND

2.1. Orthotropic Laminates

The generalized stress-strain behaviour in the case of a material having orthotropic symmetry is written in the form:

$$\begin{Bmatrix} \sigma_1 \\ \sigma_2 \\ \sigma_3 \\ \tau_{23} \\ \tau_{31} \\ \tau_{12} \end{Bmatrix} = \begin{bmatrix} C_{11} & C_{12} & C_{13} & 0 & 0 & 0 \\ C_{12} & C_{22} & C_{23} & 0 & 0 & 0 \\ C_{13} & C_{23} & C_{33} & 0 & 0 & 0 \\ 0 & 0 & 0 & C_{44} & 0 & 0 \\ 0 & 0 & 0 & 0 & C_{55} & 0 \\ 0 & 0 & 0 & 0 & 0 & C_{66} \end{bmatrix} \begin{Bmatrix} \varepsilon_1 \\ \varepsilon_2 \\ \varepsilon_3 \\ \gamma_{23} \\ \gamma_{31} \\ \gamma_{12} \end{Bmatrix} \tag{1}$$

and the strain-stress relation is given by

$$\begin{Bmatrix} \varepsilon_1 \\ \varepsilon_2 \\ \varepsilon_3 \\ \gamma_{23} \\ \gamma_{31} \\ \gamma_{12} \end{Bmatrix} = \begin{bmatrix} S_{11} & S_{12} & S_{13} & 0 & 0 & 0 \\ S_{12} & S_{22} & S_{23} & 0 & 0 & 0 \\ S_{13} & S_{23} & S_{33} & 0 & 0 & 0 \\ 0 & 0 & 0 & S_{44} & 0 & 0 \\ 0 & 0 & 0 & 0 & S_{55} & 0 \\ 0 & 0 & 0 & 0 & 0 & S_{66} \end{bmatrix} \begin{Bmatrix} \sigma_1 \\ \sigma_2 \\ \sigma_3 \\ \tau_{23} \\ \tau_{31} \\ \tau_{12} \end{Bmatrix} \tag{2}$$

where $\lfloor S \rfloor$ is the compliance matrix. The stiffness and compliance matrices are mutually inverse. For the case of an orthotropic material the relation of their components is

$$
\begin{aligned}
&S_{11}=\frac{C_{22}C_{33}-C_{23}^{2}}{C}, \quad && S_{22}=\frac{C_{33}C_{11}-C_{13}^{2}}{C}, \quad && S_{33}=\frac{C_{11}C_{22}-C_{12}^{2}}{C},\\
&S_{12}=\frac{C_{13}C_{23}-C_{12}C_{33}}{C}, && S_{13}=\frac{C_{12}C_{23}-C_{13}C_{22}}{C}, && S_{23}=\frac{C_{12}C_{13}-C_{23}C_{11}}{C},\\
&S_{44}=\frac{1}{C_{44}}, && S_{55}=\frac{1}{C_{55}}, && S_{66}=\frac{1}{C_{66}}
\end{aligned}
\tag{3}
$$

where

$$C=C_{11}C_{22}C_{33}-C_{11}C_{23}^{2}-C_{22}C_{13}^{2}-C_{22}C_{12}^{2}+2C_{12}C_{23}C_{13}$$

For the case of an orthotropic material the compliance matrix components S_{ij} in terms of the engineering constants are

$$
\lfloor S_{ij} \rfloor=\begin{bmatrix}
\frac{1}{E_1} & -\frac{\nu_{21}}{E_2} & -\frac{\nu_{31}}{E_3} & 0 & 0 & 0\\
-\frac{\nu_{12}}{E_1} & \frac{1}{E_2} & -\frac{\nu_{32}}{E_3} & 0 & 0 & 0\\
-\frac{\nu_{13}}{E_1} & -\frac{\nu_{23}}{E_2} & \frac{1}{E_3} & 0 & 0 & 0\\
0 & 0 & 0 & \frac{1}{G_{23}} & 0 & 0\\
0 & 0 & 0 & 0 & \frac{1}{G_{31}} & 0\\
0 & 0 & 0 & 0 & 0 & \frac{1}{G_{12}}
\end{bmatrix}
\tag{4}
$$

where

$$\frac{\nu_{ij}}{E_i}=\frac{\nu_{ji}}{E_{ji}} \qquad i,j=1,2,3 \tag{5}$$

2.2. *Wave propagation in composite laminates*

The wave propagation tensor in the case of an anisotropic material is described by the Christoffel equation [10]:

$$\Gamma_{ij}=C_{ijkl}n_k n_l \tag{6}$$

where C_{ijkl} are the components of the stiffness matrix and n_k (k=1,2,3) are the components of the propagation direction vector $\underline{\hat{n}}$.

The phase velocities of the three propagating waves for a given direction $\underline{\hat{n}}$ are given by the eigenvalues of the propagation tensor Γ_{ij} according to

$$\det\left(\Gamma_{ij} - \rho V^2 \delta_{ij}\right) = 0 \tag{7}$$

where the polarization directions are the corresponding eigenvectors.

In the case of an orthotropic material the components of the propagation tensor Γ_{ij} have the form:

$$\begin{aligned}
&\Gamma_{11} = C_{11}n_1^2 + C_{66}n_2^2 + C_{55}n_3^2, &&\Gamma_{22} = C_{66}n_1^2 + C_{22}n_2^2 + C_{44}n_3^2,\\
&\Gamma_{33} = C_{55}n_1^2 + C_{44}n_2^2 + C_{33}n_3^2, &&\Gamma_{23} = (C_{23} + C_{44})n_2 n_3,\\
&\Gamma_{13} = (C_{13} + C_{55})n_1 n_3, &&\Gamma_{12} = (C_{12} + C_{66})n_1 n_2
\end{aligned} \tag{8}$$

2.3. The inverse problem

The stiffness components are identified by the following successive steps.

The C_{33} stiffness component is directly computed using the velocity measured in normal incidence (i.e. $\theta = 0$)

$$C_{33} = \rho_b V^2 \tag{9}$$

where ρ_b is the bulk density in kilograms per cubic meter (kg/m^3).

The stiffness components C_{22}, C_{23} and C_{44} are identified from the velocity measurements recorded for the acquisition plane ψ =90° by minimizing the following expression:

$$F(C_{ij}) = \sum_{p=1}^{N} \{f_p(\lambda_p(n), C_{ij})\}^2 \tag{10}$$

$$f_p^{(32)}(\lambda_p(n), C_{ij}) = \lambda_p^2 - \lambda_p(\Gamma_{33} + \Gamma_{22}) + \Gamma_{11}\Gamma_{22} - \Gamma_{32}^2 \tag{11}$$

where p = 1 to N, N being the total number of measurements, and $\lambda_p = \rho V p^2$.

The stiffness components C_{11}, C_{31}, C_{55} are identified using the velocities measured in the acquisition plane ψ =0° by minimizing expression (10) with:

$$f_p^{(13)}(\lambda_p(n), C_{ij}) = \lambda_p^2 - \lambda_p(\Gamma_{11} + \Gamma_{33}) + \Gamma_{11}\Gamma_{33} - \Gamma_{13}^2 \tag{12}$$

The two remaining stiffness components, C_{12} and C_{66} are identified using the velocities measured in the non-principal plane $\psi = 45°$ and the seven stiffness components determined above. The number of modes that can experimentally be obtained in this plane, differ according to the level of material symmetry and influence the quality of the identification process.

3. EXPERIMENTAL

3.1. Materials

The material used in this study is a mullite-based matrix composite reinforced by oxide-based Nextel™ 720 fibres. The material toughness is improved by the application of a fugitive fibre/matrix interface using the sol-gel technique. The composite was manufactured using a symmetric 0°/90° fibre lay-up configuration with the polymer infiltration process (PIP) [11]. The final fibre content is 41%.

A 150x150 mm^2 was manufactured as above. Specimens were cut from the plate using a heavy duty diamond saw.

3.2. Ultrasonic measurements

As shown in figure 1, the specimen is immersed in water and placed between a transmitter (E) and a receiver (R) which are rigidly connected to each other and have two rotational degrees of freedom. A frequency of 5MHz is selected as it ensures representativeness of the measurement and provides a measurable amplitude of the Q_L wave under normal incidence $\theta_i = 0$, as well as time separation of the waves Q_L and Q_T when varying the angle of incidence θ_i (see Fig. 2).

Depending on the angle of incidence, the pulse sent by the transmitter E is refracted within the material in one, two or three bulk waves (one quasi longitudinal wave Q_L, one quasi transverse wave Q_T, or two quasi transverse waves Q_{T1}, Q_{T2}) that propagate in the solid at different velocities and in different directions. The receiver R collects one, two or three temporally delayed pulses, corresponding to each of these waves. The difference in propagation time of each of the waves and the propagation time of the emitted pulse in the coupling fluid without the specimen is measured. For each monitoring plane and θ_i, the delay δt, on the Q_L and the Q_T waves is determined by comparing the signal received in the coupling fluid alone (reference signal), and the signal received when the specimen is in the coupling fluid. The delay δt is calculated from the acquired signal using a Hilbert Transform method [12]. The evaluation procedure is based on the measurement of the time difference of the longitudinal and one or both transverse waves, and is only valid when the Q_L and Q_T waves are appropriately separated (figure 2).

For each δt the associated propagation velocity is determined as:

$$V = \frac{V_0}{\sqrt{1 + \frac{V_0 \delta t}{h}\left(\frac{V_0 \delta t}{h} - 2\cos\theta_i\right)}} \tag{13}$$

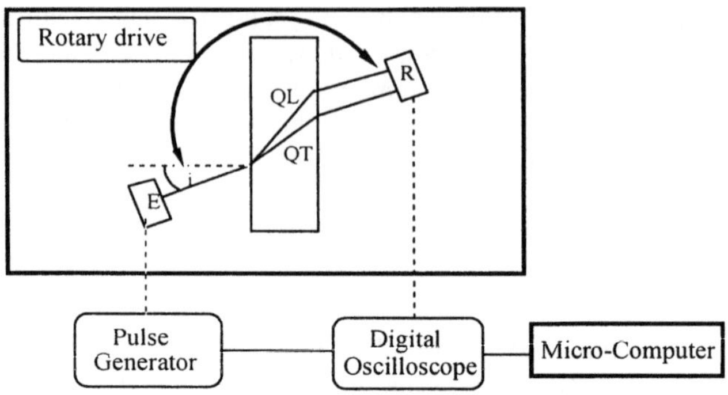

Figure 1: Ultrasonic test assembly [9]

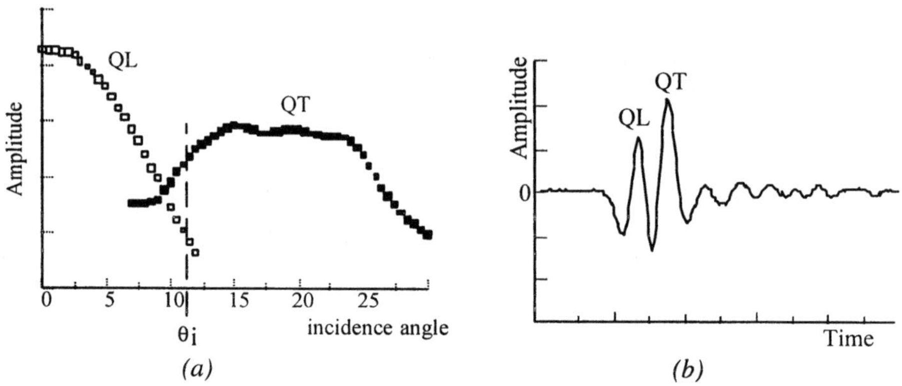

Figure 2: Q_L and Q_T waves at incidence angle θ_i, (a) amplitude (b) temporal waveform [9]

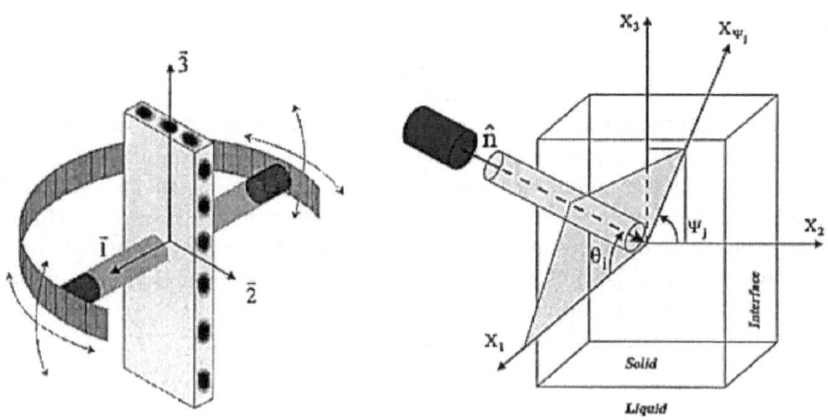

Figure 3 – Definition of the scanning planes in relation to the principal axes of the material.

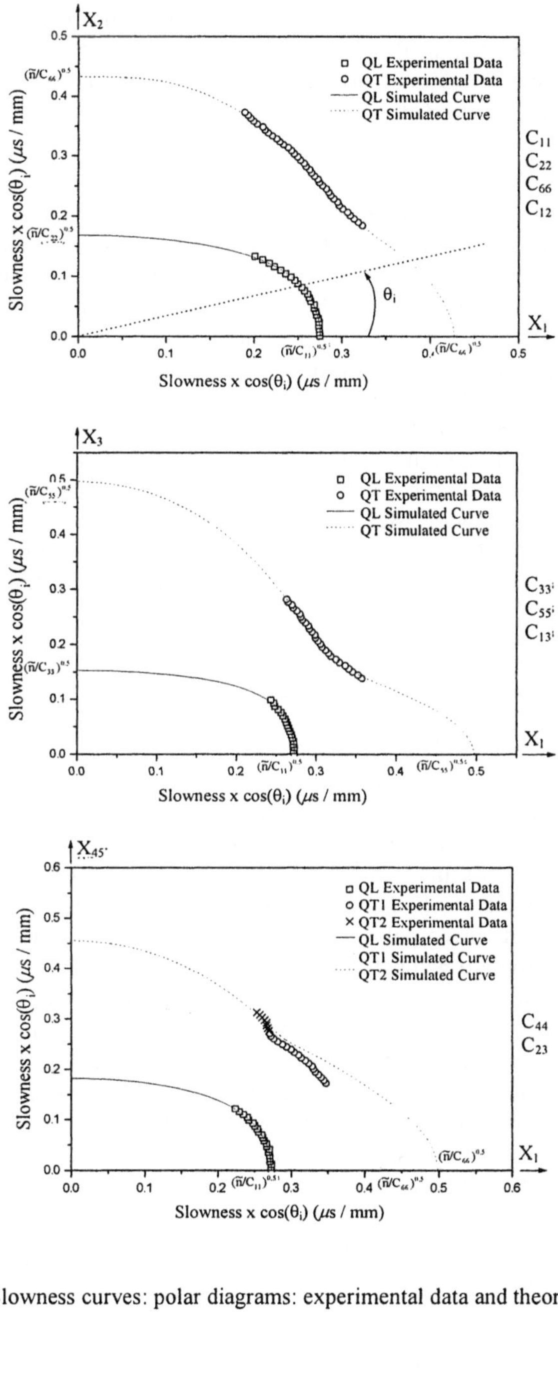

Figure 4 – Slowness curves: polar diagrams: experimental data and theoretical curves

where V is the propagation velocity in the material, in meter per second, (m/s), Vo is the propagation velocity in the coupling fluid, in meter per second, (m/s); h is the specimen thickness in meters, (m), δt is the delay time in seconds, (s) and θ_i is the angle of incidence, in degrees.

3.2. Tensile testing

Tensile tests were performed on 60 mm by 10 mm specimens according to CEN ENV 658-1. The specimens were loaded under constant crosshead velocity up to final failure. Load, displacement and strain (longitudinal and transverse) data were recorded. Material properties (modulus of elasticity, tensile strength and Poisson ratio) were calculated from the obtained data. Prior to testing, the specimens were examined using the C-Scan ultrasonic inspection Acoustic Corporation (PAC) Ultrasonic system (Pulser-Receiver ultrasonic transducers of 5 and 1 MHz) and analysis software. KYOWA strain gauge rosettes were attached to the specimens using the appropriate bonding agent. The tabs used for the gripping of the specimens were cut out of a glass/epoxy composite plate, and were bonded on the specimens using the BFD technicoll 8260 thermoset epoxy resin system. All tests were performed using a MTS Universal Testing Machine equipped with hydraulic gripping system, under displacement control loading conditions in a controlled environment of 25° C and 70% relative humidity.

4. RESULTS

4.1. Non destructive evaluation

Orthotropic materials as well as materials possessing higher symmetry (e.g. transversely isotropic or isotropic) may be fully characterised using the utrasonics non-destructive evaluation. Orthotropic materials possess three orthogonal axes of symmetry which are defined as the principal axes of the material. However, not all three wave are excited when the material is scanned perpendicular to any of the principal axes i.e. its principal planes, and as a result, a non principal plane must be chosen for the complete reconstruction of the stiffness matrix.

In figure 3, the scanning planes are defined, two of which are principal planes. The third plane which is needed for the experimental determination of the second quasi-transverse mode is the plane defined by the X_1 $X_{\psi j}$ axes, where ψ_j=45°.

In figures 4 and 5 the polar slowness curves diagrams and the slowness vs. angle of incidence are presented. The experimentally defined slowness values are depicted as points. The solid lines correspond to the theoretical curves as defined by the calculated stiffness matrix (table 1), after the stiffness components Cij, are introduced into equation 7, and the phase velocity $V^{computed}$ for the different propagation directions n is calculated as the eigenvalues of the propagation tensor Γ_{ij} using the equation :

$$V_p^{computed} = (\lambda_p/\rho_b)^{1/2} \tag{14}$$

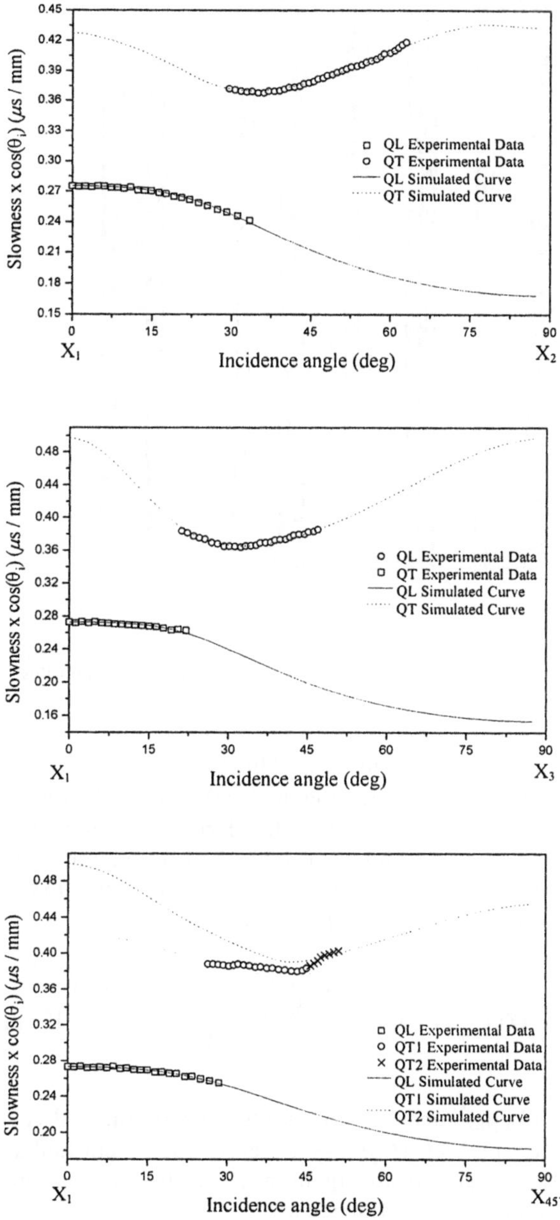

Figure 5 –Incidence angle vs. slowness: experimental data and theoretical curves

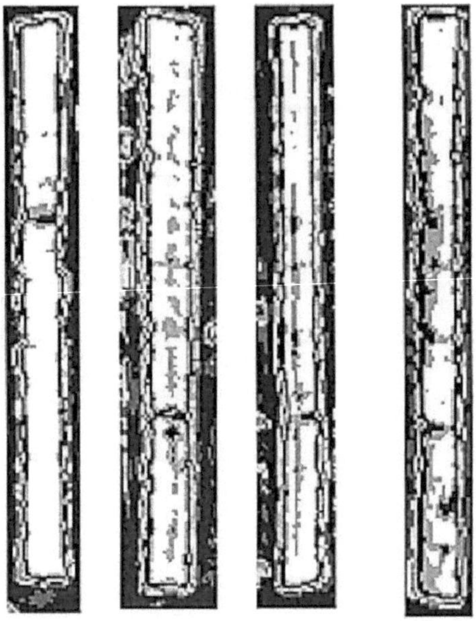

Figure 6: C scan inspection of Oxide/Oxide composites

where λ_p are the calculated eigenvalues. This back calculation of the phase velocities covers the complete range propagation angles $0° \leq \theta_i \leq 90°$ for all the used planes of incidence (ψ = 0°, 90° and 45°). The derived engineering constants and the experimental parameters are shown in table 2. The quadratic deviation (table 2) is obtained from the experimental velocity set and velocities that are computed from the reconstructed stiffness tensor as following:

$$\sigma(C_{ij},V_p)=\sqrt{\frac{1}{N}\sum_{P=1}^{N}\left(\frac{V^{computed}-V^{\exp}}{V^{computed}}\right)^2} \text{ in } \% \qquad (15)$$

where $V^{computed} = (\lambda_p/\rho_b)^{1/2}$ and V^{exp} represents the experimental velocity values.

The confidence interval $I(C_{ij})$, associated to each identified stiffness component, is determined statistically from equation 10 as the root mean square error of a normal distribution at 90% confidence level.

4.2. C-Scan inspection

The C-Scans of the specimens tested in tension is shown in figure 6. As can be seen, there are damaged regions which are aligned normal to the length of the coupons. These regions are related to areas of the initial plate where some problems were identified and may be possibly extended during the cutting phase of the samples. Although these defects are regarded as responsible for premature failure of tested samples they are not expected to macroscopically affect the elastic properties of the material as they were measured during tensile testing.

4.3. Tensile testing

Characteristic stress strain curves for the tested specimens are shown in figure 7. The respective mechanical properties of the material are shown in table 3. The values of stress and strain to failure are derived from samples previously scanned for the presence of flaws.

The modulus and Poisson's ratio values are estimated with an accuracy of ca. 1% and 2% respectively. The Poisson ratio values correspond to the in-plane Poisson ratio which corresponds to the ν_{23} in the coordinate system used for the ultrasonic inspection.

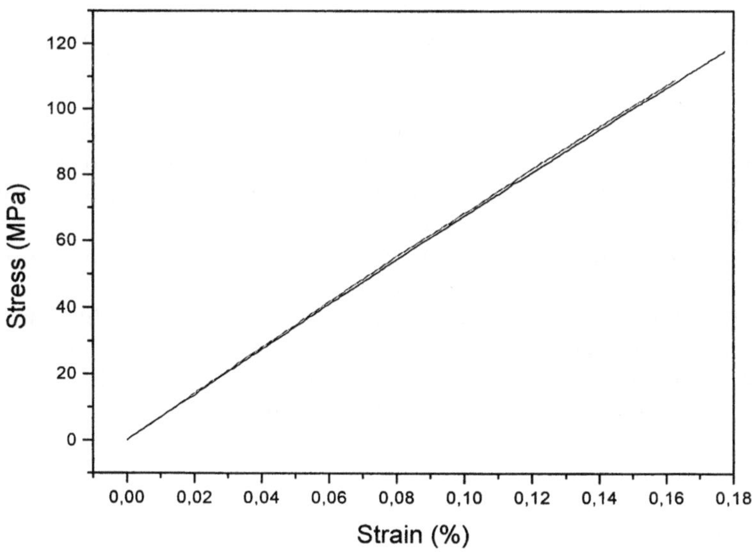

Figure 7: Characteristic stress strain curves for the Oxide/Oxide composites(2 specimens)

5. DISCUSSION AND CONSLUSIONS

The properties of the oxide / oxide composites have been successfully evaluated non-destructively using ultrasonics. The technique yielded high quality measurements for five out of nine components of the stiffness matrix. It is believed that the technique is limited by the accurate determination of the characteristic wave velocities within the material. The accuracy of the delay measurements depends on both the amplitude and the separation of the transmitted waves (figure 2).

The lamination (0°/90° symmetric) of the tested material was expected to yield similar values for the E_{22} and E_{33} engineering constants of the material. The computed difference (62.42GPa Vs. 71.23GPa) (see table 2) is attributed to the statistical nature of the ultrasonic inspection. As is expected and further enhanced by the findings of the C-scan inspection, a reasonable variation of the mechanical properties of the material is expected within the volume of the sample. The statistical variation of the material properties as far as ultrasonic inspection is concerned is more significant as material property variation is comparable to the sampling volume of the method; as the measured differences are expected to correspond to local variations, the estimation of the stiffness matrix of the material in this case would demand multiple ultrasonic sampling. The representative values would then correspond to the mean values of a series of measurements.

The comparison of the ultrasonic measurements of the material properties with the results from the destructive testing reveal very good agreement as far as the modulus of elasticity is concerned. The value obtained destructively (65.6 GPa) is close to the mean value of the computed engineering constants (circa 67GPa) which further enhances the validity of the statistical nature of the ultrasonic evaluation of the engineering constants; as is expected, the tensile test is expected to yield a macroscopically mean value of the modulus.

However, in the case of the in plane Poisson's ratio a significant difference was monitored (ν_{23}= 0.452 as compared to 0.045 measured destructively). The difference may be partly attributed to geometrical constraints imposed due to the finite specimen dimensions during the tensile test, which would account for an underestimation of Poisson's ratio. Such a discrepancy would be indicative of the need of a non conventional method for the estimation of the elastic properties of the material which is not dependent on testing geometry and conditions.

Finally, the ultrasonic technique is sensitive enough to measure differences in principal directions of the composite which are not expected to be present. Moreover, it can be a potentially invaluable tool for non-destructively monitoring the deterioration of the material throughout its service life.

Stiffness (GPa)		Matrix	Confidence Intervals (90%)	Error (%)	Measurement Quality
C_{11}	=	33.28	± 0.13		High
C_{22}	=	85.88	± 1.15	0.33	High
C_{12}	=	12.14	± 0.19		High
C_{66}	=	13.95	± 0.11		High
C_{33}	=	100.83	± 3.85		Low
C_{13}	=	16.21	± 0.64	0.29	High
C_{55}	=	9.97	± 0.47		Low
C_{23}	=	47.95	± 2.25		Low
C_{44}	=	**14.93**	± 3.58	0.46	Low

Table 1: Derived stiffness matrix for the Al_2O_3/Al_2O_3 composite

Engineering		Constants	Units
E_1	=	30.36	GPa
E_2	=	62.42	GPa
E_3	=	71.23	GPa
G_{12}	=	1.40	GPa
G_{13}	=	9.97	GPa
G_{23}	=	14.93	GPa
ν_{12}	=	0.07	-
ν_{13}	=	0.127	-
ν_{23}	=	0.452	-

Material	Al_2O_3/Al_2O_3	
Frequency (MHz)	5	
Thickness (mm)	2.56	
Density (kg/m³)	2480	
Water T (°C)	23	
Coordinate System	Principle	
	$\theta_{critical}$(deg)	
X_1-X_2	**X_1-X_3**	**X_1-X_{45}**
29.3	20.2	25.9

Table 2: Engineering constants and experimental parameters

Property	Mean Value	Standard Deviation
Modulus of Elasticity, **E** (GPa)	65.6	0.686
Poisson ratio, **ν**	0.045	0.001
Tensile Strength, σ_{ult} (MPa)	113.4	6.3
Failure strain (%)	0.17	0.01

Table 3: Tensile properties for the Al_2O_3/Al_2O_3 composite

6. ACKNOWLEDGEMENTS

Many thanks are owed to Professor S. Baste for his invaluable help and Dr Y. Pappas for his contribution to this work. The authors are indebted to the Research Committee of the University of Patras for financial support through the Karatheodoris project and to EADS/Dornier GmbH for providing the materials used in this study.

7. REFERENCES

1. Evans, G. and Cox, F.W.: The physics and mechanics of fibre reinforced brittle matrix composites, *J. Mater. Sci*, **29** (1994), 3857-3896.
2. Naslain, R.: Fibre-matrix interfaces and interphases in ceramic matrix composites processed by CVI, *Composite Interfaces*, **1** (1993), 253-286.
3. Anson, D., Sheppard, W. J. and Parks, W. P.: Impact of ceramic components in gas turbines for industrial cogeneration, ASME Paper 92-GT-393 International Gas Turbine & Aeroengine Congress & Exhibition, Cologne, Germany, June 1992.
4. Johnson, S.M., Blum, Y., Kanazawa, C., Wu, H.-J., Porter, J. R., Morgan, P.E.D., Marshall, D. B. and Wilson D.: Processing and properties of an Oxide/Oxide Composite, *Key Engineering Materials*, **127**, (1997), 231-238.
5. Norton, P.F., Frey, G.A., Bagheri, H., Flerstein, A., Twardochleb, C., Jimenez, O. and Saith, A.: Ceramic gas turbine development project – design and life assessment of ceramic components, ASME Paper 95-GT-383, International Gas Turbine & Aeroengine Congress & Exhibition, Houston, Texas, June 1995.
6. Baste, S. and Audoin, B.: On internal variables in anisotropic damage, *Eur. J. Mech., A/Solids*, **10**, (1991), 587-606.
7. Aristegui, C., and Baste, S.: Optimal recovery of the elasticity tensor of general anisotropic materials from ultrasonic velocity data, *J. Acoust. Soc. Am.* **101** (1996), 813-833.
8. Nayfeh, A. H.: *Wave propagation in layered anisotropic media*, Elsevier, Amsterdam, 1995.
9. CEN WI114: Determination of elastic properties by an ultrasonic technique, *European Technical Committee*, **184**, 2000.
10. Auld, B. A.: *Acoustic Fields and Waves in Solids*, Wiley-InterScience, New York, 1973.
11. Schafer, W., Vogel, W.D., Haug, T. and Frieß, M.: Fibre reinforced ceramics derived from composites, *High temperatures – High Pressures*, **27/28** (1995/1996), 321-324.
12. Audoin, B. and Castagnede, B.: An innovative application of the Hilbert transform to time delay estimation of overlapped ultrasonic echos, *Ultrasonics*, **34** (1996), 25-33.

INVESTIGATION OF THE MECHANICAL CHARACTERISTICS OF FIBRE REINFORCED PHOTOPOLYMERS

D. E. KARALEKAS
University of Piraeus
Karaoli & Dimitriou 80 Str.
GR-185 34 Piraeus, Greece

Abstract. The effect of the addition of fibre mat reinforcements on the mechanical properties of specimens fabricated by the laser solidification process using two different photopolymers, an acrylic- and an epoxy based photo-curable resin, has been studied. Comparisons of the mechanical properties between pure-polymer specimens and their fibre-reinforced counterparts were made, by subjecting the parts to tensile tests. It was found that the fibre-reinforced specimens yielded higher measured values of elastic modulus and ultimate tensile strength.

1. INTRODUCTION

Rapid prototyping (RP) is an advanced manufacturing technology commercialised in the middle of 1980s. At present, RP technology is widely used in engineering for conceptual models and functional models. The application of RP has been shown to greatly shorten the design-manufacturing cycle time, hence reducing the cost of product and increasing competitiveness.

There are many commercial RP systems available on the market today such as stereolithography (SLA), selective laser sintering (SLS), laminated object manufacturing (LOM), fused deposition modelling (FDM), solid ground curing (SGC) and three-dimensional printing (3DP), etc. All RP systems have a limit on the type and properties of materials that can be fabricated.

With the recent advances in computer technology and photo-sensitive polymeric materials, stereolithography has been rapidly developed, being one of the most powerful forms of rapid prototyping techniques. It uses an ultraviolet (UV) laser to selectively harden successive thin layers of photo-curable resin, building each on top of the previous layer until a three-dimensional object is formed. A complete description of the process is given by Jacobs [1], while a schematic representation of the stereolithography process is shown in Figure 1 [2].

Although stereolithography started of as a process for rapidly producing parts for visualization, has in recent years received much attention from researchers throughout the world due to problems associated with the fabrication process. Shrinkage, distortion, micro-fracture, and weak mechanical properties, which affect the accuracy and quality of the part being created, remain the major

E.E. Gdoutos and Z.P. Marioli-Riga (eds), Recent Advances in Composite Materials, 181-188

problems hindering its wider adoption. These inaccuracies are further magnified during the post-processing stage, which involves manual handling and post-curing in a UV chamber. Thus, and taking into consideration that the industry is increasingly seeking functional and semi-functional parts made by this process, understanding and characterization of the formation process is essential in building rapid prototypes of better accuracy and strength.

During the last few years, an overwhelming amount of research work has been carried out to understand and improve both the process and the photo-polymer. Various studies found in literature on photo-polymerisation [2-5] for stereolithography applications have aimed at increasing both the cure speed and degree of polymerisation. The nature and origin of shrinkage distortion in the built prototype part has been studied and documented in the past [6-11]. The recent advances in UV-curing chemistry [12] focus on developing low shrinkage resins which show high efficiency in accelerating the polymerisation process and improving the mechanical properties of the final product. Recent research work has targeted the improvement of the mechanical properties of rapid prototyping parts produced by layered manufacturing techniques utilizing long and short fibres and even micro-spheres [13-16]. The current study investigates the effects of the addition of fibre mats on the mechanical behaviour of specimens built using the stereolithography method.

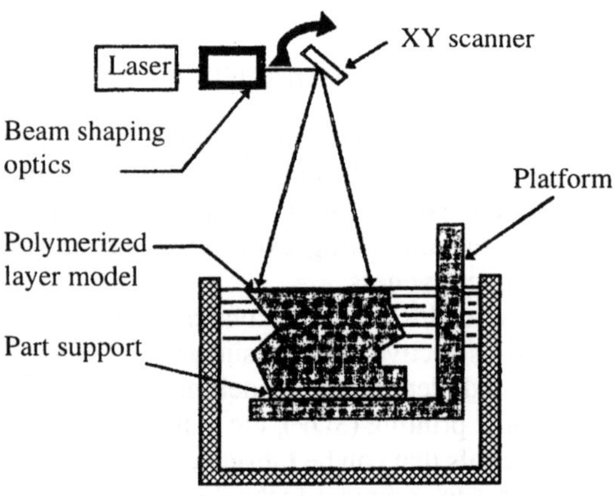

Figure 1. Schematic of stereolithography process [2]

2. EXPERIMENTAL PROCEDURE

Pure resin and reinforced test specimens were built in a stereolithography machine (Stereos Desktop S) using two kinds of photopolymers: an acrylic based resin (AlliedSignal EXACTOMER 2202SF) and an epoxy-based resin (DSM Somos 7110). The layer thickness of each specimen was controlled by pre-setting the layer pitch, which determines the step-size of each successive lowering of the stage on which the specimen was built. The cross-sectional profile was created by a laser beam using the single direction scanning pattern in which one layer of 0,15 mm thickness is scanned along the same direction as the scanning of the previous layer. A single-direction scanning pattern was used and scanning was done perpendicular to the tensile direction. The specimens were built with gauge dimensions of 5 cm x 2 cm x 0,45 cm (length x width x thickness) (Figure 3). The laser exposure density was controlled by adjusting the laser power and the scanning speed was set at 3250 m s^{-1}. Specimens were built using a laser power of 38 mW and were fabricated four at a time, at symmetrical locations within the processing vat, as shown in Figure 2. After layer fabrication, the specimens were post-cured for two hours using an ultraviolet light chamber. Tensile tests were conducted on dogbone shaped specimens using a tensile machine (Zwick Z10) with a 10kN load-cell. The specimens were tested in tension until failure, with the main parameters observed being the tensile modulus and strength. The crosshead speed of the machine was set to 5mm/mim and the force versus extension data was collected using a digital data acquisition hardware and software.

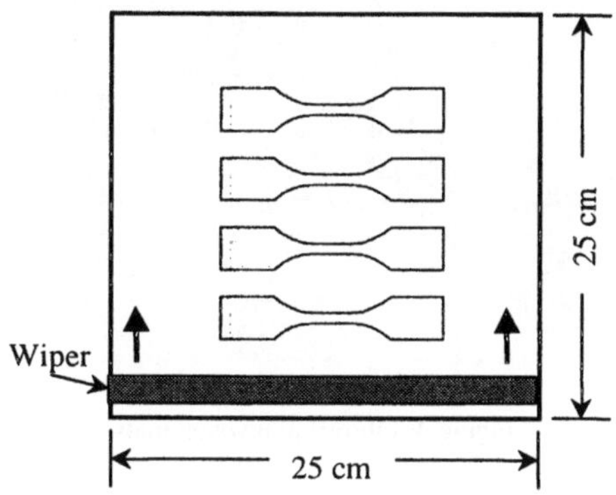

Figure 2. Location of dogbone specimens within the vat (top view)

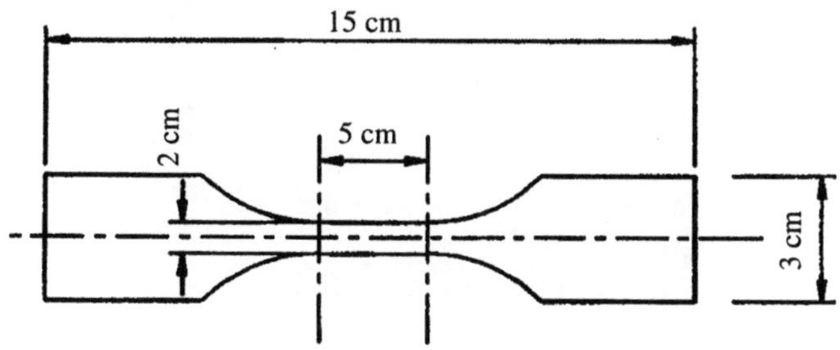

Figure 3. Dogbone-shaped tensile layered specimen

As a reinforcing material, several types of fibrous nonwoven mats were selected (Figure 4). E-Type Glass, PAN-based carbon and Para Aramid fibre mats, of different basis weight (g/m^2) and thickness, were acquired from Technical Fiber Product Inc. In the first phase of the research work, presented here, a single layer of reinforcing mat was placed in the mid-plane of the test specimens (Figure 5). Important fabrication questions that had to be solved, was firstly how to ensure adhesion of the reinforcing layer into the resin specimen and secondly how to achieve flat placing of the mat, avoiding thus phenomena of waviness or wrinkling as a result of the wiper movement placing the next layer coats to be solidified by the laser. When half the thickness of the specimen was laser built the moving platform was stopped and a precut layer of fibre mat was manually placed on top of the last cured flat surface. At the end of the fabrication process the specimens were removed from the machine platform and examined. No wrinkling of the embedded reinforcing mat was observed although in most cases a slight deviation from its original placement (parallel to the outer flat surfaces of the specimen) had taken place. A typical fibre mat reinforced specimen is shown in Figure 6.

Figure 4. Fibrous nonwoven mats

Figure 5. Single layer of glass nonwoven mat

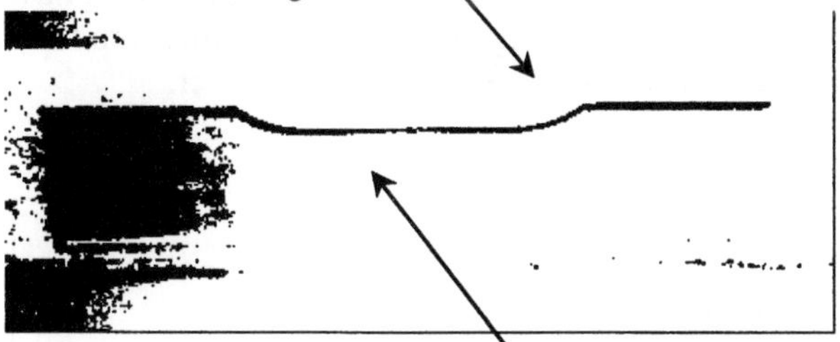

Figure 6. Fibre mat reinforced tensile specimen

3. RESULTS AND DISCUSSION

Figure 7 shows the mechanical behaviour of plain and fibre mat reinforced specimens made of acrylic-based photopolymer resin. In Figure 8, the load versus displacement curves of a plain epoxy-based resin specimens as well as two of its counterparts containing in their mid-plane embedded a single layer of glass fibre mat, of different basis weight, are presented. It is noted that the pure epoxy-based resin specimens are less ductile than the acrylic-based ones, fracturing at much higher loads. It is also seen that all reinforced specimens have become more brittle with the addition of the fibre mats.

For the acrylic-based specimens it is seen that in all cases addition of a single layer of nonwoven mat leads to an increase of the modulus of elasticity. Vast differences in the magnitude of the elastic modulus between the glass and carbon fibre reinforced specimens and their non-reinforced counterparts can be observed, indicating improvements in the stiffness of the reinforced specimens. In addition, the ultimate tensile strength increases considerably for the glass and carbon reinforced specimens while there is a significant decrease in the elongation to fracture. On the other hand, for the epoxy-based reinforced specimens it was observed that a small increase of the elastic modulus was obtained only in the case of the lower basis weight (34 g/m^2) glass reinforcement. The impregnation of the higher basis weight (50 g/m^2), and thicker, glass mat did not lead to any improvements in the modulus of elasticity. Furthermore, all reinforced specimens fractured at lower loads to those measured for the pure resin specimens.

From the experimental results obtained for the epoxy-based specimens, important considerations of degree of bonding and inter-laminar (reinforcing layer mat – solidified liquid resin layer) are raised since without strong adhesion between them it is not possible to obtain good mechanical properties. It is noted that for the reinforced epoxy-based specimens uncured resin spots close to the edges were observed at the end of the laser curing process. Post-curing in the UN chamber had to be carried out to fully cure the specimens. However, this further consolidation process presents obvious challenges to the mechanical properties of the composite, since specimens may experience significant stress concentration, shrinkage, warpage, delamination, and/or shape deformation during post-curing. Previous experimental studies on stereolithography built composites [13] have demonstrated a dependence of the resulted shrinkage in short fibre-filled reinforced specimens on processing parameters such as layer pitch and laser power.

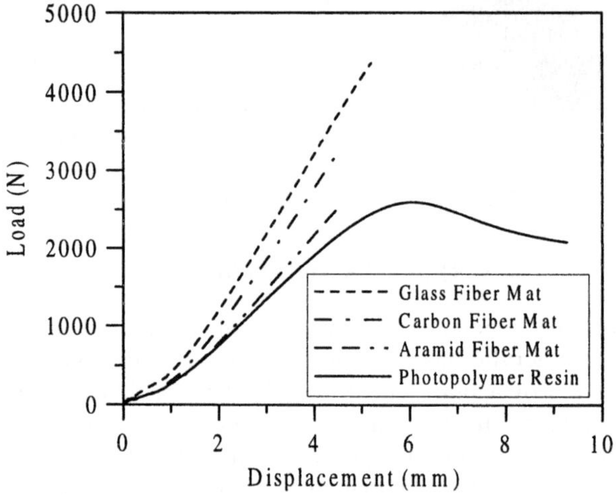

Figure 7. Typical load vs. displacement curves of acrylic-based photopolymer specimens

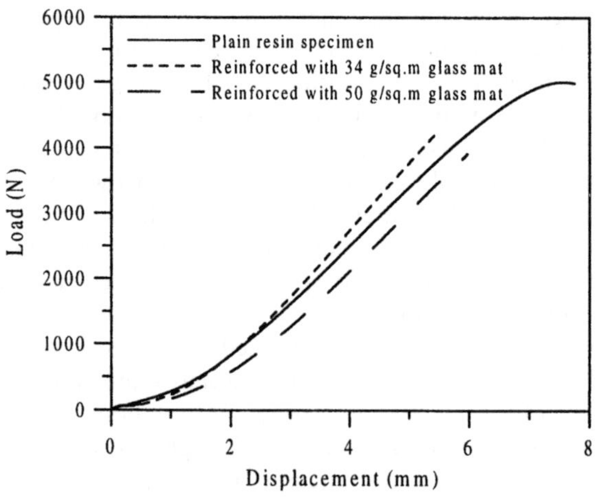

Figure 8. Typical load vs. displacement curves of epoxy-based photopolymer specimens

4. CONCLUSIONS

The aim of the undertaken work was to experimentally investigate if the embedment of commercially available fibre mats into photo-curable resins can improve the mechanical properties of the laser fabricated parts. It has been shown that the addition of just a single layer of fibrous material, in a mat/veil form, in an acrylic-based photopolymer increases significantly the elastic modulus and the mechanical strength of the stereolithography built prototype, compared to its pure resin counterpart. On the other hand, the mechanical properties did not improve considerably by the introduction of glass fibre mats to the epoxy-based photopolymer material used in this study. Future work should consider the effects of reinforcement-resin bonding problems, and of varying processing parameters (laser power, layer pitch, scan directions) on the mechanical properties of stereolithography built reinforced prototypes.

5. ACKNOWLEDGEMENTS

The author would like to acknowledge the contributions of Mr. G. Giannatsis for fabricating the test specimens and of Misses A. Adoniou and D. Karvouniari for conducting the tensile tests.

6. REFERENCES

1. Jacobs, P.F.: *Rapid Prototyping and Manufacturing: Fundamentals of Stereolithography*, Society of Manufacturing Engineers, Dearborn, MI, 1992.
2. Fuh, J.Y.H., Lu, L., Tan, C.C., Shen, Z.X., and Chew, S.: Curing characteristics of acrylic photopolymer used in stereolithography process, *Rapid Prototyping Journal*, 5(1) (1999), 27-34.
3. Fuh, J.Y.H., Lu, L., Tan, C.C., Shen, Z.X., and Chew, S.: Processing and characterising photo-sensitive polymers in the rapid prototyping process, *Journal of Materials Processing Technology*, 89-90 (1999), 211-217.
4. Ang., B.Y., Chua, C.K., and Du, Z.H.: Study of trapped material in rapid prototyping parts, *The International Journal of Advanced Manufacturing Technology*, 16 (2000), 120-130.
5. Onuh, S.O., and Hon, K.K.: Improving stereolithography part accuracy for industrial applications, *The International Journal of Advanced Manufacturing Technology*, 17 (2001), 61-68.
6. Flach, L. and Chartoff, R.P.: Laser scan rates and shrinkage in stereolithography, Proceedings of the 4th International Conference on Rapid Prototyping, June 14-17, 1993, Dayton, Ohio, 239-248.
7. Weissman P.T., Bolan, B.A., and Chartoff, R.P.: Measurement of linear shrinkage and the residual stresses developed during laser photopolymerisation, Solid Freeform Fabrication Proceedings, The University at Texas at Austin, August 8-10, 1994, Austin, Texas, 103-112.
8. Lu, L, Fuh, J.Y.H., Nee, A.Y.C., Kang, E.T., Miyazawa, T., and Cheah, C.M.: Origin of shrinkage, distortion, and fracture of photopolymerised material, *Materials Research Bulletin,* 30 (12) (1995), 1561-1569.
9. Wiedemann, B., Dusel, K.-H., and Eschl, J.: Investigation into the influence of material and process on part distortion, *Rapid Prototyping Journal*, 1(3) (1995), 17-22.
10. Aggelopoulos A., and Karalekas D.: Determination of cure shrinkage in SL layer built plates using lamination theory, *Advanced Composites Letters*, 10(1) (2001), 7-12.
11. Karalekas D., Rapti D., Gdoutos E. E, and Aggelopoulos A.: Investigation of shrinkage induced stresses in stereolithography photo-curable resins, *Experimental Mechanics*, 42 (4) (2002), 1-6.
12. Ullett, J.S., Benson-Tolle, T., Schultz, J.W., and Chartoff, R.P.: Thermal-expansion and fracture toughness properties of parts made from liquid crystal stereolithography resins, Materials & Design, 20 (1999), 91-97.

13. Cheah, C.M., Fuh, J.Y.H., Nee, A.Y.C., and Lu, L.: Mechanical characteristics of fiber-filled photopolymer used in stereolithography, Rapid Prototyping Journal, 5 (3) (1999), 112-119.
14. Zak, G.: Rapid layered manufacturing of short-fibre-reinforced parts, PhD Thesis, Department of Mechanical and Industrial Engineering, University of Toronto, 1999.
15. Zak, G., Haberer, M., Park, C.B., and Benhabib, B.: Mechanical properties of short-fibre layered composites: prediction and experiment, Rapid Prototyping Journal, 6 (2) (2000), 107-118.
16. Klosterman, D., Chartoff, R.P., Graves, G., Osborne, N., and Priore, B.: Interfacial characteristics of composites fabricated by laminated object manufacturing, Composites Part A, 29A (1998), 1165-1174.

CORRELATION BETWEEN MICROMECHANICAL AND MACROMECHANICAL PROPERTIES OF COMPOSITE MATERIALS BASED ON A NEW INTERPHASE CONCEPT

G.C. PAPANICOLAOU and N.K. ANIFANTIS
Composite Materials Group
Mechanical and Aeronautical Engineering Department
University of Patras,
GR-265 00 Patras, Greece

Abstract. In the present investigation, a novel approach of the interphase concept is introduced according which the interphase volume fraction represents the percentage of the bulk matrix surrounding the inclusions in which a specific matrix property is strongly affected by the existence of the reinforcement, while the interphase thickness represents the maximum radial distance from the inclusion boundary at which this property varies. This means that both the interphase volume fraction and the interphase thickness are not simple geometrical/structural concepts but they are property-dependent, that is, their values depend on the property considered at the time. Such an interphase is the so-called *"hybrid interphase"*. Based on this new interphase concept, a micromechanics semi-analytical model has been developed for the determination of stresses developed in the interfacl area of unidirectional fiber - reinforced composites incorporating a hybrid interphase region. .Macromechanical material characterization and predictions of interfacial stress distribution indicate the effect of the hybrid interphase, while a satisfactory agreement between theoretical predictions and respective experimental results was found.

1. INTRODUCTION

The need for high strength, light weight, damage tolerant structural materials in the aerospace industry has led to the development of various high performance composite material systems. Towards this end, high strength, high modulus inclusions and tough, damage tolerant matrices have been developed for use in structural applications. During the manufacturing of these materials, the presence of inclusions in the matrix material affects the polymerization of the polymeric matrix and the mobility of the macromolecules at the interfaces. As a result, reduced conformations of molecules in boundaries occur and denser interfacial layers are formed in comparison to the bulk matrix [1-2]. In semicrystalline thermoplastic polymer matrices, crystallization in the presence of fibers may result in the development on the inclusion surface of a transcrystalline region, which acts as an interphase. This has been shown for a number of polymers. Whereas the thermal treatment controls the type and degree of crystallinity, the type of inclusion determines the ability to form a transcrystalline layer. Like the interphase in thermoset matrix composites, the transcrystalline layer is expected to have an effect

E.E. Gdoutos and Z.P. Marioli-Riga (eds), Recent Advances in Composite Materials, 189-202

on the mechanical performance of the composite. It has been demonstrated, for example, that transcrystallinity enhances the fibre/matrix interfacial bond strength and increases the transverse strength of the composite [3]. This analysis enables the fact that in the matrix area surrounding embedded inclusions a complex situation occurs revealing imperfect bonding, material gradients and stress concentrations. In such a situation, the composite may be considered as not consisting from two phases, but during manufacturing a third phase is developed due to component interaction. This extra phase is commonly called *interphase*.

The role of interphase in composite materials is the focus of an increasing number of studies. An interphase is a third, relatively thick, intermediate phase present between the constituents. Its elastic and mechanical properties effect the overall performance of the composite material. For example, a gradient modulus interphase approach has been proposed to promote the mechanical properties and the fatigue life by reducing the modulus mismatch of the reinforcement and the matrix, while a soft, elastic interphase has been advocated for high fracture toughness, and for reduced stress concentrations in the matrix around the reinforcements. A number of theoretical models link the properties of the interphase and its relative thickness to the performance of the composite material.

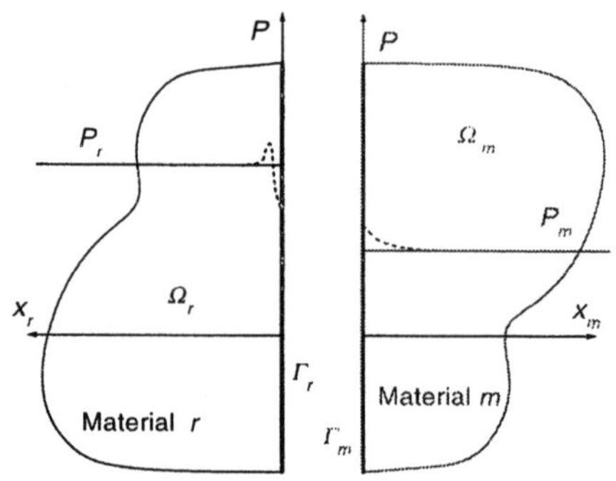

Figure 1. Two opposing semi-infinite media with material property mismatch.

According to the IUPAC *Compendium of Chemical Terminology* [4], an interfacial layer is defined as the inhomogeneous space region intermediate between the primary phases in contact, and where properties are significantly different from, but related to, the properties of the bulk phases. Nowadays, the existence of the interphase is not an assumption. Its existence has already been experimentally proved by a great number of researchers and its properties have already been experimentally measured and theoretically modeled. Researchers dealing with the micromechanics of reinforced composite materials have long recognized the need for including an interphase separating the reinforcement inclusion phase from the host matrix phase. According the previous discussion, the interphase may represent: a physical coating on the inclusion, a degraded material zone, a weak interfacial layer due to imperfect bonding between primary phases, a diffusive or chemical interaction zone, and an inelastic region in metal-matrix composites. This study addresses the first four types of interphase but not the fifth. If the properties of the interphase are distinct from those of the primary phases, their influence on the effective properties of the composite needs to be modeled explicitly. Several researchers have contributed to an improved understanding of the different aspects of the interphase region as evident from the abundance of experimental and analytical work reported in literature. Various models have been considered for the physical properties of the interphase in an effort to study the effect of the interphase on residual thermal stresses [5-11], stress distribution [12-15], characterization [16-21], and so forth, in a composite. Models have also been proposed [22-27] for the formation of the interphase and the resulting interphasial properties and substantiated in a limited manner through experiments. These models range from hypothetical to semi-empirical and provide approximate representations of the interphase. Although the representations provided by both experiments and models are approximate, the models take into account some of the complexities of the interphase that are usually not considered in the interpretation of the experiments.

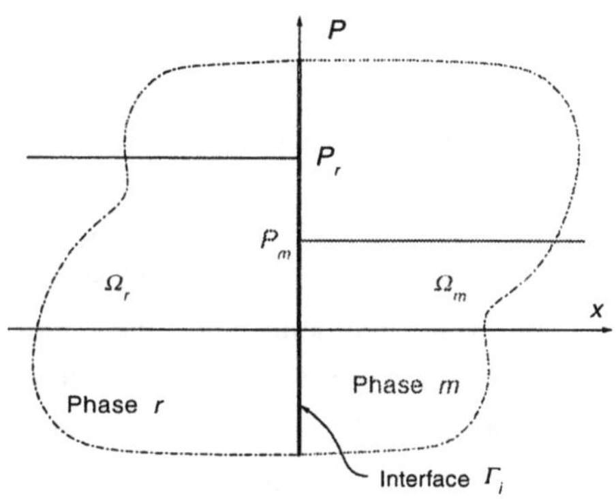

Figure 2. Development of an interface from the bonding of two semi-infinite media.

In a recent publication [28], the thickness of the interphase formed between a glass fibre and a polymer resin has been measured by applying three different experimental techniques, namely, nanoindentation, nanoscratch and thermal capacity jump measurements. By comparing the results extracted from these techniques, great differences in the interphase thickness values were found, which depended on both the experimental technique used and the particular property measured at the time. These differences were in some cases on the order of 30% and in one case more than a 100%. The interphase thickness has been measured by means of the interphase elastic modulus variation applying nanoindentation. The coefficient of friction has been predicted by applying scratch tests and the thermal capacity by applying DSC-tests. The authors of [28] attributed these differences to the different sensitivity of each one of the methods used. If this explanation is adopted, then the question, which arises, is: what is the real interphase thickness value? Our opinion is that there is no a unique value for the interphase thickness. Each experimental method is based on a principal which, in turn, is based on the measurement of a specific property. Thus, in that sense, all of the interphase thickness values thus taken are correct. As a result, what is called interphase thickness must be reconsidered. This article presents a survey of the models available for the characterization of the physical properties of the interphase and addresses a novel model referred in the sequel as *hybrid interphase model*. According to this approach, the interphase volume fraction represents the percentage of the bulk matrix surrounding the inclusions in which a specific matrix property is strongly affected by the existence of the reinforcement, while the interphase thickness represents the maximum radial distance from the inclusion boundary at which this property varies. This means that both the interphase volume fraction and the interphase thickness are not simple geometrical/structural concepts but they are property-dependent, that is, their values depend on the property considered at the time. Such an interphase is the so-called *hybrid interphase*.

2. COMPLIANT INTERPHASE MODELS

Consider two semi-infinite subdomains, occupied by different media, in a two-dimensional space (x_a, y_a), as shown in Figure 1. Subscripts $a = r, m$ denote the separate media i.e. the reinforcement and the matrix, respectively. These homogeneous or nearly homogeneous media are defined on the subdomains, Ω_r, Ω_m having the boundary surfaces Γ_r, Γ_m, respectively. Manufacturing operations or processing conditions may arise local inhomogeneities or residual stresses near a very narrow surface layer of each medium. Excluding the presence of residual stresses, these materials are stress free and the boundary surfaces are subjected to the same condition. The letters P_r, P_m denote a specified material property, P_a, such as the modulus of elasticity. Obviously, $P_r \neq P_m$ and the homogeneity condition requires that $P_r = const.$ and $P_m = const.$ Figure 1 illustrates the variation of these properties on each subdomain.

If by an appropriate manufacturing procedure these materials come and join together, as illustrated in Figure 2, then a composite structure results in occupied by an infinite medium having piecewise variation of the material property, P_a. The new domain is $\Omega = \Omega_r \cup \Omega_m$, and the boundary surfaces Γ_a form a surface called *interface*, Γ_i. Due to the problem definition, the normalized volumetric composition of each constituent in the composite structure is $\upsilon_r = \upsilon_m = 1/2$. Along this interface, a jump in the material property P is occurring. A question then arises, regarding the micromechanics of the interface, i.e. what is the real variation of the property P_a on the position $x = 0$, how it is affected on the manufacturing procedure, the preexisting surface inhomogeneities, the residual stress, and the bonding conditions between the primary material phases?

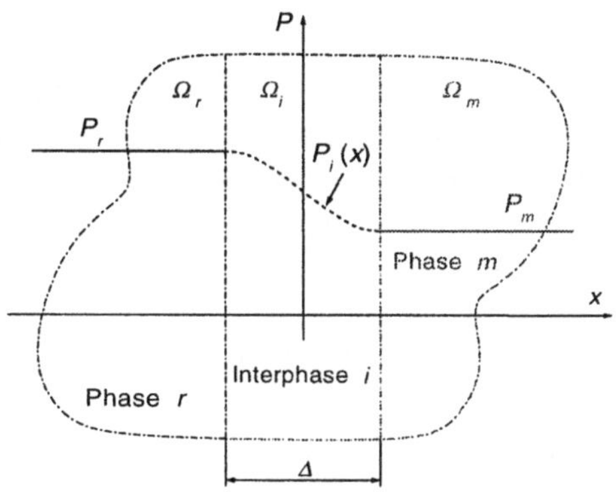

Figure 3. Development of an interphasial layer between two primary material phases.

A simplified way to treat this question reveals to the concept of the interphase. This layer extends between the primary phases assuming expansion of the interface to this layer, Figure 3. In this manner, a new phase is defined occupying the subdomain Ω_i, and having volume fraction υ_i, very small compared with the volume fraction of the primary phases. The conventional definition of the interphase considers constant thickness Δ, degradation of material property P, and compliance according to the physics of the problem. The gradient of the function P follows the material phase sinteraction, i.e. changes from P_r to P_m. Inefficient bonding conditions may be included in this degradation by setting $P_{r-}(x=-\Delta/2) \neq P_{r+}(x=-\Delta/2)$. This type of material discontinuity prevents displacement or traction singularities and describes as well the two-phase composite structure if the difference between properties is set to be $P_{r-}(x=-\Delta/2) - P_{r+}(x=-\Delta/2) = P_r - P_m$. The requirement for a compliant interphase demands for continuity of P at $x = \Delta/2$. Various laws of variation of property P within the interphase $P_i = P_i(x)$ have been proposed in literature, ranging from linear, to exponential or more complicated [14-16, 22, 24]. The most efficient model involves a compliant interphase with exponential material degradation involving imperfect adhesion conditions. According to this model, the property P within the interphase varies as

$$P_i(\xi) = P_m\,[1 + (\lambda_P\, q_P - 1)]\, R(\xi) \tag{1}$$

The function $R(\xi)$ describes the exponential material degradation

$$R(\xi) = (1 - \xi\, e^{1-\xi})/(1 - \xi_r\, e^{1-\xi_r}) \tag{2}$$

The composite material structure being characterized by the parameters

$$\lambda_P = P_r / P_o, \qquad q_P = P_r / P_m. \tag{3}$$

In above equations, the variables ξ, ξ_r express the normalized abscissa x, and its value on the interface with respect to some reference length L, respectively. This model has been applied extensively in the past to study the response of composite materials including both fibrous and particulates. Some outcomes of these investigations will be discussed later on.

3. HYBRID INTERPHASE MODEL

Considering the composite material construction illustrated in Figure 3, the model involves the reinforcing phase, the matrix phase and the interphase, assigned by the subscripts r, m and i, respectively. The hybrid interphase model follows some of the ideas set forth in the corresponding models available in literature and involves some additional aspects as described in the following. An inhomogeneous material layer lying between the reinforcement and the matrix interprets this. The material properties of the interphase region differ from the respective ones of primary phases. Although the model may exhibit overall anisotropy, the reinforcing and matrix phases are assumed to be homogeneous and isotropic. The property mismatch index $q_P = P_r / P_m$ determines the overall degree of inhomogeneity between the fiber and matrix, where the subscript P denotes the property under consideration. Sometimes, the material phases are non-isotropic. Then, the anisotropy index $s_P = P_L / P_T \geq 1$ defines the overall anisotropy of the composite material structure, the subscripts T, L being denoting the principal values of material properties along transverse and longitudinal directions, respectively.

Improper adhesion between the reinforcing member and the matrix may release bonding conditions and thus causes changes in the stiffness of the construction. The perfect bonding implies continuity of tractions and displacements at the boundaries of interphase, and reveals for the formulation of a compliant interphase. An equivalent way to model the localized loss of stiffness, depending on the degree of debonding or damage in the interphase, lies on the formulation of a simultaneously compliant and weak interphase having lower properties than those of the reinforcement, but preserves continuity of tractions and displacements [22, 29]. On the side of matrix, the interphase is compliant. This condition requires that

$P(\xi_P)=1$ and $P'(\xi_P)=0$, where primes indicate differentiation in respect to ξ, and $\xi_P=\Delta_P/L$. On the side of reinforcement, the designation of reduced properties simulates the presence of imperfect adhesion conditions. Then, at the reinforcement surface exists a discontinuity on derivative $P'(\xi_{r-})\neq P'(\xi_{r+})$, and a jump in the property of the normalized form $P(\xi_r)=\lambda_P\, q_P$. The parameter λ_P describes the efficiency of bonding for property P at $\xi=\xi_r$, where ξ_r is the position of the reinforcement interphase bounding surface. To be the model compliant with the properties of primary constituents, this parameter must lie in the range $1/q_P \leq \lambda_P \leq 1$. The value $\lambda_P=1$ corresponds to a three-phase model with a distinct property $P(\xi)$, at $\xi=\xi_o$. The marginal value $\lambda_P=1/q_P$ yields a two-phase model, and maintains minimum adhesion conditions, the interphase being disappeared. When the parameter λ_P varies between $1/q_P \leq \lambda_P \leq 1$, intermediate three-phase composite materials result, which present a discontinuity at $\xi=\xi_r$. This general formulation describes imperfect adhesion as an extension of the perfect one, by assigning proper values to the parameter λ_P.

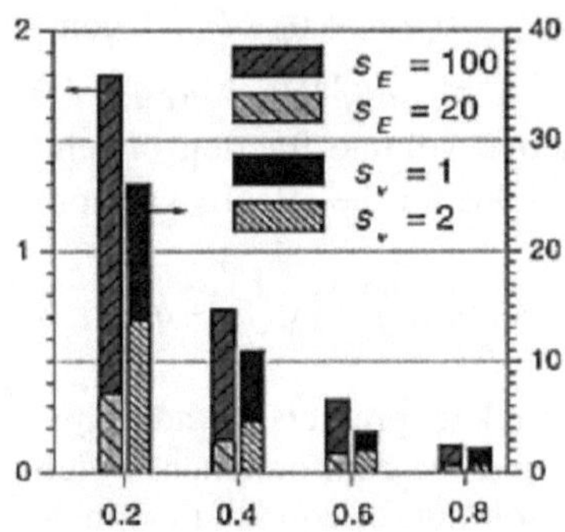

Figure 4. Thickness of hybrid interphase for a fibrous composite.

In order to derive an equation describing the variation of property $P(\xi)$ within the interphase, some assumptions are required. The first assumption states that each property of the interphase $P(\xi)$ has a rate of variation, which is proportional to the property. There is a great difference between the values of the properties of the matrix and the respective values of the fiber. The thinner is the interphase; the more abrupt is the transition from the properties of the fiber to the respective properties of the matrix. Assuming that the rate of variation of each material property within this region is proportional of the property strength, one can write the equation

$$\mathrm{d}P(\xi)/\mathrm{d}\xi = P(\xi)/\Delta_P \tag{4}$$

The second assumption presumes that the interphase thickness depends on the bonding efficiency, the property under consideration, and the nature of properties of composite constituents. Thus, the interphase thickness must be written in terms of these parameters. When perfect adhesion conditions exist, the interphase thickness diminishes. On the other hand, if no adhesion exists, the interphase thickness is infinite. The thickness of the interphase is a function of the respective property under consideration and of the overall aspects of the model. These simple conditions may be written down in the form

$$\Delta_P = s_P(1-\lambda_P)/(\lambda_P - 1/q_P) \tag{5}$$

Based on above observations, equation (5) relates the bonding efficiency parameter λ_P to the hybrid interphase thickness that corresponds to property P. When perfect bonding exists, i.e. for $\lambda_P \to 1$, then the normalized thickness Δ_P tends to zero, while the limit $\lambda_P \to 1/q_P$, yields a thickness dependent on the property mismatch index q_P. Assuming that υ_r, υ_m are the volume fractions of the reinforcement and matrix, respectively, the law of mixtures yields the conditions $P_L = \upsilon_r P_r + \upsilon_m P_m$ and $1/P_T = \upsilon_r / P_r + \upsilon_m / P_m$. Upon substitution, the anisotropy index takes the form $s_P = (\upsilon_r + \upsilon_m / q_P)(\upsilon_r + \upsilon_m q_P)$. Accounting on the existence of the interphase, the effective volume fraction of reinforcement is assumed to be $\upsilon_r{*} = \lambda_P \upsilon_r$. This analysis yields for the effective anisotropy the following equation

$$s_P{*} = [\lambda_P \upsilon_r + (1-\lambda_P \upsilon_r)/q_P]\,[\lambda_P \upsilon_r + q_P(1-\lambda_P \upsilon_r)] \tag{6}$$

Equation (6) may be used to predict the adhesion efficiency in terms of the effective anisotropy index and the reinforcement volume fraction, which both are measurable quantities. The solution of equation (4) in conjunction with equation (5) that satisfies the above stated conditions, has the general form

$$P = P_m \begin{cases} 1 + (\lambda_P q_P - 1)\dfrac{\xi\xi_P e^{1-\xi\xi_P} - 1}{\xi_P e^{1-\xi_P} - 1} & \text{when} \quad \xi_r \le \xi < \xi_P \\ 1 & \text{when} \quad \xi = \xi_P \end{cases} \tag{7}$$

Equation (7) expresses the material degradation within a compliant hybrid interphase, in terms of material inhomogeneity, volumetric composition, and imperfect adhesion.

4. PIECEWISE HOMOGENIZATION PROCEDURE

In order to elucidate the role of the overall features and the micromechanics aspects of composite structures on the material response, an appropriate method must be developed. The presence of inhomogeneities render the problem unresolved by pure analytical procedures, leading to the development of approximate ones. Among these, the method of successive continuously varying subdomains with piecewise homogeneous materials defined on them is adopted for the analysis of the imposed problem. This approach is axially an extension of the CCM [11, 30-31], and it is implemented briefly in this section to explain how it is possible to predict the response of composite structures incorporating a hybrid interphase. To reduce the size and the complexity of the approach, the analysis is performed on an appropriate unit volume accounting the symmetry conditions and the packing sequence of the composite. The domain of the problem consists from three primary subregions i.e., the inclusion, the interphase and the matrix. Materials defined on inclusion and matrix subregions are isotropic and homogeneous ones. Hybrid interphase material is non-homogeneous and its spatial dependence is discussed in previous section. The coordinate system depends on the spatial, cylindrical or spherical geometry of the problem.

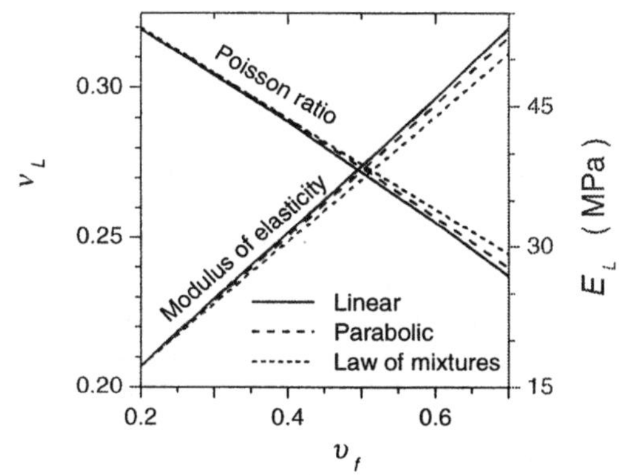

Figure 5. Prediction of longitudinal material properties of a fibrous composite.

To treat material inhomogeneity, this method defines a set of subdomains with equivalent homogeneous material properties. Subdivision of the maximum interphase thickness $\Delta_P = \xi_P - \xi_r$ that corresponds to some property P, into n segments of equal length enables discretization of the problem domain into $n+2$ subdomains. Each subdomain is uniquely defined in the region $[\xi_e, \xi_{e+1}]$, where $e = 1, 2, \ldots n+2$. This discretization enables piecewise approach also of the field functions $u = u(\xi)$, by a sequence of functions $u_e(\xi)$ each one defined within the segment $[\xi_e, \xi_{e+1}]$. A particular material property $P = P(\xi)$ within the interphase varies according to equation (7). This function is approximated over the segment $[\xi_e, \xi_{e+1}]$ by the average value $P_e = [P(\xi_e) + P(\xi_{e+1})]/2$. Then, over each subsegment of hybrid interphase all material properties are defined uniquely, though some of the outermost may occupy the matrix region and thus keep the homogeneous properties of the matrix.

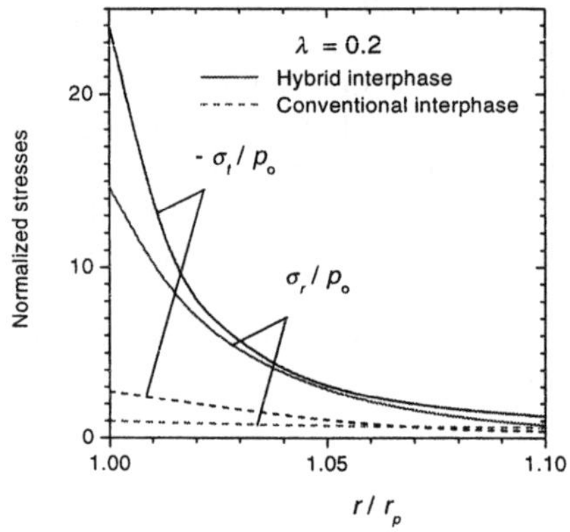

Figure 6. Stresses developed within the interphase region of a particulate material.

This methodology allows for analysis only on a particular subsegment $[\xi_e, \xi_{e+1}]$ and handling of appropriate equations with the aid of numerical procedures. Beyond usual boundary conditions, interfacial conditions accounting on continuity and compatibility between adjacent elements must be imposed to the model. The piecewise homogenization approach described in the previous section may be formulated in such a way that allows automatic evaluation of matrix coefficients assembly of equations, and construction of solution assemblage. A regenerative numerical procedure can be formulated, which enables solution of any particular composite construction, defined in terms of packing sequence, volumetric composition, materials and imperfect adhesion.

5. FIBROUS MATERIAL CHARACTERIZATION

The characterization of a fiber-reinforced material needs a great number of parameters. Despite the volumetric composition and microscopic structure, the physical behavior of the composite system depends on the particular properties of primary material phases as well as on their interaction. The proposed hybrid interphase model involves such interactions of parameters and thus it is expected to yield more realistic material characterizations to composites than the conventional interphase. The prediction of the macroscopic material moduli in terms of both the material properties of the components and the aspects of the composite is one of the major tasks in scientists involved with such materials. In this section, the effect of interphase aspects on the material longitudinal material properties of fibrous composites is studied. With reference to a unidirectional fiber-reinforced epoxy, the piecewise homogenization procedure developed in previous section, allows for the approximation of the stress equilibrium problem and computation of principal material properties by means of energy balance principles [16]. Equation (7) in terms of polar coordinates yields the variation of the interphasial modulus of elasticity and Poisson ratio, respectively

$$E_i(r) = E_m(\lambda_E\, E_f - E_m)\,\mathrm{EXP}\left(-\lambda_E\, s_E\, \Delta_E / r_f\, (1-\lambda_E)\right) \tag{8}$$

$$v_i(r) = v_m(\lambda_v\, v_f - v_m)\,\mathrm{EXP}\left(-\lambda_v\, s_v\, \Delta_v / r_f\, (1-\lambda_v)\right) \tag{9}$$

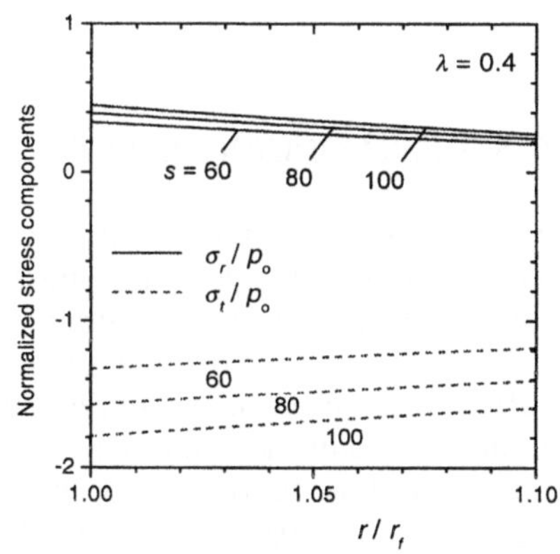

Figure 7. Strains developed within the interphase region of a fibrous material.

In the above equations, r_f is the fiber radius, E, v being the modulus of elasticity and the Poisson's ratio, respectively. These equations define a hybrid interphase, being not compliant on the matrix side, and having different thickness with respect to these properties. Considering hexagonal fiber packing sequence, the inter-fiber spacing is

$$d = 2\left(\sqrt{\pi / 2\sqrt{3}\,\upsilon_f} - 1\right) r_f \tag{10}$$

When $2\Delta_P > d$, then the interphase is stiffer than the matrix at the position $r = r_i = \Delta + r_f$. This happens for high volumetric compositions, where the fibers come close. Figure 4 illustrates the thickness of interphase for various values of the adhesion efficiency, assuming that $\lambda = \lambda_E = \lambda_\nu$. The modulus of elasticity being the most dominant property has very small domain of variation, though the Poisson ratio being the less dominate property has very large domain of variation. Depending on the adhesion efficiency, these variations may differ one order of magnitude with respect to the fiber radius. The overall anisotropy seems to play a less important role in the extent of the interphase. Figure 5 illustrates predictions of the longitudinal material properties for the fibrous composite data given in reference [16]. For comparison reasons, results are presented for three different approaches of the conventional interphase, i.e. linear and parabolic laws of variation plus one more accounting on the law of mixtures. The application of the hybrid interphase to this composite yields the same approximately results as those predicted by the parabolic law of variation. These predictions show that the interphase as involving phase interactions, affects the macroscopic material properties, especially for higher fiber volume fractions. The influence on the macroscopic property may be of the order of 5%.

6. STRESS CONCENTRATION

Recent investigations have shown that the boundary layer surrounding the inclusion in a composite material plays an important role in the stress transfer. Due to steep material gradients, within this layer high stress concentration is developed and thus may be an area of failure initiation. The existence of a degraded even hybrid interphase is expected to affect these mechanisms definitely. In this section some numerical results are presented to study the influence of the interphase on stress states developed in particulate as well as fibrous composites. Consider a particulate with $E_m = 3\,\text{GPa}$, $\nu_m = 0.35$, $E_p = 70\,\text{GPa}$ and $\nu_p = 0.35$, where the subscript p denotes the particle properties. In this case, the overall material anisotropy is meaningless ($s = 1$), and equation (7) is simplified accordingly. Application of the piecewise homogenization procedure outlined above yields the elastic response of such a composite structure. Figure 6 illustrates a comparison between radial and tangential stress components developed within the interphase region. In this Figure, numerical results are normalized with respect to the pressure p_o applied equivalently at the inclusion region. These results correspond to efficiency of adhesion $\lambda = 0.4$. In comparison to the conventional interphase, the existence of a sophisticated interphase region seems to increase significantly the stress state. Very intense stress concentration occurs near the skin of the inclusion. This explains the failure initiation process from this region. Hybrid interphase affects the stress states developed also in unidirectional fibrous composites. For such a composite with $E_m = 4.3\,\text{GPa}$, $\nu_m = 0.34$, $E_f = 220\,\text{GPa}$ and $\nu_f = 0.2$, Figure 7 illustrates the stress state developed in the neighbourhood of the interphase region. In this case, very small stress concentration takes place, depending on the overall anisotropy characterizing the material. When the parameter s takes large values, then smoother stresses occur. Numerical results show that strain and stress intensifications appearing within the interphase region, strongly suggest that interphase aspects play a significant role in the elastic response of composite materials.

7. CONCLUSIONS

The interphase region determines the residual stress state and the interface adhesion in composites. In the present investigation, an overview on *interfaces* and *interphases* in composite materials is undertaken while the new *"hybrid interphase"* concept is introduced by the authors Based on this new interphase concept a micromechanics semi-analytical method has been developed for the determination of the mechanical interfacial stresses of composites incorporating a hybrid interphase region. Interrelations between macro and micro-mechanical behaviour have been elucidated.

8. REFERENCES

1. Lipatov, Yu.S.: Physical chemistry of filled polymers, Translated from the Russian by R.J. Moseley, *International Polymer and Technology Monograph* No. 2, Originally Published in *Khimiya*, Moscow, 1977.
2. Papanicolaou, G.C., Paipetis, S.A. and Theocaris, P.S.: The concept of the boundary interphase in composite mechanics, *Colloid and Polymer Science,* **256**, (1978) 625-630.
3. Incardona, S., Migliaresi, C., Wagner, H. D., Gilbert, A.H. and Marom, G.: The mechanical role of the fibre/matrix transcrystalline interphase in carbon fibre reinforced J-polymer microcomposites, Composites Science and Technology, 47, (1993) 43-50.
4. McNaught, A.D. and Wilkinson, A.: *The UIPAC Compentium of Chemical Terminology*, 2nd Edition, The Royal Society of Chemistry, Cambridge, UK, 1997.
5. Papanicolaou, G.C. and Theocaris, P.S.: Thermal properties and volume fraction of the boundary interphase in metal-filled epoxies, *Colloid and Polymer Science,* **257**, (1979) 239-246.
6. Theocaris, P.S. and Papanicolaou, G.C.: The effect of the boundary interphase on the thermomechanical behaviour of composites reinforced with short fibres, *Fibre Science and Technology,* **12**, (1979) 421-433.
7. Jayaraman, K. and Reifsnider, K.L.: Residual stresses in a composite with continuously varying Young's modulus in the fiber/matrix interphase, *Journal of Composite Materials*, **26**, (1992) 770-791.
8. Sotos, N.R. and McCullough, R.L.: The influence of interphase regions on local thermal displacements in composites, *Composites Science and Technology*, **44**, (1992) 319-332.
9. Jayaraman, K. and Reifsnider, K.L.: The interphase in unidirectional fiber-reinforced epoxies: Effect on residual thermal stresses, *Composites Science and Technology*, **47**, (1993) 119-129.
10. Anifantis, N.K., Kakavas, P.A. and Papanicolaou, G.C.: Thermal stress concentration in fiber reinforced composites due to imperfect adhesion, *Composites Science and Technology*, **57**, (1997) 687-696.
11. Papanicolaou, G.C., Michalopoulou, M.V. and Anifantis, N.K.: Thermal stresses in fibrous composites incorporating hybrid interphase regions, *Composites Science and Technology*, **62**, (2002) 1881-1894.
12. Waas, A.M.: Effect of interphase on compressive strength of unidirectional composites, *Journal of Applied Mechanics*, **59**, (1992) 183-188.

13. Hiemstra, D.L. and Sottos, N.R.: Thermally induced microcracking in polymer matrix composites, *Journal of Composite Materials*, **27**, (1993) 1030-1051.
14. Kakavas, P.A., Anifantis, N.K., Baxevanakis, K., Katsareas, D.E. and Papanicolaou, G.C.: The effect of interfacial imperfections on the micromechanical stress and strain distribution in fiber-reinforced composites, Journal of Materials Science, 30, (1995) 4541-4548.
15. Kiritsi, C.C. and Anifantis, N.K.: Load carrying characteristics of short fiber composites containing a heterogeneous interphase region, Computational Materials Science, 20, (2001) 86-97.
16. Theocaris, P.S., Sideridis, E.P. and Papanicolaou, G.C.: The elastic longitudinal modulus and Poisson's ratio of fibre composites, Journal of Reinforced Plastics and Composites, 4, (1985) 396-418.
17. Sideridis, E.P., Theocaris, P.S. and Papanicolaou, G.C.: The elastic modulus of particulates as defined by the concept of mesophase, Acta Rheologica, 25, (1986) 350-357.
18. Dasgupta, A. and Bhandarkar, S.M.: A generalized self-consistent Mori-Tanaka scheme for fiber-composites with multiple interphases, Mechanics of Materials, 14, (1992), 67-82.
19. Low, B.Y., Gardner, S.D., Pittman Jr, C.U. and Hackett, R.M.: A micromechanical characterization of graphite-fiber/epoxy composites containing a heterogeneous interphase region, Composites Science and Technology, 52, (1994) 589-606.
20. Subramanian, S., Lesko, J. and Reifsneider, K.L.: Characterization of the fiber-matrix interphase and its influence on mechanical properties of unidirectional composites, Journal of Composite Materials, 30, (1996) 309-324.
21. Chouchaoui, C.S. and Benzeggah, M.L.: The effect of interphase on the elastic behavior of a glass/epoxy bundle, Composites Science and Technology, 57, (1997) 17-622.
22. Papanicolaou, G.C., Theocaris, P.S. and Spathis G.D.: Adhesion efficiency between phases in fibre-reinforced polymers by means of the concept of boundary interphase, Colloid and Polymer Science, 258 (1980), 1231-1237.
23. Palmese, G.R.: Origin and influence of interphase material property gradients in thermosetting composites, Ph.D. dissertation, Department of Chemical Engineering, University of Delaware, Newark, DE, 1991.
24. Jayaraman, K., Reifsnider, K.L. and Swain, R.E.: Elastic and thermal effects in the interphase: Part II. Comments on modeling studies, Journal of Composites Technology and Research, 15, (1993) 14-22.
25. Pompe, G. and Mäder, E.: Experimental detection of a transcrystalline interphase in glass - fibre/polypropylene composites, Composites Science and Technology, 60, (2000) 2159-2167.
26. Yang, F. and Pitchumani, R.: Modeling of interphase formation on unsized fibers in thermosetting composites, Proceedings of the ASME Heat Transfer Division, HTD 366, (2000) 329-337.
27. Fisher, F.T. and Brinson, L.C.: Viscoelastic interphases in polymer-matrix composites: theoretical models and finite-element analysis, Composites Science and Technology, 61, (2001) 731-748.
28. Kim, J.K., Sham, M.L. and Wu, J.: Nanoscale characterization of interphase in silane treated glass fibre composites, Composites Part A, 32, (2001) 607-618.
29. Kakavas, P.A., Anifantis, N.K. and Papanicolaou, G.C.: The role of imperfect adhesion on thermal expansivities of transversely isotropic composites with inhomogeneous interphase, Composites, Part A, 29A, (1998) 1021-1026.
30. Tandon, G.P.: Use of composite cylinder model as representative volume element for unidirectional fiber composites, Journal of Composite Materials, 29, (1995) 388-409.
31. Davies, G.C. and Bruce, D.M.: A stress analysis model for composite coaxial cylinders, Journal of Materials Science, 32, (1997) 5425-5437.

MICROFIBRILLAR REINFORCED COMPOSITES – A NEW CONCEPT FOR THE PREPARATION OF STRONGER, BIODEGRADABLE IMPLANTS

K. FRIEDRICH*, J. HOFFMANN*, M. EVSTATIEV** and S. FAKIROV**

**Institute for Composite Materials (IVW GmbH)*
University of Kaiserslautern, Erwin-Schrödinger-Strasse, 67663 Kaiserslautern, Germany
***Laboratory for Structure and Properties of Polymers*
University of Sofia, 1 J. Bouchier, 1126 Sofia, Bulgaria

Abstract. In contrast to the classical composites, microfibrillar reinforced composites are not prepared via melt blending of the matrix and the reinforcing material. In fact, the reinforcing elements of this composite, the microfibrils, are created during processing. This advantage allows the manufacturing of a full biodegradable composite material with improved mechanical properties. Basing on a blend of two common biodegradable polymers, polylactide and polyglycolide, this new composite shows promising values under ambient conditions.

1. INTRODUCTION

1.1 The MFC-Approach

During the last decade a new type of polymer-polymer composites, originally called "microfibrillar reinforced composites" (MFC), was developed [1-9]. In contrast to the classical composites, the new type of composite can not be manufactured using the common approach via melt-blending of the two starting components, the matrix and the reinforcing material. Since the reinforcing elements in the MFC, the microfibrils, do not exist as a separate material, a special technique must be applied for their creation.

The manufacturing of the microfibrillar polymer composites starts with blending of immiscible polymer partners that should have different melting points, T_m. The essential stages of MFC preparation are as follows: (i) blending and extrusion, (ii) drawing (with good orientation of all components), and (iii) thermal treatment at a temperature being above T_m of the lower–melting component, but below T_m of the higher–melting one. During the drawing step, the blend components are oriented and microfibrils are created (fibrillization step). In the subsequent step, when melting of the lower–melting component occurs (isotropization step), it is necessary to ensure that the fibrillar structure of the higher–melting component is preserved (Figure 1a).

E.E. Gdoutos and Z.P. Marioli-Riga (eds), Recent Advances in Composite Materials, 203-214

Based on lab-scale experiments, recently commercially relevant investigations were carried out [5]. While the drawing under lab conditions has been carried out off-line by means of a tensile testing machine, here the drawing was performed continuously using a prescribed drawing ratio. In addition, isotropization in a lab ambient, i.e. the conversion of the drawn blend into a proper composite material comprising an isotropic matrix, has generally been done by annealing the sample with fixed ends at a temperature between T_m of the two components.

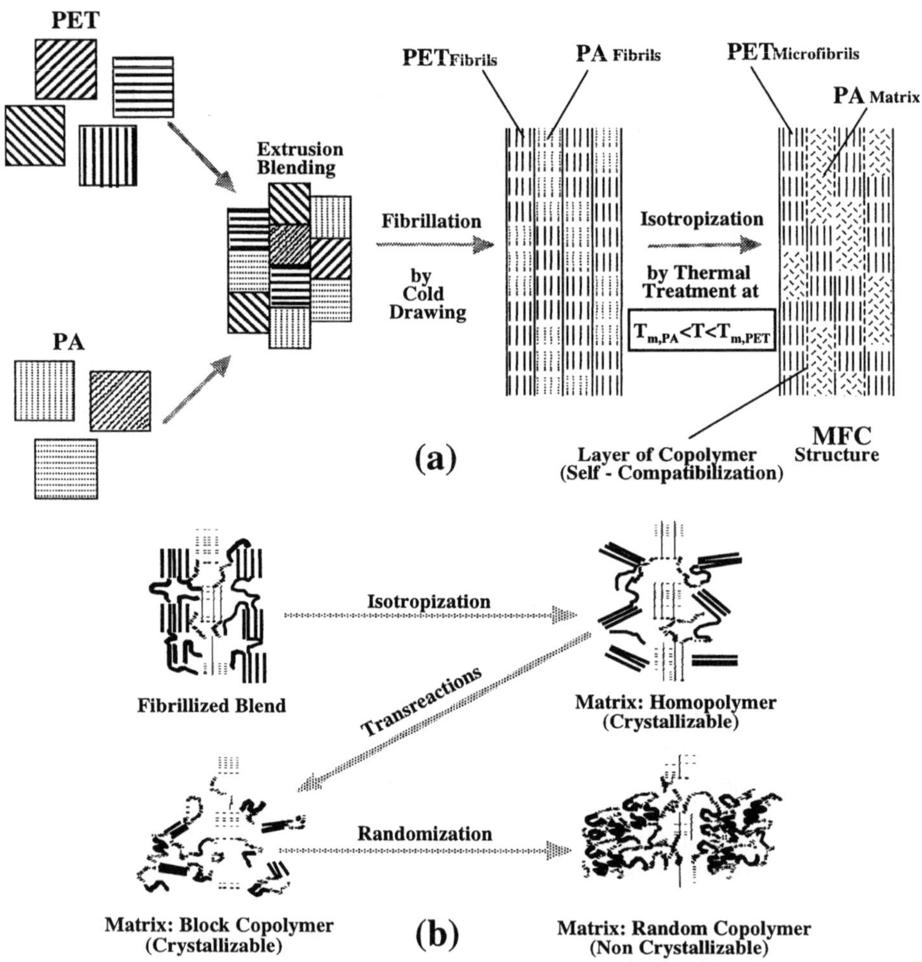

Figure1. Schematic representation of MFC preparation (a) and molecular interactions at the fibrils/matrix interface (b).

In the commercially relevant experiments, the isotropisation is carried out during processing (e.g. injection molding), i.e. one obtains final goods with a composite structure, in which an isotropic matrix is reinforced by microfibrils. If two polycondensates are used for this process (e.g. polyethyleneterephthalate PET and polyamide PA), transreactions can occur at the fibril/matrix interface, leading to the formation of a copolymeric phase and self-compatibilization phenomena (Figure 1b).

For carrying out the commercially relevant experiments, low density polyethylene (LDPE) (matrix) and recycled PET (reinforcing material) from soft-drink bottles were melt blended (in 30/70 and 50/50 PET/LDPE wt ratio) and extruded; it was followed by continuous drawing, pelletizing and injection molding of dog bone samples. Samples of each stage of MFC manufacturing and processing were characterized by means of SEM, WAXS and mechanical (tensile, wear and impact) testing. SEM and WAXS showed that the extruded blend is isotropic, but becomes highly oriented after drawing, being converted into a polymer-polymer composite upon injection molding at a temperature below the melting temperature of the extruded PET phase. This polymer-polymer composite is characterized by an isotropic LDPE matrix reinforced by randomly distributed PET microfibrils, as concluded from the WAXS patterns and SEM observations (Figure 2).

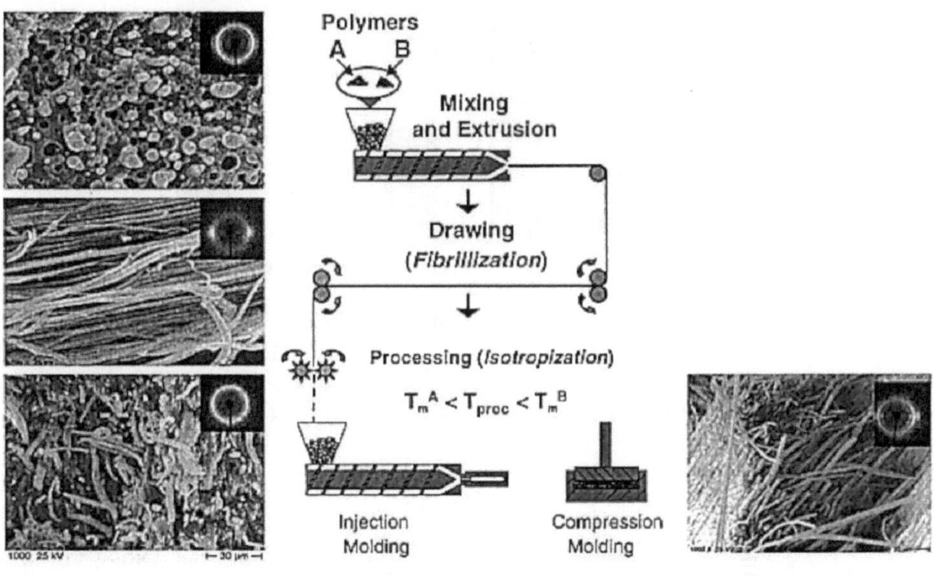

Figure 2. Schematic presentation of the essential steps during manufacturing (mixing with extrusion and drawing) of MFC as well as their processing via injection molding or compression molding, together with the respective morphological peculiarities according to SEM and WAXS (inserted patterns) studies for the blend PET(recycled)/LDPE

The dog bone samples with an MFC structure show impressive mechanical properties - the elastic modulus is about 10 times higher than that of the homo-LDPE and about 3 times higher than LDPE reinforced with glass spheres (Figure 3), approaching the modulus of LDPE reinforced with 30 % short glass fibers (GF). The tensile strength is at least two times higher than that of the homo-LDPE or of LDPE reinforced with glass spheres, approaching that of LDPE reinforced with 30 wt % GF (Figure 3).

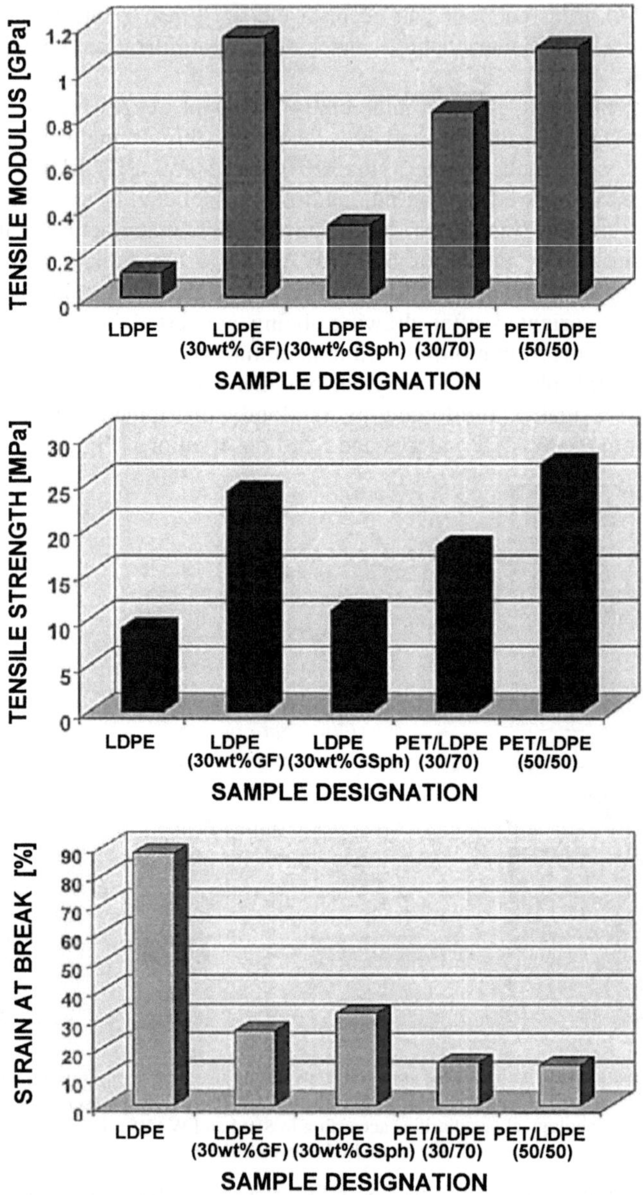

Figure 3.Tensile modulus (a), tensile strength (b) and strain at break (c) of MFC based LDPE blends

The impact strength of LDPE increases by 50 % after being reinforced with PET [6]. It is concluded that: (i) the MFC-approach can be applied under industrially relevant conditions using various blend partners, and (ii) the MFC concept represents an attractive alternative for recycling of PET as well as other polymers.

1.2 Biodegradable Implants, Objectives

There is an increasing interest in the use of resorbable polymers for the preparation of internal fixation devices. Many such devices produced from various polyhydroxyacids are available on the market at present, mainly used for the treatment of small fragments. The advantage of such resorbable implants is that they do not have to be removed from the implantation site after fracture healing is completed. One limitation of the broad use of these resorbable implants in bone fracture treatment relates to their relatively poor mechanical properties. While the latter still seem to be sufficient when the implants are used in areas of restricted load, they are inadequate, however, to allow for the use of implants in high-load situations [10].

Various techniques are applied to produce resorbable internal fixation devices with enhanced mechanical properties [11-13]. For example, compression moulding of reinforcing fibres embedded in a matrix of the same or a different polymer is such a technique. Orientrusion, a cold- or hot-drawing process above the glass transition temperature and below the melting temperature, is another technique for enhancing the mechanical properties. However, many problems of these techniques, like delamination between the matrix and the reinforcing component or high crystallinity of the polymer, have not been solved yet.

The aim is to develop a strong, completely biodegradable material under the aspect of free shaping. The difference to other techniques is the in situ processing of the composite (according to the MFC procedure) and the nearly short fibre reinforcement through one of the blend components, which can be of advantage regarding the degradation behaviour. In particular, it is planned in this study to produce small rods and small bone nails with complex shapes and to enhance the mechanical properties of such implants.

2. EXPERIMENTAL

2.1 Materials

Different types of Polylactides (PLA) and a type of Polyglycolide (PGA) from Boehringer Ingelheim KG (Ingelheim, Germany) were used. The raw materials were blended and injection moulded into cylindrical bars. The samples were then dried and stored in a desiccator prior to the beginning of the MFC-processing procedure.

2.2 Sample Preparation

According to the MFC technology, the blends were drawn between the glass transition temperature and the melting temperature of the lower melting phase (fibrillization step), so as to create at first a fibrillar structure in both phases. The last step was to anneal and isotropize the stretched samples within a window between the melting points of the two components (isotropization step). The samples obtained by compression molding were rectangular in shape and had a geometry of approximately 50 mm in length and 4 mm in width. The samples produced by injection molding had the shape of a bone nail according to the geometry of a commercially available implant.

2.3 Thermal Analysis

A Mettler-Toledo differential scanning calorimeter (Type "DSC 821") calibrated with indium was used for evaluation of the samples' thermal characteristics. The specimens had an average weight of roughly 10 mg and were scanned in aluminium pans at a heating rate of 10°C/min under dry nitrogen flowing at a rate of 30 ml/min.

2.4 Mechanical Tests

To measure tensile modulus and tensile strength of the samples, a Zwick universal testing machine (Type 1474) was used. The measurements were carried out at room temperature and ambient condition (approximately 50% humidity). The cross-head speed amounted to 1 mm/min.

2.5 Light Microscopy and Scanning Electron Microscopy (SEM)

For light microscopy,a Leitz model "Diaplan" was operated at different magnifications. Additionally, Joel models "JSM 5400" and "JSM 6300" SEM allowed to characterize the morphology of the samples. The instrument was operated at 20 kV.

3. RESULTS AND DISCUSSION

3.1 Thermal Characteristics by DSC-Study

Figure 4 shows typical DSC thermograms for the blended samples, the samples after orientation (fibrillization step) and the annealed samples (isotropization step). The samples show four major phase transitions; two glass transitions in the lower temperature range which are close together and fall at one point and two melting peaks in the upper temperature range.

The difference of the melting peaks is about 40°C, enough for preparing the microfibrillar reinforced composite. The tagged region (A) refers to the processing window for the final isotropization step.

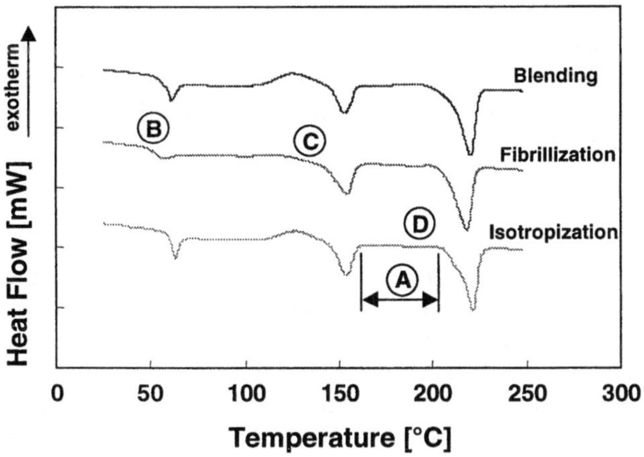

Figure 4.Representative DSC thermograms (obtained during heating) for the samples subjected to various processing steps.

Table 1 displays the measured thermal properties of the neat materials. Polyglycolide is the strongest material with the highest melting point and the reinforcing component in the composite. Polylactide is less strong, lower in melting temperature and it is the matrix material in the composite.

Table 1.Properties of the biodegradable polymers.

Polymer	Tensile Modulus [MPa]	Glass-Transition Temp. T_G [°C]	Melting Temp. T_M [°C]	Processing Temp. T_P [°C]
Polyglycolide	4420	20-35	215-228	~230
Polylactide	2600	45-60	145-155	160-180

This figure shows also effects of the fibrillization step. The glass transition of the oriented and fully fibrillized material is shifted to the left, as shown by mark (B). The blended and the annealed samples show a glass transition superimposed by an endothermic peak as a result of the additionally crystallization of both blends components.

Another difference can be seen by the absence of the exothermic peak (C) in the thermogram of the fibrillized sample. In this case, crystallization cannot be observed during the first run due to the high orientation of the molecules. Both samples, the blended and the annealed one, show a cold crystallization during the DSC-measurement. It is a result of the not fully completed crystallization of the processed samples. This effect can be proofed with a second measurement to exclude prehistory-influence on the measurement.

In addition orientation effects of the higher melting component can also be observed. The thermogram of the blended sample shows a small exothermic peak at about 190°C before melting of the second component. It is the cold cristallization of the higher melting component. In the thermograms of the fibrillized and the isotropized samples, this peak is missing (D). As a result of this observations it can be stated, that especially the higher melting component has to be preserved during the annealing process.

3.2 Sample Morphology

MFC-processing of the biodegradable blends leads to the formation of a typical fibrillar structure, as pointed out in the following series of microscopy pictures. At first, the MFC process starts with blending the raw material components. Figure 5a shows a photograph of the blended material. It can be seen that both components are not miscible, and that the phases are arranged separately. In particular, the blending resulted in a system of small spheres of the higher melting component, embedded in a matrix of the lower melting one.

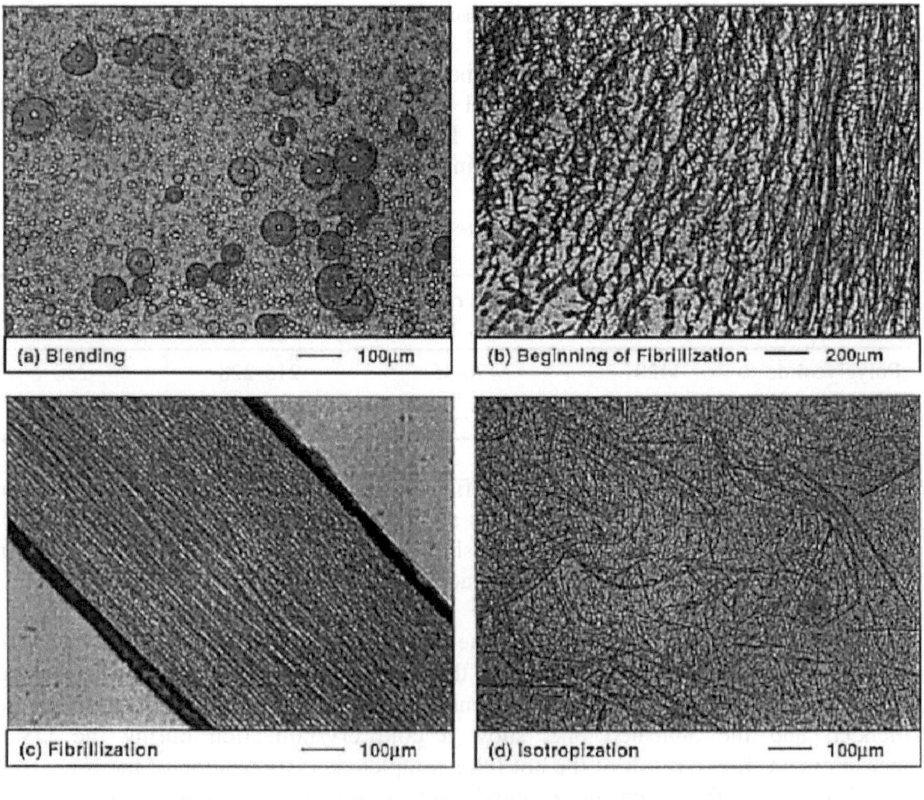

Figure 5.Sequences of the MFC-process for the biodegradable PGA/PLA-composite.

A different structure can be seen after the second processing step, the fibrillization. The beginning of fibrillization is shown in Figure 5b, where the polyglycolide component starts with deformation. Figure 5c illustrates the changes in the material which have lead to an increase in strength; both components have been transferred into an oriented state. The fibrils of the higher melting component are uniformly stretched within the surrounding matrix. The latter has also experienced a certain degree of orientation, as evidenced by the oriented blend structure transverse to the fibrillization direction in the photograph.

The structure after annealing the samples is displayed in Figure 5d. The fibrils with an average diameter of less than 5 μm, formed from the high-melting component, are embedded in the low- melting component. Although an isotropization of the lower melting component has taken place, the fibrillar structure of the higher melting component was preserved. The same can be observed on the fracture surfaces in the micrographs of Figure 6.

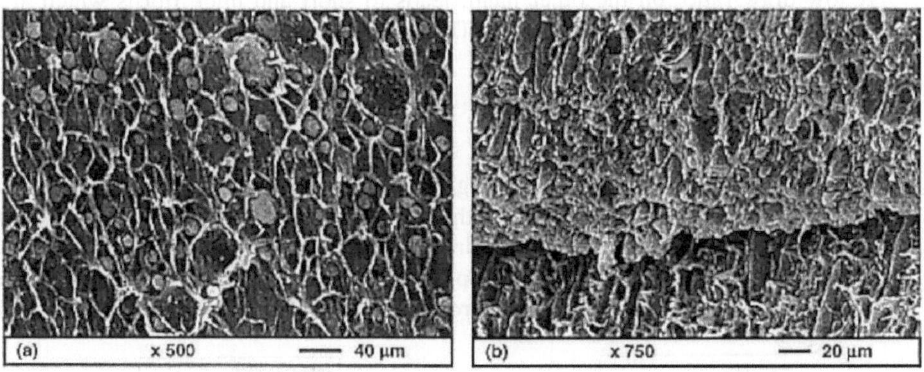

Figure 6. Micrographs on the fracture surfaces a.) blended material, b.) MFC-material

The following Figure 7 illustrates the difference in the material as a function of draw ratio. For high draw ratios the formation of long fibrils can be detected like in photograph 7a.

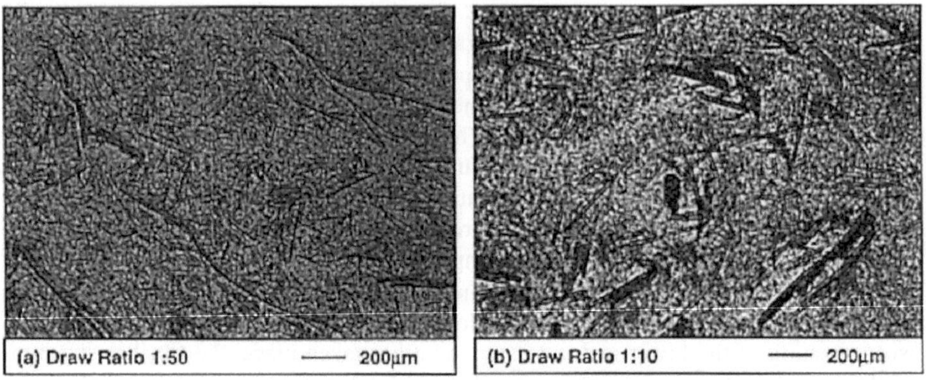

Figure 7. Correlation of draw-ration and fibril-size in isotropized materials

A low draw ratio, on the other hand, leads to short and thick fibrils which are lower in strength than the long fibrils. The latter statement can be verified by the mechanical test results.

3.3 Mechanical Properties

Better mechanical properties can strongly be achieved by MFC processing. As shown in Figure 8, the tensile strength and the tensile modulus of the 50/50 PGA/PLA-blend can be increased after drawing by up to three times, in comparison to the isotropic blend. After the isotropization step, carried out by compression molding, more than a two fold value in strength and stiffness is still remaining.

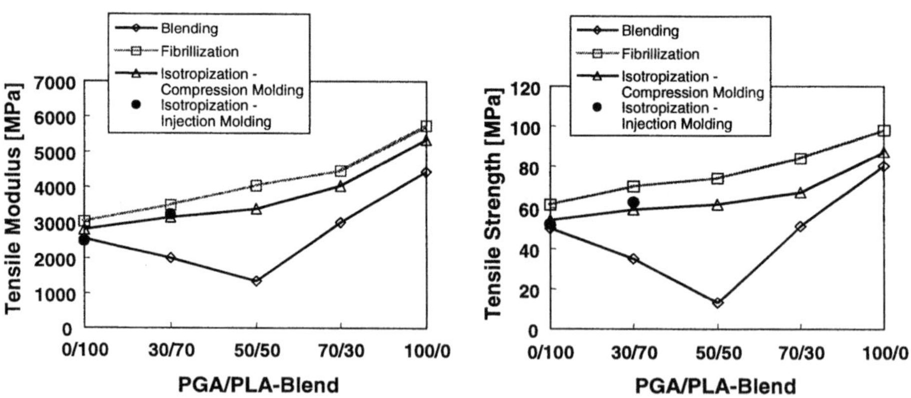

Figure 8. Mechanical properties of the MFC-processing steps from the PGA/PLA-blend.

The injection-molded samples bring comparable results to the compression molded ones, as shown in Figure 8 for a 30/70-blend. This blend is the upper limit for the injection molding process window, since increasing the polyglycolide content of the blend leads to an unprocessable material, due to its high melt viscosity.

According to Figure 8, it is obvious that the mechanical properties of blends can be adjusted or improved by the MFC-process in two ways: One way is the ratio of the blend components, e.g. adding more of the stiffer component to the blend will increase the stiffness of the composite. This is standing in contrast to the isotropic and untreated blend where a 50/50-blend will result in a minimum of the mechanical properties. The second way to affect the mechanical values is to control the fibril size by the processing parameters in the fibrillization step. That means, if one is controlling the fibril size and the fibril geometry, every mechanical value between the isotropic blend and the "MFC-maximum" can be set without changing the component ratio.

4. CONCLUSION

High-strength and high-modulus resorbable composites can be produced by the MFC technique. The initial tensile strength and the tensile modulus of the samples produced are two times higher than that of the unmodified blend of the two components used. The injection molding process can be used for the preparation of biodegradable PGA/PLA microfibrillar composites; it allows the processing into more complex shapes such as surface structured bone fixation implant nails. The first step, to increase the initial mechanical properties of the material in the dry state, was achieved. Tests in the future will clarify and evaluate the long-term medical potential of these biodegradable implants. Hence, the processing of microfibrillar reinforced composites from resorbable polymers is a promising method for the preparation of medical implant parts with enhanced mechanical properties. One of the first prototypes made of these materials is shown in Figure 9.

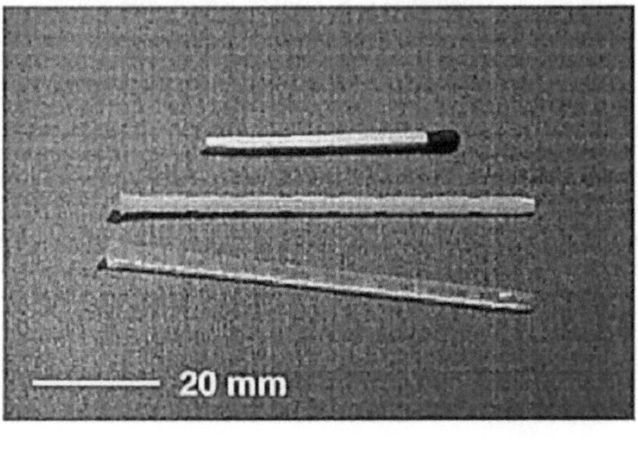

Figure 9. Commercial bone nail (in the front) and MFC-prototype.

5. ACKNOWLEDGEMENT

The authors acknowledge the financial support of the foundation of Rhineland-Palatinate for Innovation (project 495) and the material supply from Boehringer Ingelheim KG (Ingelheim, Germany).

6. REFERENCES

1. Fakirov, S., Evstatiev, M. and Friedrich, K. (2000). From Polymer Blends to Microfibrillar Reinforced Composites. in: Polymer Blends, Vol. 2 Performance,. Paul, D. R. and Bucknall, C. B., (eds.), Wiley J & Sons, Inc., New York, Chapter 33, 2000, 455.
2. Evstatiev, M., Schultz, J., M., Fakirov, S. and Friedrich, K. (2001). In Situ Fibrillar Reinforced PET/PA6/PA66 Blends. Polym. Eng. Sci., Vol. 41, 192.
3. Evstatiev, M., Fakirov, S., Bechtold, G., Friedrich, K (2000). Structure-Property Relationships of Injection- and Compression Molded Microfibrillar Reinforced PET/PA6 Composites. Adv. Polym. Technonology. Vol. 19, 249.
4. Evstatiev, M., Fakirov, S., Krasteva, B., Friedrich, K., Covas, J., Cunha, A. (2002). Recycling of PET as Polymer-Polymer Composites. Polym. Eng.Sci., Vol. 42, 826.
5. Friedrich, K., Ueda, E., Kamo, H., Evstatiev, M., Krasteva, B., Fakirov, S. (2002). Direct Electron Microscopic Observation of Transcrystalline Layers in Microfibrillar Reinforced Polymer-Polymer Composites. J. Mater. Sci. (in press)
6. Evstatiev, M., Fakirov, S., Friedrich, K. (2000). Microfibrillar Reinforced Composites - Another Approach to Polymer Blends Processing. in: Structure Development during Polymer Processing, A.M.Cunha, S. Fakirov (Eds), Kluwer, Dordrecht, 344
7. Fakirov, S., Evstatiev, M., Friedrich, K. (2001). Nanostructured Polymer Composites from Polymer Blends - an Alternative Approach for Polymer Recycling. in: Proceeding of the International Conference on "Polymeric Blends `01", Merserburg, Germany, March 2001, 17
8. Friedrich, K., Evstatiev, M., Fakirov, S., Evstatiev, O. (2002). Effect of Blend Composition on the Structure-Properties Relationship of Nanostructured Polymer Composites from Polycondensate/Polyolefine Blends. Int. J. Polym. Mater. (in press)
9. Wintermantel, E., Mayer, J., Ruffieux, K., Bruinink, A., Eckert, K.-L. (1999). Biomaterialien. Der Chirurg, Vol. 70, 847-857
10. Nordström, P., Pihlajamäki, H., Toivonen, T., Törmälä, P., Rokkanen, P. (1998). Tissue Response to Polyglycolide and Polylactide Pins in Cancellous Bone. Arch. Orthop. Trauma. Surg., Vol. 117, 197
11. Mainil-Varlet, P., Cordey, J., Gogolewski, S. (1997). Resorbable Polymeric Inserts as a Means of Enhancing Fixation of Fractures of Porotic Bones. Biomaterials, Vol. 18, 289
12. Pihlajamäki, H., Böstman, O., Hirvensalo, E., Törmälä, P., Rokannen, P. (1992). Absorbable Pins of Self-Reinforced Poly-L-lactic Acid for Fixation of Fractures and Osteotomies, The journal of bone and joint surgery, Vol.6, 853-857
13. Hoffmann, J., Friedrich, K., Evstatiev, M., Fink, U. (2001). A Totally Bioresorbable Fibrillar Reinforced Composite System: Structure and Properties. Intern. J. Polymeric Mater., Vol. 50, 469
14. Hoffmann, J., Friedrich, K., Fink, U. (2002). Rapid Manufacturing of Bone Nails with Improved Mechanical Properties. in: Proceedings of the IV World Congress Biomechanics, Calgary, Canada, August 4-9, 2002, Index H

LONG TERM BEHAVIOUR OF CONTINUOUS FIBER OXIDE/OXIDE COMPOSITES UNDER THERMAL EXPOSURE

V. KOSTOPOULOS and D. E. VLACHOS
Applied Mechanics Laboratory
University of Patras
GR-265 00 Patras, Greece

Abstract. The present work deals with the mechanical characterization of a new Al_2O_3/Al_2O_3 composite material and its durability performance during the exposure at high temperature environment that simulates the working conditions of a gas turbine. Stiffness and strength variation versus the exposure duration at 1100 °C are identified by means of tensile tests and ultrasonic stiffness matrix measurements. More precisely, a new characterization technique based on the inversion of phase velocity data gained by immersive ultrasonic tests, was used for the determination of the complete stiffness matrix of the oxide/oxide composite, and anisotropic damage functions was introduced in the case the material experiences long term high temperature exposure.

1. INTRODUCTION

Continuous Fibre Ceramic Matrix Composites (CFCCs) have been identified as the most promising class of materials for future high temperature industrial applications, where high structural efficiency under mechanical and thermal loading is required. Up to now CFCCs have been used in applications ranging from high performance braking systems, to heat shields for re-entry vehicles. One of the main disadvantages of conventional CFCC materials is their vulnerability to oxidation at high temperature environment. Although a number of oxidation protective top layer gradient coating systems have been developed and liquid silica infiltration techniques have been used, the problem is far to be solved.

A remedy to this, which has been proposed, is the use of oxide/oxide composite systems that will allow the material to withstand high temperatures without oxidation protective layers or cooling systems. One of the most active research and development efforts in the field of oxide/oxide composites is underway by the gas turbine industry. The main goal is to increase the working temperature of the machines, reducing the cooling air supply and thus to increase the plan efficiency (and reduce the fuel consumption) enabling a consistent reduction in pollutant emission and meeting this way the ever more stringent emission limits of NO_x, CO and CO_2.

E.E. Gdoutos and Z.P. Marioli-Riga (eds), Recent Advances in Composite Materials, 215-226

Ceramic composites of oxide matrix materials reinforced with continuous oxide fibres potentially combine increased fracture resistance, having the ability to retain strength in the presence of holes and notches as a result of the built in active stress redistribution mechanisms, and reliable performance under severe thermo-mechanical conditions.

To this direction, there are a number of activities worldwide, mainly at laboratory scale for the manufacturing of oxide/oxide composites using Nextel™ type fibres. In addition, some working prototypes, e.g. gas turbine liners and exhaust tubes for aero-engines, have already been produced and successfully tested at moderately high temperature level [1-5]. To prevent brittleness and increase fracture toughness, different material tailoring approaches are under development.

The material under investigation in the present study is an improved, compared to former attempts [6], mullite matrix NEXTEL 720 (3000 denier) fibre [7-8] reinforced composite with a fugitive fibre/matrix carbon interface applied by Sol/Gel technique [9]. The composite laminate was made using a symmetric 0°/90° fibre lay-up orientation, with a total fibre volume fraction of 41%. The fabrication process followed is the polymer infiltration process (PIP) [10-11].

2. ULTRASONIC STIFFNESS MEASUREMENTS

In the case of an isotropic material only two engineering constants (Young's modulus and Poisson ratio) are sufficient to fully describe its elastic behavior. These constants can be easily determined from the results of a simple tensile test.

When dealing with anisotropic materials, such as composites, a larger number of independent constants are required. These constants are the coefficients of the elasticity tensor that fully describes the elastic behavior of the material. In this case conventional mechanical testing can determine directly only a part of the elements of the elasticity tensor.

However since elastic wave propagation is directly related to the elastic properties of the medium of propagation, ultrasonic testing can allow for the complete determination of the elasticity tensor of the anisotropic materials.

The determination of the elastic properties of the anisotropic material (assumed to be orthotropic in the case of oxide/oxide composites) can be achieved by calculating the coefficients of the propagation equation of an elastic plane wave, based on a set of properly chosen velocity measurements along known material directions. These measurements are carried out using a special experimental setup that is described below.

A thin specimen with plan parallel faces is immersed in an acoustically coupling fluid (water). The specimen is placed between a transmitter (E) and a receiver (R), which are rigidly connected to each other through a supporting arm. Both the specimen and the arm, holding the two probes, can rotate as seen in Figure 1.

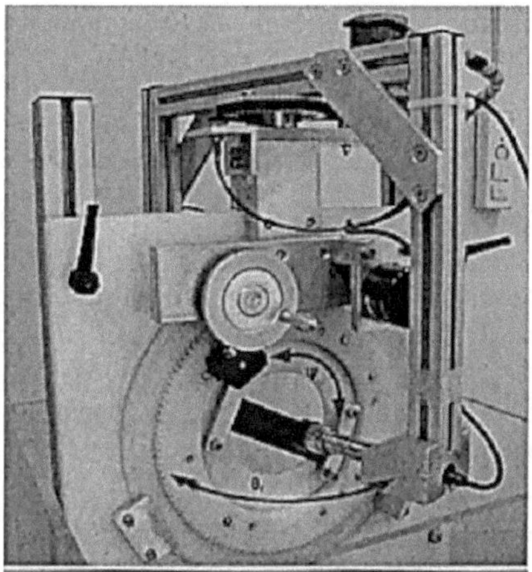

Figure 1: Photo of the experimental setup illustrating the angle movements

Depending on the angle of incidence, the pulse emitted by the transmitter E is refracted within the material in one, two or three bulk waves (one quasi longitudinal wave QL, one quasi transverse wave QT, or two quasi transverse waves QT_1, QT_2) that propagate in the anisotropic solid at different phase velocities and having different polarization vectors. Using appropriate signal processing instrumentation, the phase velocities of each of the propagated waves within the specimen are calculated for a number of specimen/probes relative orientations [12-14].

The calculation of the phase velocity is based on the difference between the flight time of each the waves in the presence of the anisotropic material and the flight time of the emitted signal within the coupling fluid without the specimen.

The procedure is based on the measurement of the time difference for the longitudinal and one or both transverse waves, and is only valid when the QL and the QT waves are appropriately separated [14].

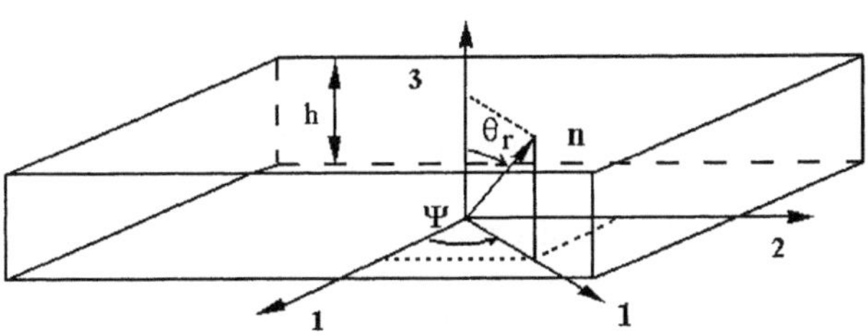

Figure 2. Definition of angles

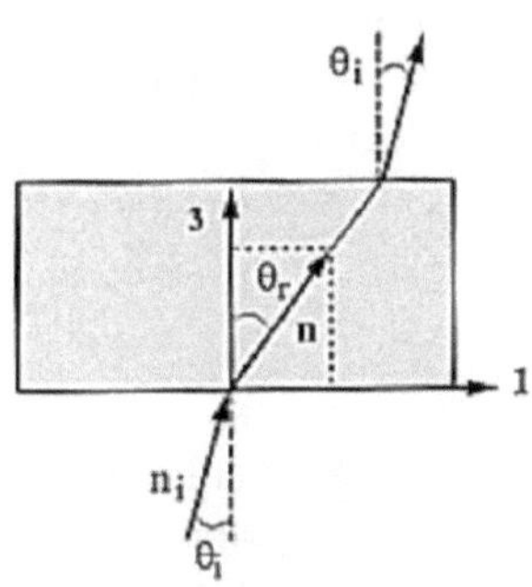

Figure 3. Propagation in the plane of incidence

The wave propagation tensor in the general case of an anisotropic material has the following form [15]:

$$\Gamma_{ij} = C_{ijkl}\, n_k\, n_l \qquad (1)$$

where C_{ijkl} are the elements of the elasticity tensor/stiffness matrix of the anisotropic elastic solid under investigation and n_k (k = 1, 2, 3) are the components of the propagation direction vector **n**, as it is shown in Figures 2 and 3, where n_1, n_2 and n_3 are the direction cosines ($n_1 = \sin\theta \sin\psi$, $n_2 = \sin\theta \cos\psi$, $n_3 = \cos\theta$).

The phase velocities of the three propagated waves for a given propagation direction **n** are given by the eigenvalues of the propagation tensor Γ_{ij} according to the wave propagation equation:

$$\det(\Gamma ij - \rho V2\, \delta ij) = 0 \qquad (2)$$

The components of the elasticity tensor are obtained from the propagation velocities through a least square regression analysis, which minimizes the residuals of the wave propagation equations for the complete set of measurements.

$$F(C_{ij}) = \mathring{a}_{p=1}^{N} \left\{ f_p \left(\lambda_p (\mathbf{n}), C_{ij} \right) \right\}^2 \qquad (3)$$

where p = 1 to N, N being the total number of measurements of a range of incident angles θ_i, each corresponding to a different propagation direction **n**, and $\lambda_p = \rho_b V_p^{\,2}$

After the calculation of the components C_{ij} of the elasticity tensor, the engineering constants of the laminate structure (orthotropic in the present case) are determined.

This non-destructive testing procedure is very efficient, since it yields the complete stiffness matrix of the material, with adequate accuracy.

Following this process for the determination of the elements of the stiffness matrix of a given solid at the different stages of its loading life, the evolution of the anisotropic damage formed within the material structure can be traced. This is still possible even in the most extreme case where the ascertained damage alters the initial degree of anisotropy of the loaded material.

3. CHARACTERIZATION PROCEDURE

Based on the common practice for the design of composite components, designing using CFCCs requires detail characterization of the mechanical properties of the material used, knowledge of the damage modes evolution and their effect on the mechanical performance of the material plus information on the intrinsic damage characteristic of the composite. Since CFCCs are intended to work at High Temperature (HT) environment, oxidation and properties degradation in general due to HT exposure, is of critical importance for any design attempt, and should be identified through testing. In particular, for working temperature over 1000 °C, oxide/oxide composites, exhibit a reduction of their stiffness and strength increasing the exposure duration. This is due to instigation of a number of transformation processes that affect the material structure and have a detrimental effect on its properties. In order to obtain an exact figure of the material degradation, both stiffness and strength have to be determined as a function of exposure time at a given temperature.

More precisely within the frame of the present work, the strength measurements were obtained by standard tensile tests, while stiffness measurements across the loading direction were taken both using tensile tests and a non-destructive ultrasonic technique, where the latter one yields the complete stiffness matrix of the composite material under investigation.

For the purposes of the present study, a detailed testing procedure was planned. A number of test coupons were prepared according to CEN ENV 658-1 and subjected to high temperature exposure for different exposure periods, in the absent of any mechanical loading, into a specially designed chamber that simulates the working environment of a gas turbine. This chamber was actually a furnace producing temperatures up to 1200°C, and it was equipped with a system that supplied a gas compound having the same chemical composition as the combustion gases, thus simulating the conditions inside a gas turbine combustion chamber.

The conditioning characteristics (temperature, duration) of the specimens used are shown in Table 1. Each set of specimens corresponding to different exposure duration was subjected to a series of tests. Firstly, all the samples were passed through C-scan examination in order to obtain their reference state concerning the quality of the material for each specimen. This was a prerequisite for the initial evaluation of the results of the following tests.

At this stage the ultrasonic stiffness matrix measurements were performed for all the specimens, using the experimental set-up described above. Two 5 MHz, 15mm diameter panametric immersion transducers were used. The frequency was carefully chosen in order to assure wavelengths at least 5 times the dimension of the representative volume element of the material under investigation in the propagation direction and simultaneously to achieve proper separation between the QL and QT waves [14].

An in house built up optimisation scheme permits the transformation of the velocity measurements and yields the stiffness matrix of the material. Knowing the stiffness coefficients, the engineering constants of the material can be easily obtained for the given exposure time.

For the needs of the present analysis it was assumed that the damage developed during HT exposure does not affect the degree of anisotropy of the material. This means that assuming the starting state of the material is orthotropic, the induced damage does not change its degree of anisotropy, keeping the same principal axes of symmetry. Following the non-destructive ultrasonic tests, the specimens were tested in tension until fracture. The results of these tests are used to derive both modulus of elasticity in the loading direction and the ultimate tensile strength.

The comparison of the results gained by the tensile tests for the modulus of elasticity against the relative results for the same quantity measured by the ultrasonic technique permits the evaluation of the validity of the ultrasonic technique as a tool for the determination of the damage function of composite materials and provides a measure of the accuracy of the ultrasonic non-destructive methodology.

Three samples were used for each exposure duration (plus six samples which were not exposed and used as reference material) and six exposure duration steps were applied as it is clearly indicated in Table 1. The tested coupons have been properly mixed in order to avoid the correlation of the damage to the part of the plate from where the sample has been cut out.

Table 1: Analytical presentation of the High Temperature exposure conditions for all the tested samples used in the present study

Specimen	Temperature (^oC)	Duration (hours)	Specimen	Temperature (^oC)	Duration (hours)
T1	-	0	T13	-	0
T2	-	0	T14	-	0
T3	-	0	T15	-	0
T4	1100	50	T16	1100	408
T5	1100	50	T17	1100	50
T6	1100	100	T18	1100	100
T7	1100	150	T19	1100	100
T8	1100	150	T20	1100	150
T9	1100	216	T21	1100	216
T10	1100	216	T22	1100	300
T11	1100	300	T23	1100	300
T12	1100	408	T24	1100	408

4. RESULTS

The above described characterization process was applied in the case of an oxide/oxide composite manufactured by EADS/Dornier GmbH for the needs of a EU supported research project (BRPR-CT98-0791). The material consists of sol-gel carbon fugitive coated 3M Nextel 720 fibers, with a mullite-based matrix. The specimens were 12-layer symmetric cross-ply laminates produced by pyrolysis and infiltration of polymer performs [9].

All 24 specimens were tested non-destructively and the ultrasonic stiffness matrix measurements were performed in their "as-received" condition. For each specimen the stiffness measurement was repeated three times at three different well-determined regions. Six specimens were kept out of the high temperature exposure process in order to serve as reference material and to be used for the mechanical characterization of the composite at its initial/original state.

Following the high temperature exposure into the special chamber, the same specimens were again tested non-destructively and ultrasonic stiffness matrix measurements were performed once more at the same regions noted above. This way it was possible to identify the change in stiffness components for each individual specimen. This approach assists to compensate the inaccuracies of the method and the properties variation within the same test sample.

As it can be seen in the following figures, where the results are presented, the differences of the actual stiffness values between the specimens could lead to a misinterpretation of the results, since in some cases the stiffness appears to increase versus exposure time.

The results of the ultrasonic measurements for the three moduli of elasticity E_1, E_2 and E_3 (1 and 2 are the in plane principle directions, with direction 1 associated to the length of the specimen) are shown in figures 4, 5 and 6 respectively. Since the material is a 0/90 symmetric laminate E_1 and E_2 are expected to be equal. However a difference of about 10% is shown in the results. This can be attributed to edge effects, since the width of the specimen is lower than the length and these results to lower values for the E_2 modulus. In square plates the method yields consistent results, i.e. almost equal E_1 and E_2 values.

The characterization process shows a drop in modulus of up to 16 % for the E_1 modulus and 14 % for the E_2 and E_3 moduli, increasing the exposure duration. It must be stressed that more than a 50% of the stiffness reduction occurs at the first 50 hours of exposure.

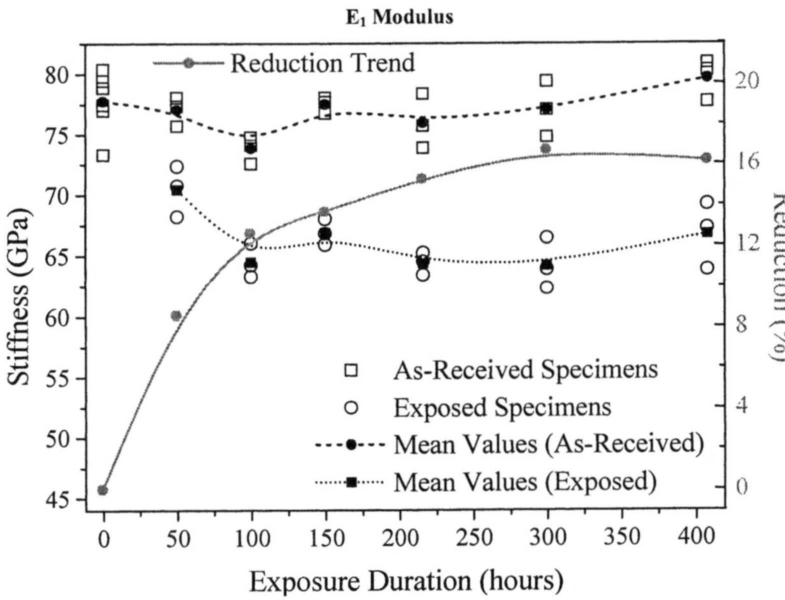

Figure 4: Variation of modulus of Elasticity E_1 versus the exposure time at 1100 °C

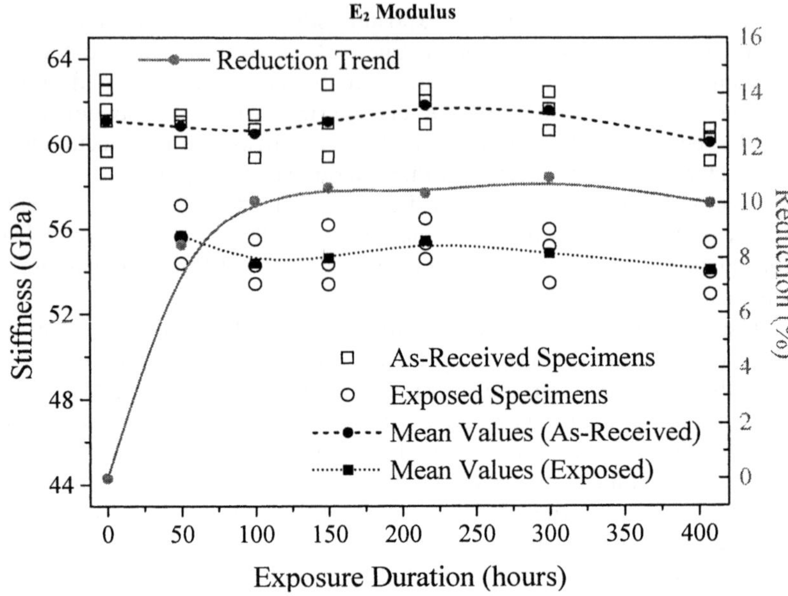

Figure 5: Variation of modulus of Elasticity E_2 versus the exposure time at 1100 °C

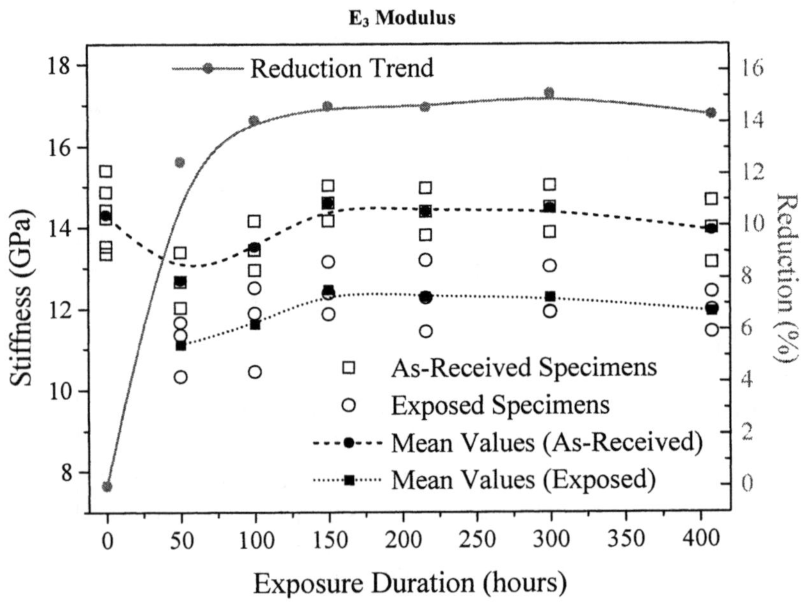

Figure 6: Variation of modulus of Elasticity E_3 versus the exposure time at 1100 °C

As already stated, the exposed specimens and a number of "virgin" specimens were subjected to tensile tests up to fracture. A set of indicative stress-strain curves is presented in figure 9. Tensile strength and strain to failure results are also presented in figures 7 and 8 respectively. Both quantities show an abrupt drop, at the early stages of exposure (50 hours) and an almost constant value thereafter. The obtained results for the Young's modulus are compared against the non-destructive ultrasonic measurements as seen in figure 10. The ultrasonic stiffness measurements are in very good agreement with the mechanical test results. Minor differences are present however, the overall reduction trend is fairly accurate.

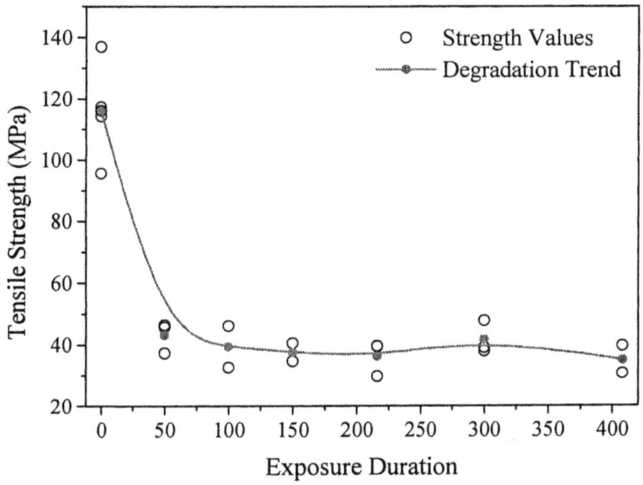

Figure 7: Variation of tensile strength in the loading direction versus the exposure time at 1100 °C

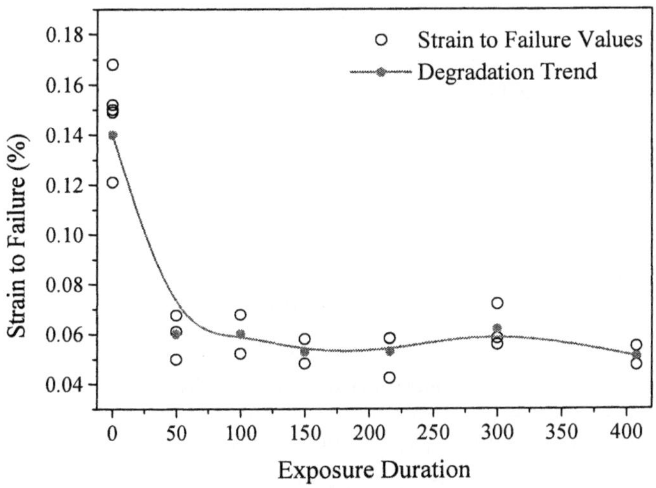

Figure 8: Variation of strain to failure in the loading direction versus the exposure time at 1100 °C

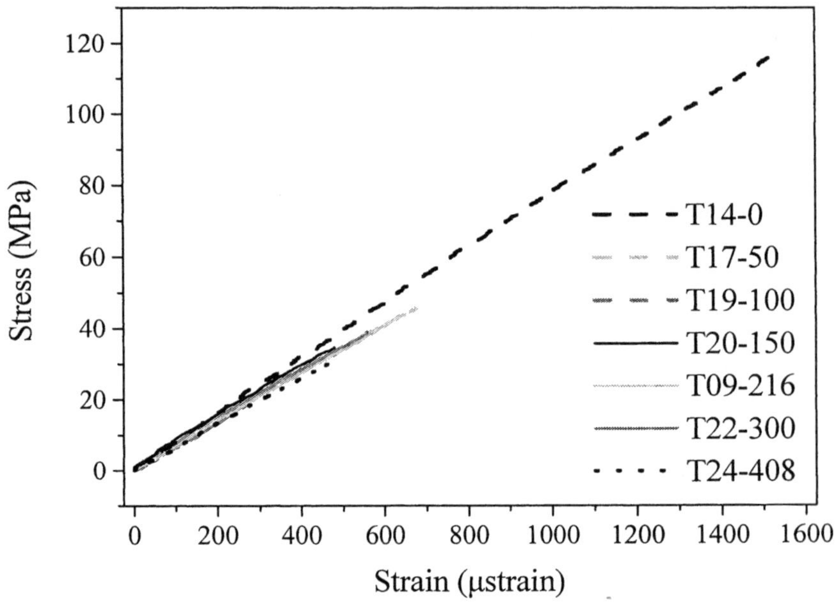

Figure 9: Typical Stress-Strain curves for exposed and as-received specimens

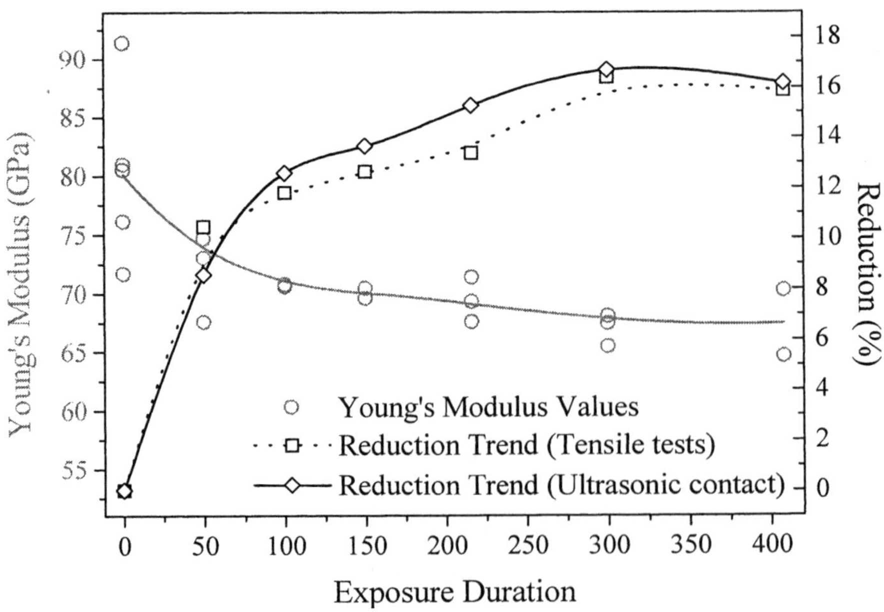

Figure 10: Comparison of the variation of the modulus of elasticity E_1 measured using the ultrasonic stiffness measurement technique and tensile tests, versus the exposure duration at 1100 °C

5. DISCUSSION AND CONCLUSIONS

Characterisation of an oxide/oxide material using mechanical and non-destructive ultrasonic test has identified a significant drop of stiffness and strength, occurring at an early stage of the thermal exposure process at 1100°C. The stiffness reduction is primarily attributed to degradation of the matrix properties. It was shown that there are changes in the matrix structure and the scale of porosity, which will alter the crack propagation paths within the material structure during loading. In particular, coarsening of porosity have severely reduced the ability of the matrix to diffuse cracks decreasing that way the crack resistance of the material and diminishing the ultimate strength of the composite. Grain growth of the fibres has been also observed, which leads to lower strength and stiffness values. The performance of the fibre/matrix interface is affected too [16-17]. Cracking of the fugitive coating allows for sintering of the fibre and matrix across the interface gaps, which also reduces the ability for fibre debonding and pullout. This is consistent to experimental results on the Nextel 720 found in literature. [7]

The ultrasonic stiffness measurements were in good agreement to the mechanical tests. Although the values of Young moduli, shear moduli and Poisson ratios determined by the two methods are not comparable in principle, particularly for materials which exhibit highly non-linear stress-strain behaviour, the ultrasonic method can provide a fairly accurate mean for quantitative stiffness characterization. In particular, the applied methodology can be considered as a very effective tool during material development stage.

The main advantage of the proposed approach, which is based on the developed methodology of the ultrasonic stiffness measurements, is its capacity to follow the evolution of the anisotropic damage and its direct effect on the material properties at various stages of a progressive fermentation such as oxidation, mechanical and thermal fatigue etc, using a limited number of specimens. Since results are obtained in the form of the complete stiffness matrix of the composite laminate, it is possible to derive equivalent lamina properties, a valuable input for the structural design phase of any components made of laminated materials.

Identification of changes in the material's anisotropy is also possible using a more complicated formulation for the inversion of the phase velocity data.

6. ACKNOWLEDGEMENTS

The present work has been performed within the frame of R&D programme BRITE III, 'Ceramic Components for Industrial Gas Turbines - CERCO' of the European Commission. The authors wish to thank EADS/Dornier GmbH for supplying the used materials and Dr. Y. Z. Pappas for his valuable contribution in this work.

7. REFERENCES

1. Marshall, D.B. and Davis, J. B.: Ceramics for future power generation: fiber reinforced oxide composites, *Current Opinion in Solid State & Materials Science*, **5**, (2001), 283-289.

2. Hay, R. S., Boakeye, E., Petry, M.D., Berta, Y., Von Lehmden, K. and Welch, K.: Grain growth and tensile strength of 3M Nextel 720 after thermal exposure, *Ceram. Eng. Sci. Proc.*, **20**, (1999), 165-172.
3. Evans, G., Marshall, D. B., Zok, F. and Levi, C.: Recent advances in oxide-oxide composite technology, *Adv. Composite Mater.*, **8**, (1999), 17-23.
4. John, R., Zawada, L. and Buchanan, D.: Fracture and creep rupture behavior of a notched woven oxide/oxide CMC, 24th Annual Conference on Composites, Materials and Structures, Coco Beach, Florida, January 2000.
5. Szweda, et al.: Exposure of CFCC to high temperature, high pressure combustor environments, presented at the 24th Annual Conference on Composites, Materials and Structures, Coco Beach, Florida, January, 2000.
6. Paipetis, A. and Kostopoulos, V.: Ultrasonic Stiffness Matrix Measurements Of Oxide/Oxide Composites, submitted for publication in the book: *Recent Advances in Composite Materials*, Kluwer Academic Publishers, The Netherlands 2002.
7. Dassios, G., Filiou, C., and Steen, M.: The Statistical Properties of Alumina Fibres to 1200°C, accepted for publication at *Journal of Engineering Materials*, (2002).
8. 3M Nextel Ceramic Fiber Technical Notebook, 3M Ceramic Fiber Products, 2001.
9. Innocenti, M., Del Puglia, P., Pappas, Y.Z., Dassios, C.G., Steen, M., Kostopoulos V. and Vlachos, D.E.: Properties of Oxide/Oxide CMCs for High Temperature Applications in Gas Turbines, European conference on Gas Turbines, Liege, Belgium, 2002.
10. Duran, A., Aparicio, M., Rebstock, K. and Vogel, W. D.: Reinfiltration processes for polymer-derived fibre reinforced ceramics, *Key Engineering Materials*, (1997), 127-131, 287-294.
11. Knabe, H., Haug, T , Schafer, W. and Waldenmaier, T.: Oxide Ceramic Matrix Composites for Aerospace Applications. Report Dornier GmbH und Dornier Luftfahrt GmbH, Friedrichshafen, (2001).
12. Baste, S. and Audoin, B.: On internal variables in anisotropic damage, *Eur. J. Mech., A/Solids*, **10**, (1991), 587-606.
13. Aristegui, C., and Baste, S.: Optimal recovery of the elasticity tensor of general anisotropic materials from ultrasonic velocity data, *J. Acoust. Soc. Am.* **101** (1996), 813-833.
14. CEN Technical Committee 184, SubCommittee 1: Advanced Technical Ceramics,- Ceramic Matrix Composites - Mechanical Properties at Room Temperature - Determination of Elastic Properties by an Ultrasonic Technique, WI 114, March 2001.
15. Nayfeh, A. H.: *Wave propagation in layered anisotropic media*, Elsevier, Amsterdam, 1995.
16. Peters, P.W.M., Daniels, B., Clemens, F. and Vogel, W.D.: Mechanical characterisation of mullite-based ceramic matrix composites at test temperatures up to 1200 °C, *Journal of the European Ceramic Society*, **20**, (2000), 531-535.
17. Holmquist, M., Lundberg, R., Sudre, O., Razzell, A.G., Molliex, L., Benoit, J. and Adlerborn, J.: Alumina/alumina composite with a porous zirconia interphase - Processing, properties and component testing, *Journal of the European Ceramic Society*, **20**, (2000), 599-606.

4. Metal Matrix Composites

ON A RELATIONSHIP BETWEEN COD AND STRAIN ENERGY DENSITY IN A MODERN MMC

N. P. ANDRIANOPOULOS
Dept. of Mechanics,
National Technical University of Athens
GR-157 73 Athens, Greece

Abstract. A relationship between the critical value of Crack Opening Displacement and the mean value of dilatational strain energy density, T_V, computed along crack-lips, is indicated in the present work. Namely, T_V takes a maximum value at the moment of crack initiation. This relationship does not hold when the crack tends to be parallel to loading axis and shear stresses dominate over normal ones. Experimental evidence from a metal matrix composite supports the present conclusions.

1. INTRODUCTION

Probably, the most attractive for practical applications crack initiation criterion is that of Crack Opening Displacement (COD), stating that a crack initiates when the opening of the crack-lips measured near its tip takes a critical value. However, and in spite of four decades been passed since Wells [1] introduced COD, doubts are still expressed regarding both the physical meaning and the accurate definition of COD [2]. In addition, limited experimental information is available concerning the behavior of COD versus crack inclination [3]. This information indicates that COD is quite sensitive to crack-inclination angle, β, deteriorating its service as a critical failure parameter.

Thus, it seems interesting an attempt to connect COD with other purely geometrical characteristics of the crack like its surface or volume and compare their behavior with mechanical quantities playing role in crack initiation.

2. SOME THEORETICAL CONSIDERATIONS

Consider a linear elastic plate containing a straight crack of length $2a$, inclined by an angle, β, with respect to the uniaxial loading axis, Fig.1. When the external stress, σ_∞, is applied the crack becomes a thin elliptical hole with semiaxes, a and b (Fig.2). A convenient technique to obtain COD is to measure it at a certain distance, c, from the crack tip. Usually, c is a portion (0.1~0.5) of the half crack length, a. Assume,

E.E. Gdoutos and Z.P. Marioli-Riga (eds), Recent Advances in Composite Materials, 229-235

then, that for various crack inclinations, β, a set of critical COD-values, δ, and fracture stress, σ_{∞}, is obtained.

For an elliptical crack with major semiaxis, a, and passing through the point (x, y) = (ca, δ), elementary calculations indicate that the minor semiaxis equals to:

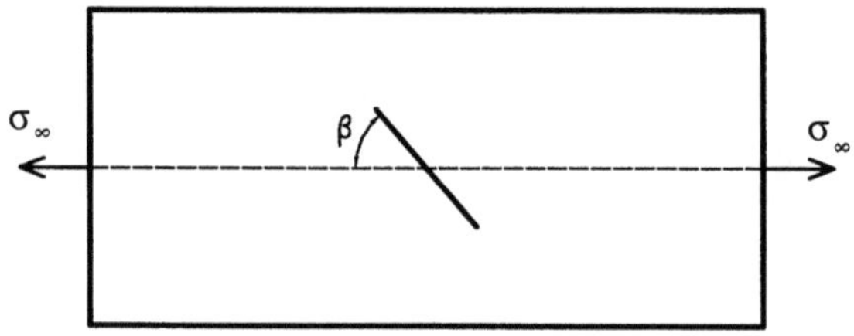

Figure 1. A plate with an inclined crack of length 2a.

$$b = \frac{\delta}{\sqrt{c(2-c)}} \tag{1}$$

and the surface, S, of the crack in the xy-plane is:

$$S = \pi ab = \frac{\pi a \delta}{\sqrt{c(2-c)}} \tag{2}$$

In order to evaluate the crack volume, V, it is necessary to estimate the plate-thickness along crack-lips. It is expressed as:

$$\bar{w} = w_0(1+\bar{\varepsilon}_{zz}) \tag{3}$$

where w_0 is the initial plate thickness, $\bar{w}$ is the mean value of thickness along crack-lips and ε_{zz} is the mean value of the strain normal to the plate thickness. Plate thickness being known, the volume of the crack is:

$$V = S\bar{w} = Sw_0(1+\bar{\varepsilon}_{zz}) \tag{4}$$

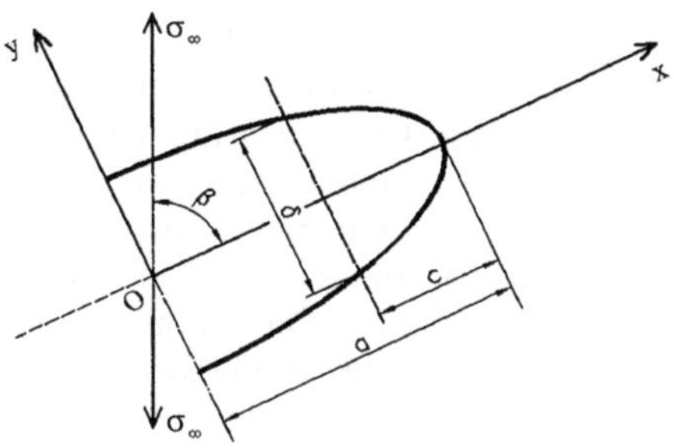

Figure 2. Crack tip geometry.

The only available exact method to obtain stress/strain fields (including ε_{zz}) at points not only near crack-tip but, also, far away from it, is that of complex potentials [4]. A detailed description of the application of this classical method has minor importance for the present case, but, anyway, it can be found in [5], where an identical to the present one geometry was studied.

3. APPLICATION

Experimental results from the 8090 Al-Li Metal Matrix Composite [3] were used for the examination of the behavior of the geometrical quantities described in the previous section. In these experiments six different crack-inclination angles, β, were used in plates with dimensions (length, width, thickness)=(0.12, 0.04, 0.001)m^3 and initial crack length a=0.01m. It must be noticed that the specimens in those experiments contained an edge crack instead of a central one. It implies that in a quantitative manner, the present analysis may be not fully accurate. But, the qualitative behavior of critical COD, δ, is expected to be similar in both geometries. This assumption accompanied by the observation that both surface and volume of the crack depend linearly on δ, allows for the conclusion that the present results may be scaled by an unknown but constant scaling factor, which is expected to have minor influence (if at all) on the validity of the present study. So, in Table 1 experimental results obtained from [3] are given:

Table 1. Experimental data

β $^{(0)}$	**8**	**10**	**26**	**50**	**70**	**90**
σ_∞ **(MPa)**	244.4	185.7	159.9	140.9	129.4	138.0
δ **(mX10^{-3})**	0.415	0.490	0.550	0.605	0.640	0.620

In Fig.(3) values of δ, S and V, reduced to their respective value for β=90^0, are plotted versus angle, β.. As it can be seen, not significant difference in the behavior of the three quantities can be located. It implies that these quantities can be interchanged in the role of a crack initiation index. It is obvious from Eq.(2) where S/δ=const but not clear from Eq.(3). Furthermore, the "equivalence" between δ, S and V allows for the replacement of COD-measurements by surface-measurements. Crack surface is a much more stable quantity compared to COD, as it balances local irregularities along the crack-lips, which may strongly affect point measurements required by COD. This may resolve long lasting doubts on inherent difficulties and high scattering in COD-measurements [6]. On the other hand, from a practical point of view, surface measurements are roughly of the same difficulty to be performed as COD measurements.

At this point, it is useful to recall that neither δ nor S or V remain constant versus crack-inclination angle, β.. Typically speaking, none of these three purely

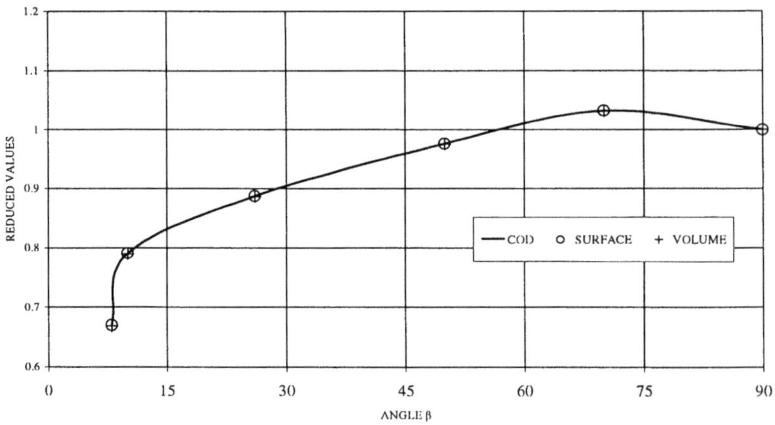

Figure 3. Variation of reduced values of δ, S and V vs crack inclination.

geometrical quantities can be considered as a critical one for the prediction of crack initiation, as long as it varies with specimen geometry (angle β).

An attempt to enhance the physical basis of COD and the other purely geometrical characteristics of a crack, by connecting them with mechanical quantities, like stresses and strain energy densities, crashes on the fact that crack-lips (excluding the close vicinity of the crack-tip) are practically unloaded in comparison to the stresses and strains near tips. However, if the near the singularities points are excluded, then some interesting results can be obtained.

In fact, stress and strain fields were computed along the crack-lips in the interval $-0.99a \leq x \leq 0.99a$. Then, the mean value of dilatational strain energy density, T_V, was computed. As it is known, T_V is given by:

$$T_V = \frac{1-2\nu}{6E}(\sigma_{xx} + \sigma_{YY})^2 \tag{5}$$

where (E, ν) are the Young's modulus and Poisson ratio of the material. The results obtained are plotted in Fig.(4) versus stress at infinity, σ_∞, and for the six crack inclinations, β, studied here. In Fig.(4a) showing results for β=50^0, 70^0 and 90^0, mean T_V presents a local maximum at a value of external stress, σ_∞, roughly equal to the experimentally obtained values (see Table 1). On the contrary, results given in Fig.(4b) for β=8^0, 10^0 and 26^0, present a monotonous increase of mean T_V. An additional observation in Fig.(4b) is that the curve for β=26^0 shows a change in its curvature from concave to convex at a value of external stress roughly equal to the respective experimental value $\sigma_\infty \approx$160MPa, although the other two curves for β=8^0 and 10^0 do not show such a curvature change.

The above observations may allow for the conclusion that when normal stresses dominate (high values of angle β) the critical value, δ, of COD is achieved when mean T_V along crack-lips takes a maximum value. When crack tends to be parallel to the loading axis (low values of angle β) and shear stresses play a more important role than the normal ones, δ does not relate to T_V. In this case, rotation (slip) along crack-lips instead of translation seems to be the main factor related to δ [7], the separation line between the two states being near β=26^0, for the present material.

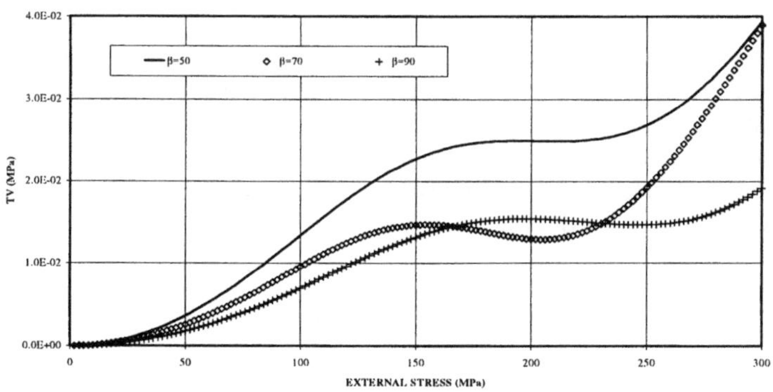

Figure 4a. The mean value of T_V vs external stress, for high crack inclinations.

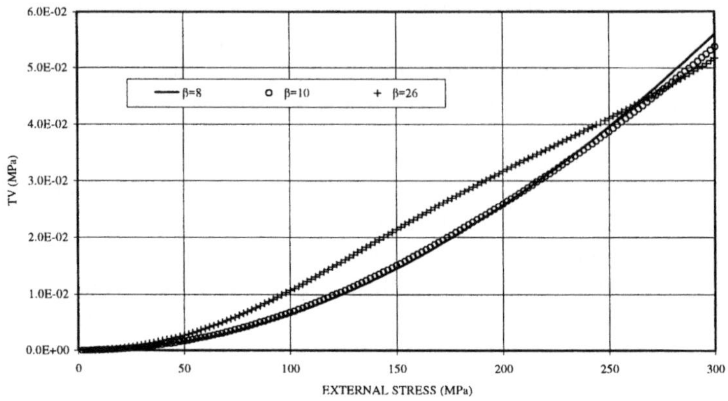

Figure 4b. The mean value of T_V vs external stress, for low crack inclinations.

However, the as above-described relationship between mean dilatational strain energy density, T_V and critical crack opening displacement, δ, may reflect the fundamental behavior of materials at failure. Critical geometrical quantities, like δ, representing linear displacement (separation of the material) are related to normal stresses, represented by T_V, as far as these stresses dominate over the shear ones. When this balance is changed in favor of shear stresses (low β-values) δ and T_V loose their relationship and failure is caused mainly by slip (rotation). In the latter case, it could be interesting to relate distortional strain energy density with the slip of the crack-lips. Unfortunately, experimental data on slip are not available. Should they existed, a reasonable generalization of the COD-concept could be the pair of a Crack Opening Surface (COS) index for brittle fracture and a Crack Opening Rotation (COR) index for ductile failure.

4. CONCLUSIONS

In the present work it was indicated that:

1. An equivalence between critical values of COD, crack surface and crack volume exists, permitting a free interchange between these three geometrical quantities in the description of critical conditions for a crack to initiate. Surface measurements, although a little more difficult to be performed, may be preferable. Their expected lower scattering, as they are based on multi-point information may be a considerable advantage over single-point COD measurements.
2. The mean value of dilatational strain energy density, computed along the deformed crack lips (excluding the close vicinity of crack-tip) shows a maximum value directly related with the critical value of COD. This relationship vanishes when shear stresses (small values of crack inclination angle β) dominate over normal

stresses. Under these conditions the validity of COD-concept is questioned and may be necessary to be replaced by another geometrical index related with crack-lips slip.
3. Experimental evidence from a metal matrix composite material indicates good agreement with the present analysis.

5. REFERENCES

1. Wells, A.A., in *Symposium on Crack Propagation*, Vol. 1, Cranfield, England (1961) 210-230.
2. Dawes, M.G., in Elastic-Plastic Fracture, ASTM STP 668, J.D. Landes, J.A. Begley and G.A. Clark (eds.) (1979) 307-333.
3. Saragas, S., Kourkoulis S.K. and Andrianopoulos N. P., Int. Journal of Fracture 79 (1996) 179-187.
4. Muskhelishvili, N.I.: Some Basic Problems in the Mathematical Theory of Elasticity, Noordhoff International Publishers, Leyden (1975).
5. Andrianopoulos, N.P., Theocaris P.S., Int. Journal of Mechanical Sciences 27 (1985) 793-801.
6. Knauf, G. Riedel H., in Advances in Fracture Research, Vol. 5, D. Francois (ed.) Pergamon Press (1981).
7. Theocaris, P.S., Pazis D. and Konstantellos B.D., Int. Journal of Fracture 30 (1986) 135-153.

DEPENDENCE OF FRACTURE CHARACTERISTICS ON THE THICKNESS OF THE SPECIMEN IN CASE OF METAL MATRIX COMPOSITES: AN EXPERIMENTAL STUDY

S.K. KOURKOULIS
National Technical University of Athens, School of Applied Sciences, Department of Mechanics, Theocaris Building, Zografou Campus, GR 157 73 Athens, Greece.

Abstract. The dependence of various fracture characteristics on the thickness of the specimen is studied experimentally in case of a particulate Metal Matrix Composite. Both intact and cracked specimens are considered. The Double Edge Notched Tensile specimen is used for the experimental program, in the case of pre-cracked specimens. The ductility and fracture strength appear to be increasing functions of the thickness in case of intact specimens. On the other hand, the critical J-integral and the critical Crack Tip Opening Displacement are found to increase in an almost linear manner with thickness for relatively small specimen thickness. However, the dependence becomes non-linear as the thickness increases further, leading finally to a maximum value. Dependence of the above quantities on the direction of the specimen with respect to the rolling axis is also detected, indicating the importance of the plastically induced anisotropy due to the manufacturing process.

1. INTRODUCTION

The problem of defining a mechanical quantity in the form of a single-parameter that could serve as a failure criterion (or equivalently as an index of the structural integrity) in case of pre-cracked specimens or structural members has been the subject of an enormous number of research papers already from the very early steps of the development of Fracture Mechanics. The introduction of the concept of the stress intensity factor, K, by Westergaard [1] and Irwin [2] as well as of the concept of the critical stress intensity factor or fracture toughness, K_C, gave the impression that an acceptable solution of the problem was achieved.

The initial enthusiasm was, however, shadowed by a long series of experimental findings, which indicated that K_C depends on the thickness of the specimen as it is shown in Fig.1 [3]. It is concluded from this figure that only after a certain specimen thickness, beyond which the material is under predominantly plane strain, K_C tends to a constant value, which is usually considered as a material property. Below this limit, K_C depends in a sigmoid manner on the thickness of the specimen exhibiting a well-defined maximum value. For very thin specimens the dependence is not exactly known (dotted portion of the graph). The above behaviour is attributed to the fact that for very thin plates the fracture toughness is a result of the plastic energy spent in the neck in front of the crack tip which in general is significantly larger in comparison to the energy spent for the overall damage of the material. It is,

E.E. Gdoutos and Z.P. Marioli-Riga (eds), Recent Advances in Composite Materials, 237-250

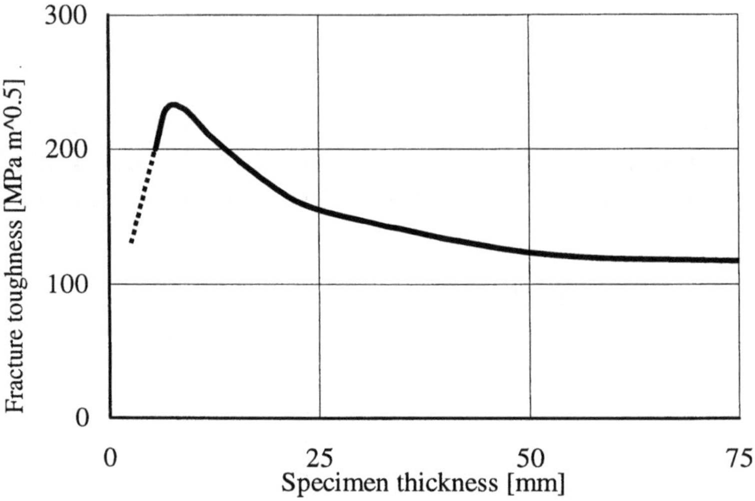

Figure 1. The typical dependence of the fracture toughness on the thickness of the specimen.

however, obvious that the work of necking depends directly on the thickness of the specimen and as a consequence the fracture toughness varies, also, with thickness as it was pointed out by Bluhm [4] and Swedlow [5] already from the early sixties.

It is thus concluded that for purely plane stress the fracture toughness depends almost linearly on the thickness of the specimen. On the other hand, however, as long as the thickness increases and the stress state deviates from the plane stress one the increase of fracture toughness is found to slow down reaching finally a maximum value for specimen thickness about 10 mm. From this point on the tests indicate that the fracture toughness decreases until it reaches a constant value corresponding to the pure plane strain state.

A different explanation of the above dependence was provided by Krafft et al. [6] in terms of the influence of the specimen thickness on the critical value of the energy release rate, G_C, which is directly related to the fracture toughness by the equation, $K_C=[EG_C/(1-v^2)]^{1/2}$. Knott [7] assumed that the thickness of the shear lips is more or less constant, which means that the fraction, S, of the crack front that is under plane stress decreases with increasing specimen thickness. Analyzing the experimental data provided by Krafft et al. [6] for the 7075/T6 aluminum alloy Knott arrived to the empirical equation $G_C=20(1-S)+200S^2$, which explains in a very satisfactory manner the dependence of the fracture toughness on the specimen thickness for all but the thinnest specimens. Unfortunately, experimental results concerning the dependence of fracture toughness on thickness for very thin specimens are rather scarce and as a result the relation between the resistance to cracking initiation and plate thickness remains insufficiently understood until today [8].

Similar observations were made very recently by Asserin-Lebert et al. [9] for the fracture toughness and the ductility of Aluminum 6065 plates. They conclude that for increasing thickness (from 1.4 mm to 6.0 mm) the ductility of the specimens increases by an amount of about 75% (!). They observed, also, that reducing the thickness did not cause significant hardening and the influence on the stress - strain curve up to the point of maximum load was negligible. They noticed, however, that the deformation was anisotropic and the size reduction along the short transverse direction (along the thickness of the specimens) was always larger than along the transverse one. Similar conclusions were recently drawn, also, by Kourkoulis and Andrianopoulos [10] for metal matrix composites.

The main purpose of the present paper is to study further experimentally the dependence of some fracture characteristics on the thickness of the specimens for the case of composite materials. Attention is focused on the initial portion of the respective curves, i.e. that of very thin specimens, for which the respective data are very few. The characteristics studied include the ductility, the fracture stress, the critical J-integral and the critical Crack Tip Opening Displacement.

2. THEORETICAL CONSIDERATIONS

In Linear Elastic Fracture Mechanics (LEFM) a single parameter characterizes the singular stress and strain fields around a crack tip. It can be obtained in terms of a balance of energy rates which leads to the energy release rate concept, uniquely related to the stress intensity factor. Following a rather similar procedure for the case of non-linear materials the J-contour integral (or simply called the J-integral) was introduced [11]. It describes the flow of energy into the tip region and it is proved that the dominant term of the stress and strain singularities around the crack tip can be written in terms of it.

The J-Integral with respect to a point s on a crack front in a non-linear elastic material is defined by the formula:

$$J(s) = \int_{\Gamma}\left(W\delta_{1j} - P_{ji}\frac{\partial u_i}{\partial X_1}\right)n_j d\Gamma - \int_{A_\Gamma}\frac{\partial}{\partial X_3}\left(P_{3i}\frac{\partial u_i}{\partial X_1}\right)dA \tag{1}$$

Here Γ is an arbitrary curve enclosing the crack front at the position given by s, in the plane $X_3=0$, A_Γ is the surface area defined by Γ, W is the deformation work per unit volume, u is the displacement vector, P_{ij} denotes the Piola-Kirchhoff stress tensor, X_i is a component of the position vector of a material point in the undeformed configuration and n_j is a component of a unit vector perpendicular to Γ.

For the Double Edge Notched Tensile (DENT) specimen, used in the present work, Rice [12] introduced a simple formula for the determination of the J-integral:

$$J_{Rice} = \frac{K_I^2}{E} + \frac{1}{\ell_0 t_0}\left(2\int P du_P - P u_P\right) \tag{2}$$

where K_I is the stress intensity factor, E is the Young's modulus, P is the applied load and u_p is the plastic displacement. The stress intensity factor, K_I, for this type of specimen is given by the following formula [13]:

$$K_I = \frac{P}{t_0}\sqrt{\frac{\pi\lambda}{2w_0(1-\lambda)}}\left(1.122 - 0.561\lambda - 0.205\lambda^2 + 0.471\lambda^3 + 0190\lambda^4\right) \tag{3}$$

In the above Eq.(3) w_0 is the half width of the plate. The parameter λ is defined as $\lambda = a_0/w_0$ where a_0 is the initial crack length.

The validity of the J-integral concept was the subject of a long series of papers. The point most intensively criticized is related to the fact that the whole concept is strictly founded for non-linear elastic materials, which are of reversible nature concerning the loading-unloading path. Thus using the J-integral for elasto-plastic materials appears illegitimate and the concept seems to be applicable only until no unloading occurs in any part of the material, in other words until crack extension starts. Other points related to the difficulties encountered during the practical determination of the critical value of J-integral and the necessity to combine the experimental techniques with finite element analysis are of minor importance.

The consensus reached today is that the J-integral can be calculated at the tip of a crack even for an elasto-plastic solid until crack starts propagating subject to the additional condition that the crack tip finite strain zone in which damage occurs is sufficiently confined. Indeed, the loss of constraint, appearing with large scale yielding, is directly related to the deviation of the relationship between the J-Integral and the Crack Tip Opening Displacement (δ_t) from the one proposed by Shih [14], even for a stationary crack:

$$J = \frac{\sigma_0 \delta_t}{d_n} \tag{4}$$

The above relationship was obtained by Shih using the HRR solution for the stress and strain fields. The latter is valid within the framework of the deformation theory of plasticity, assuming however that no unloading occurs at any point. The flow behaviour of the material is simulated by a power law, generalized as:

$$\varepsilon_{ij} = \frac{3}{2}\alpha\left(\frac{\sigma_e}{\sigma_0}\right)^{n-1}\frac{S_{ij}}{E} \tag{5}$$

where ε_{ij} and S_{ij} are the components of the strain- and of the deviatoric stress-tensor, respectively, σ_e is the effective stress, defined as $\sigma_e=[3/2(S_{ij}S_{ij})]^{1/2}$, E is the modulus of elasticity, σ_0 is a reference value for the stress (usually defined as the 0.2% offset yield strength) and α, n are experimentally defined constants. Assuming then that the above restrictions are fulfilled, the stress field components, σ_{ij}, can be described according to Shih's approach by the following equation:

$$\sigma_{ij} = \sigma_0 \left(\frac{EJ}{\alpha \sigma_0^2 I_n r} \right)^{1/(n+1)} \tilde{\sigma}_{ij}(n, \theta) \tag{6}$$

where, r is the radial distance from the crack tip, I_n is a dimensionless constant which depends on strain hardening and $\tilde{\sigma}_{ij}$ is, also, a dimensionless constant depending on strain hardening and angle θ from the crack plane. Concerning the function $d_n = d_n(n, \varepsilon_0)$ of Eq.(4), which was given by Shih as a nomogramme, Omidvar's et al. [15] approximate solution can be adopted, according to which it holds that:

$$d_n(n, \varepsilon_0) = \frac{\sigma_0}{E}^{1.05/(n-0.1)} (1 + \frac{3}{n}) \tag{7}$$

However, it is to be emphasized at this point that the previous analysis is valid as long as each material point experiences proportional loading. For cracked bodies this is not the case, since an intense strain region exists, within $\sim 2\delta_t$ of the tip, that experiences highly non-proportional loading. In such a case the analysis is only valid if the intense strain region is surrounded by a region in which the HRR model assumptions still prevail. This is true as long as δ_t is relatively small compared to both the crack size and the uncracked ligament length. Otherwise fracture toughness becomes strongly geometry dependent [16].

For the material under study it has been determined experimentally in a previous work [17] that the intensively damaged zone surrounding the tip is restricted to less than 5% of the initial crack length. Hence the HRR model appears to be valid, at least for the classes of specimens tested, although the failure mode detected using Scanning Electron Microscopy is a rather ductile one. It means that the above Eqs.(4-7) can be safely used for the description of the stress field according to the HRR assumptions and thus the critical value of the J-integral can be considered as a failure parameter.

3. EXPERIMENTAL PROCEDURE AND NUMERICAL ANALYSIS

3.1. The material and specimens

The material used in the present study is the BP-2124 Al-Cu particulate Metal Matrix Composite. It is produced by a novel powder metallurgy process, which reinforces aluminum alloys with extremely fine silicon carbide (SiC) particles. It combines a series of excellent mechanical properties with relatively low specific weight (high specific stiffness and high specific strength). The chemical composition of the matrix alloy is listed in Table 1. The respective composite material is obtained by adding to the matrix alloy about 20% wt of SiC particles of average diameter 3 μm [18].

According to the production process introduced by the manufacturing company the raw materials, i.e. the atomized aluminum alloy powder and the silicon carbide microgrit, are bought-in to an internal specification covering chemistry and size. The powders are then processed and blended to a homogeneous mixture using technology developed at BP's research centre. The mixed powders are then canned, degassed and hot isostatically pressed to full density using conventional aluminum powder metallurgy practices. The hot pressed billets are decanned and converted to wrought product using standard metal working equipment.

The material was available in rolled plates of thickness equal to t_o=12 mm. After the specimens were cut from the plate and machined in their standardized shape (by slowly removing away material in the thickness direction by mechanical milling) they were subjected to heat treatment as it was suggested by the manufacturer. The procedure included solution treatment for ninety minutes at 505 ^oC followed by immediately cold-water quench. No visible distortion or surface cracking of the specimens was observed due to the quenching. Scanning Electron Microscopy study did not reveal internal stress concentration points also.

Table 1. The composition of the material

Element	Wt%
Copper	4.0-4.4
Magnesium	1.3-1.6
Manganese	0.4-0.7
Iron	0.3 max
Zinc	0.25
Silicon	0.2
Chromium	0.1
Titanium & Zirconium	0.2
Others (each)	0.05 max
Others (total)	0.2 max
Aluminum	Balance

Two different types of specimens were used: The first one included prismatic specimens with orthogonal cross section of constant width w=9 mm and thickness varying from 1.2 mm to 9 mm, for the determination of the dependence of fracture strength and ductility on the specimen thickness. The second type included DENT specimens of length 150 mm and width 60 mm, the thickness of which varied again between 1.2 and 9 mm. The width of the ligament was kept constant equal to 20 mm since it was not among the purposes of the paper to study the influence of the length of the ligament on the process. During the numerical analysis of the experimental results it was verified that path independence at cracking initiation was secured and no case was detected for which the whole ligament length underwent finite strains. The edge cracks of initial length a_0=20 mm were machined by means of a slowly rotating diamond disc of thickness 100 μm. The specific cutting procedure was chosen since it creates cracks with minimal mechanical damage at their root.

The DENT specimens were cut from the rolled plate along the rolling direction and the load was induced along the same direction. On the other hand the prismatic specimens were cut either along the rolling direction or perpendicular to it or inclined 45° with respect to it, in order to study the influence of the plastically induced anisotropy on the ductility and fracture stress.

The prismatic specimens were prepared very carefully to ensure parallel gauge length and faces that were orthogonal to the plane of the sheet. A system of 8 strain gauges bonded in the mid-section of the specimens was used to measure in real time both the transverse and the through-thickness strains as well as the longitudinal ones, which are necessary for the determination of the constitutive law. Gauges were bonded on all four sides of the specimens to monitor possible bending parasitic effects. A series of successive loading-unloading loops, was executed in each experiment, in order to measure the pure elastic constants of the material as well as the plastic portion of strains. It is to be mentioned at this point that considerable difficulties were encountered during the measurement of strain in the case of the very thin specimens, the maximum thickness of which did not exceed in any case 1.2 mm. Bonding gauges on so thin surfaces is not possible and uncertainties appear concerning the validity of the results. To ensure that no parasitic effects influenced the experimental results, a number of experiments was carried out using two additional strain measurement methods, namely one based on the laser technology and a purely mechanical one based on conventional extensometers. The differences recorded between the two methods were insignificant.

For the DENT specimens the displacements were measured by means of very sensitive extensometers. The onset of fracture was detected by the appearance of thumbnails on the crack front at the center of the plates. Concerning the measurement of Crack Tip Opening Displacement a procedure similar to the one described in ref. [19] was adopted, based on continuous monitoring and successive photograph of the specimen at various stages of the testing procedure. The measurements were carried out by means of a Gruman-type visioscopic fringe analyser, equipped with a digital electronic micropositioning system, of maximum error ±0.5 μm.

All specimens were subjected to static tensile loading with the aid of a very stiff hydraulic loading frame. The crosshead speed did not exceed 0.2 mm/min.

3.2. Experimental results: Intact specimens

As a first experimental observation it should be mentioned that the axial strains measured in the rolling plane, $\varepsilon_{axial,rp}$, diverge slightly, but systematically, from the axial strains measured in the thickness plane, $\varepsilon_{axial,tp}$, at the same load level. The divergence becomes higher with increasing width to thickness ratio, as it can be seen from Fig.2, in which the ratio $\varepsilon_{axial,rp}/\varepsilon_{axial,tp}$ has been plotted versus the axial stress, for a series of characteristic experiments with prismatic specimens cut and loaded along the rolling direction. Also, it is observed from the same figure that while $\varepsilon_{axial,rp}$ exceeds $\varepsilon_{axial,tp}$ during the initial loading steps, however, as the load increases the phenomenon is inversed and tends to be eliminated as the load approaches the failure limit. The above observations indicate a non-uniformity of the deformation along the thickness of the plates, which becomes more pronounced as the thickness of the specimens increases. The conclusions for the other two types of prismatic

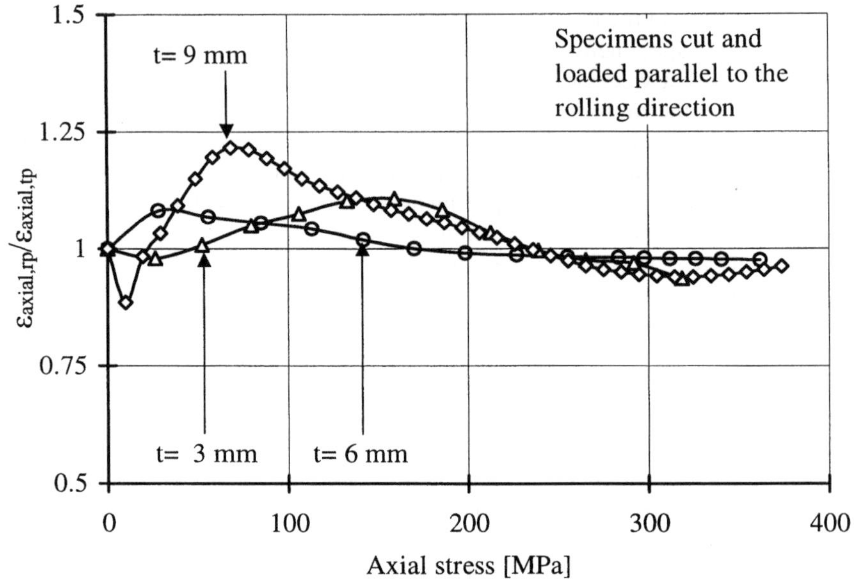

Figure 2. The divergence between the axial strains measured in the rolling- and in the thickness-plane for specimens cut and loaded along the rolling direction.

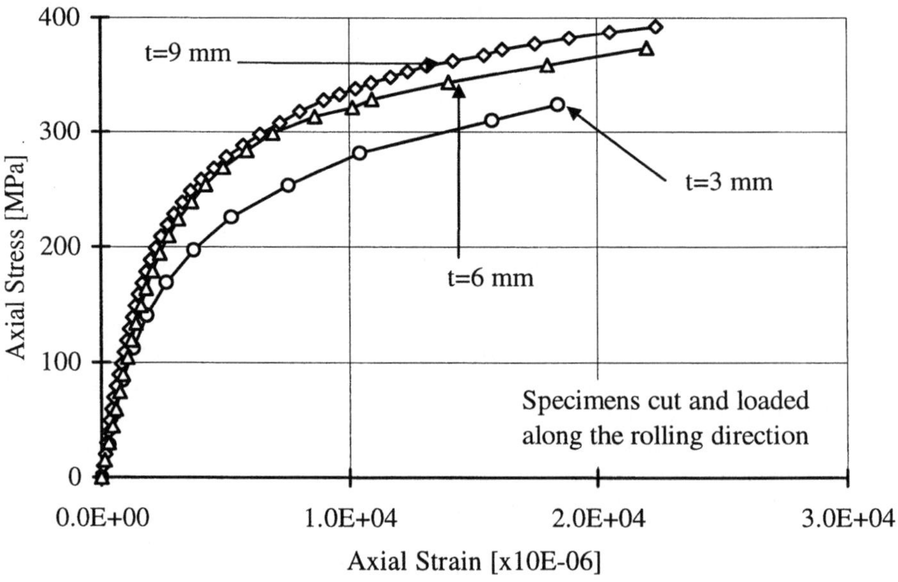

Figure 3. The axial stress versus the axial strain measured in the rolling plane for prismatic specimens cut and loaded along the rolling direction.

specimens, namely the ones cut perpendicular to the rolling direction and diagonal to it are of similar qualitative nature.

As far as it concerns the variation of the fracture strength and ductility with specimen thickness it was concluded that both quantities increase with increasing thickness. In Fig.3 characteristic axial stress - axial strain curves are plotted for the specimens cut and loaded parallel to the rolling direction. The axial strains were measured in the rolling plane. It can be seen from this figure that, contrary to the observations by Asserin-Lebert et al. [9], the deviation between the graphs starts relatively early, namely almost immediately after surpassing the linearity limit. Such a behaviour seems to contradict common sense, however it can be explained if one takes into account that the material studied here is not a homogeneous continuum medium due to the presence of the reinforcing dispersed phase of the silicon carbide particles. The conclusions for the specimens cut and loaded perpendicular and inclined 45° with respect to the rolling axis are almost identical. The respective results are summarized in next Figures 4 and 5 for all types of specimens.

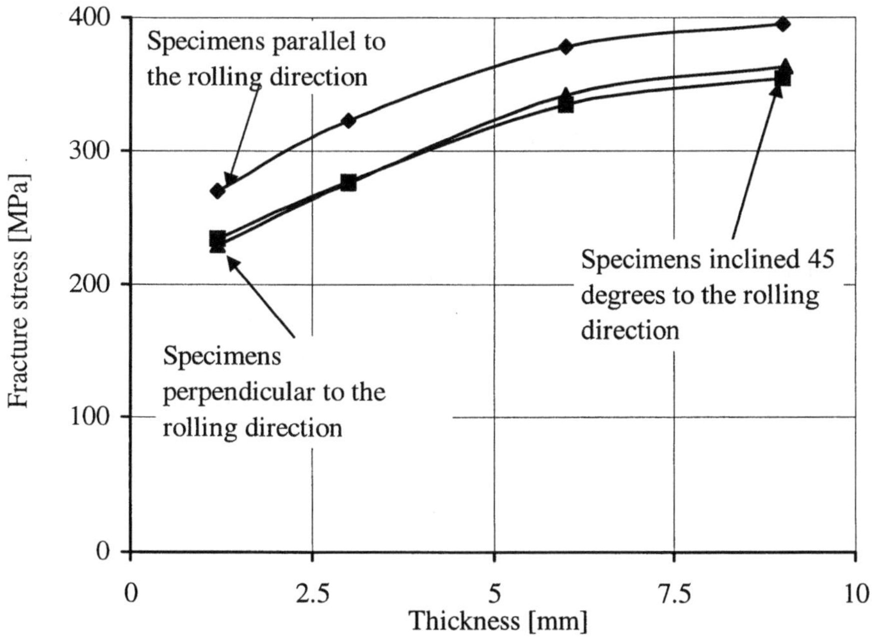

Figure 4. The dependence of fracture stress on the specimen thickness.

In Figure 4 the fracture stress is plotted versus the specimen thickness for all three types of prismatic specimens. It is seen from this figure that the fracture stress–specimen thickness curve is almost linearly increasing with increasing thickness for thicknesses between 1.2 mm and 5 mm. From this point on the linearity is lost and the curve reaches its maximum value for the specimens with thickness about 9 mm.

The absolute differences between the fracture strength of the thinnest and thickest specimens was about 40% for all three types of specimens.

Concerning the influence of plastically induced anisotropy it is seen from Fig.4 that the rolling direction is the strongest one while the differences between the other two directions are negligible no matter what the thickness of the specimen is.

In next Figure 5 the dependence of the fracture strain on the specimen thickness is plotted for all types of prismatic specimens. Although the qualitative behaviour of the graph is similar to that of the fracture stress it is worth to be mentioned here that the specimens cut along the direction inclined 45° with respect to the rolling one appear to be the most ductile, especially for small thickness and they reach their respective maximum value of ductility already for a thickness equal to about 6 mm. Concerning the absolute differences they vary from about 80% for the specimens cut perpendicular to the rolling direction to 50% for the ones cut parallel to it.

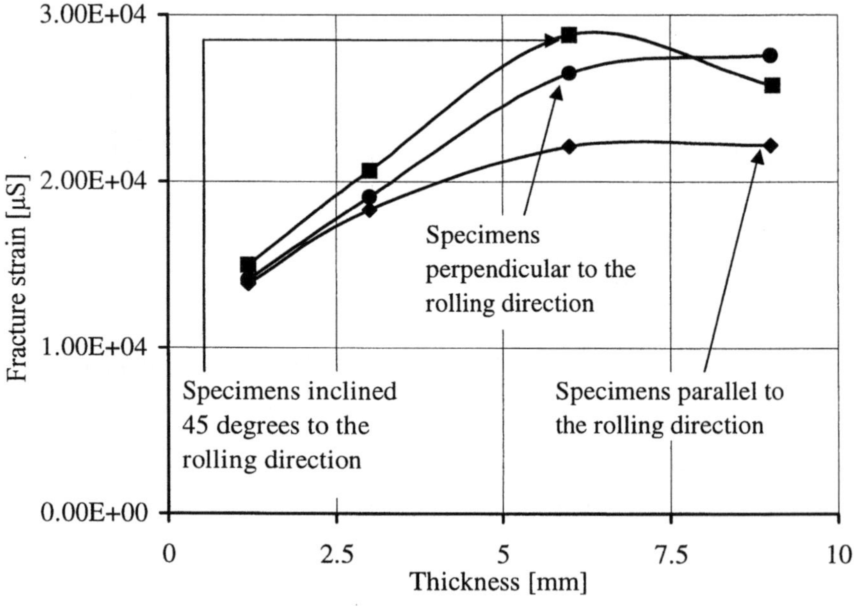

Figure 5. The dependence of the ductility on the specimen thickness.

3.3. Experimental results: DENT specimens

The values of the critical Crack Tip Opening Displacement are plotted in Figure 6 versus the thickness of the specimen. Similar conclusions with the previously analysed quantities can be drawn from this figure concerning both the initially linearly increasing behaviour of the graph as well as the attainment of a maximum value after this portion. The difference, however, is that in this case the increasing portion of the graph is much more abrupt and the maximum value is already attained for a specimen thickness equal to about 6 mm.

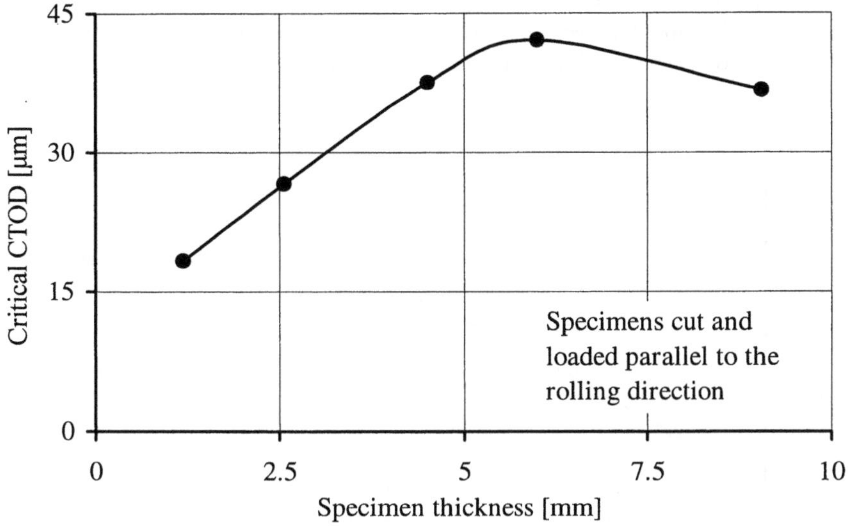

Figure 6. The dependence of the critical Crack Tip Opening Displacement on the thickness of the specimen.

3.4. Numerical analysis and the critical J-Integral

In order to determne the critical values of the J -Integral a numerical analysis similar to the one followed by Pardoen et al. [8] was adopted, based on the 3D domain integration technique introduced by Shih et al. [20]. According to it J(s) is computed from various contours in order to check the path independence and a mean value is obtained by averaging the local values on the entire thickness. However, it is to be mentioned that the differences detected between these local values were relatively small for all specimens. As it was mentioned earlier path independence at the moment of crack propagation onset was observed for all geometries tested.

The results of the analysis are plotted in Figure 7. As it can be seen from this figure the critical values of J-integral vary in a similar manner with Crack Tip Opening Displacement. However, contrary to other fracture characteristics studied in the previous paragraph, the values of the critical J-integral do not reach their maximum even for specimens with thickness equal to about 9 mm although the increasing tendency is almost zeroed. Such a behaviour is rather unexpected since the material studied here is less ductile to the one studied by Pardoen et al. [8] for which the maximum value was reached for a thickness of about 5-6 mm. However, it could be explained on the basis of the different failure modes activated in case of pure metallic materials and composite ones, as it will be discussed in the next paragraph. In any case it is obvious that additional analysis is required (including tests with thicker specimens also) in order to determine the exact form of the relation between J_C and thickness.

On the other it is encouraging for the analysis and it should be mentioned that the differences detected between the values obtained numerically and the ones obtained using the J_{Rice} formula from Eq.(2) together with the experimental data concerning the load – displacement curve are extremely close to each other for all experiments. The maximum discrepancy did not exceed 4% in any case.

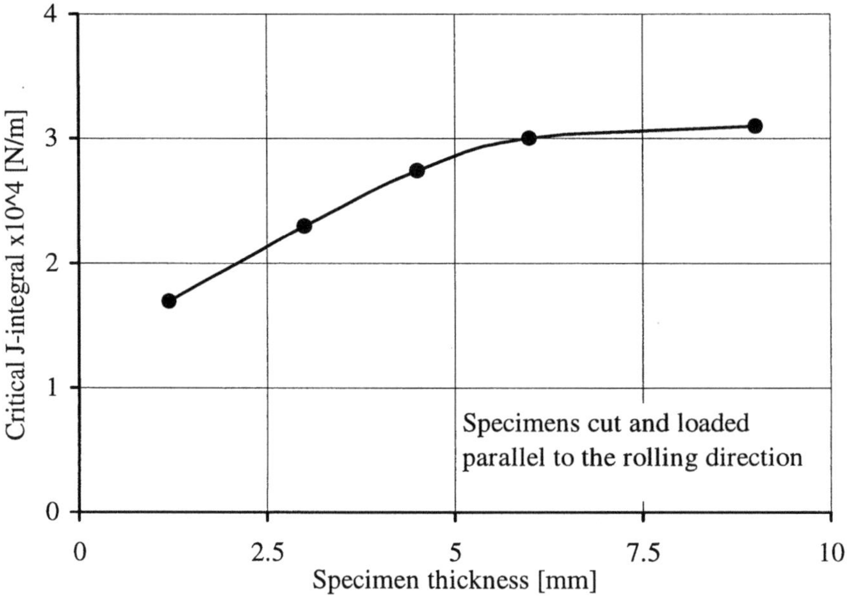

Figure 7. Numerically calculated critical values of the J-integral versus the thickness of the specimen.

4. THE FRACTURE SURFACES AND THE FRACTURE MODE

The study of the fracture surface of the specimens using Scanning Electron Microscopy revealed that the plane-stress “macromode” of fracture ahead of the crack tip is clearly activated. This failure mode is a combination of a shear component (distinguished by the shear lips) and a normal one (normal cone). The shear component prevails close to the surface layers whereas the normal one prevails at the mid-thickness layers of the specimen. It was concluded that the overall failure process is a ductile plane-stress one with an additional feature: The “dimple microfailure mode” is active. This is attributed to the microvoid coalescence mechanism, on which the ductile macrofracture of metals [21] is based.

Additionally the signal differentiation processing technique was used, by which contour enhancement or edge sharpening of the dimples is achieved permitting one to distinguish better the differences in the morphologie between the two areas of the failure surface (shear lips and normal cone). It was observed that the shear prevailing area is characterized by an elongated dimple pattern, in contrast to the normal cone area which is characterized by a more or less “quasi-equiaxed” dimple pattern. Such a behaviour is attributed to the loss of severity of the triaxiality of the stress state near the surface layers of the material.

The analysis of the SEM images pointed out, also, that the fracture process is still controlled by the void coalescence mechanism despite the presence of the reinforcing SiC particles. On the other hand, SEM analysis of the fracture surface of uncracked specimens indicates that the micro-cracks are initiated in the matrix rather than at the particle-matrix interface. This peculiar, at least for Metal Matrix Composites, phenomenon can be explained by the fineness of the reinforcement phase as well as by the process used for the production of the specific Metal Matrix Composite (powder metallurgy) which is responsible for some porosity. Thus, in case of uncracked specimens void nucleation rather than void coalescence is the critical step for failure. When the local plastic relaxation, that relieves stress concentration becomes difficult, as in case of pre-cracked specimens, where it is concentrated in the vicinity of the tip of the macroscopic crack, then the final failure occurs at rather low strains. It seems thus that the presence of macro-cracks changes the failure mechanism, explaining in this way the deterioration of the mechanical properties of cracked Metal Matrix Composites [19].

5. CONCLUSIONS

The dependence of some characteristic mechanical quantities related to the fracture of either intact or pre-cracked materials on the thickness of the specimens was studied experimentally in the present work. The specimens were made form the BP-2124 Al-Cu Metal Matrix Composite, made form aluminium matrix and reinforced with very fine SiC particles. The quantities studied were the fracture strength, the ductility, the critical Crack Tip Opening Displacement and the critical J-integral. The results of the study can be shortly summarized as follows:

All four quantities depend on the specimen thickness in the same qualitative way: Namely they increase almost linearly with increasing thickness for very thin specimens and then the increasing ratio decreases and finally disappears.

A maximum value is attained for a specimen thickness, which varies from 6 mm for the Crack Tip Opening Displacement to 9 mm for the fracture strength. From this point on a decreasing tendency appears, which should be studied further using thicker specimens, in order to determine the respective plane strain value, which should be considered as the respective material property.

The values of the critical J-integral do not attain their maximum value even for the thickest specimens used, although the increasing tendency of the respective curve is almost eliminated.

Finally the Scanning Electron Microscopy study indicated that the failure mode of the cracked specimens was a combination of a shear component (distinguished by the shear lips) and a normal one (normal cone) and the micro-cracks are initiated in the matrix rather than at the particle-matrix interface.

REFERENCES

1. Westergaard, H. M.: Bearing pressures and cracks, *Transactions of the ASME*, **61**(1939), A49-A53.
2. Irwin, G. R.: Analysis of stresses and strains near the end of a crack transversing a plate, *J. Appl. Mech.*, **24**(1957), 361-364.
3. Ewalds, H. L. and Wanhill, R. J. H.: *Fracture Mechanics*, Edward Arnold Publishers, London, 1985.

4. Bluhm, J. L.: A model for the effect of thickness on fracture toughness, *ASTM Proc.*, **61**(1961), 1324-1331.
5. Swedlow, J. L.: The thickness effect and plastic flow in cracked plates. Aerospace Research Laboratories, Office of Aerospace Research, U.S. Air Force, Wright-Patterson Air Force Base, OH, 1965.
6. Krafft, J. M., Sullivan, A. M. and Boyle, R. W.: Effect of dimensions on fast fracture instability of notched sheets, *Proc. of Crack Propagation Symposium*, The College of Aeronautics, Cranfield, England, **1**(1962), 8-28.
7. Knott, J. F.: *Fundamentals of Fracture Mechanics*, Butterworths, London, 1973.
8. Pardoen, T., Marchall, Y. and Dellannay, F.: Thickness dependence of cracking resistance in thin aluminium plates, *J. Mech. Phys. Solids*, **47**(1999), 2093-2123.
9. Asserin-Lebert, A., Bron, F., Besson, J. and Gourgues, A. F.: Rupture of 6056 aluminum sheet materials: Effect of sheet thickness on strain localization and toughness, *Proc. ECF-14: Fracture Mechanics Beyond 2000*, EMAS Publishing, Sheffield, England, **1**(2002), 97-103.
10. Kourkoulis S. K. and Andrianopoulos, N. P.: Plastically induced anisotropy on metal matrix composites, *Mechanics of Composite Materials and Structures*, **7**(2000), 1-18.
11. Rice, J. R.: A path independent integral and approximate analysis of strain concentration by notches and cracks, *J. of Appl. Mech.*, **35**(1968), 379-386.
12. Rice, J. R.: Elastic – plastic models for stable crack growth, *Proc. of the Conference on Mechanics and Mechanisms of Crack Growth*, Churchill College, Cambridge, England, (1973), 14-39.
13. Anderson, T. L.: *Fracture Mechanics – Fundamentals and Applications*, CRC Press, Boca Raton, 1995.
14. Shih, C. F.: Relationships between the J-integral and the COD for stationary and extending cracks, *J. Mech. Phys. of Solids*, **29**(1981), 305-326.
15. Omidvar, B., Wnuk, M. P. and Choroszynski, M.: Relationship between the CTOD and the J-integral for stationary and growing cracks, *Int. J. of Fracture*, **87**(1997), 331-343.
16. Schapery, A.: Report MM5762-88-1, Texas A&M University, College Station TX, 1988.
17. Kourkoulis, S. K. and Kytopoulos, V.: Experimental quantification of crack tip parameters for particulate metal matrix composites, *Proc. of the 6th Congress of the Hellenic Association of Theoretical and Applied Mechanics*, Thessaloniki, Hellas, **1**(2001), 339-347.
18. Particulate Metal Matrix Composite Processing Handbook, Issue 1, BP Metal Composites, Farnborough, England, 1991.
19. Kourkoulis, S. K.: The influence of cracks on the mechanical behaviour of MMCs: An experimental study, *Archives of Mechanics*, **53**(2001), 439-456.
20. Shih, C. F., Moran, B. and Nakamura, T.: Energy release rate along a three dimensional crack front in a thermally stressed body, *Int. J. Fract.*, 30(1986), 79-102.
21. Fractography and materials science, ASTM-STP 733, 1981.

THEORETICAL EVALUATION OF STRENGTH DEGRADATION DURING MONOTONIC AND CYCLIC LOADING IN UNIDIRECTIONAL METAL MATRIX COMPOSITES

C. A. RODOPOULOS
Division of Aeronautical Applications, Structural Integrity Research Institute of the University of Sheffield (SIRIUS)
Department of Mechanical Engineering
The University of Sheffield, Mappin Street, Sheffield, S1 3JD
United Kingdom

Abstract. By implementing the concept of critical fibre length into a modified version of the Law of Mixtures and considering the effect of the thermal residual stresses, the monotonic strength of unidirectional MMCs is predicted. The above is also used as a reference value for strength degradation in the case of cyclic loading. Herein, the conditions of constraint effect and flow resistance degradation are taken as the major, strength controlling, damage mechanisms. The approach is then executed for two kinds of titanium matrices and the commonly used SCS-6 fibre. The approach allows the study of the above composite system in terms of the mechanical properties of the constituent phases and the fibre volume fraction.

1. INTRODUCTION

In recent years, it was experimentally shown that fatigue damage in unidirectional metal matrix composites (uMMCs) is the product of various damage and crack shielding mechanisms [1-3]. Fibre bridging and fibre constraint effect (also known as elastic shielding effect), are cited as mechanisms that confined the development of high crack tip plastic displacements [4-7]. On the other hand, matrix cracking and fibre failure, either ahead or behind the crack tip, are reported to promote crack growth [8,9]. It is now known that the above mechanisms are exhibiting different degrees of efficiency which depend on the mechanical properties of the fibre, matrix, interface, the distribution of inherent defects and the loading history of the material [8].

To incorporate uMMCs into real-world applications, the designer should have sufficient knowledge on the residual strength behaviour of the material. Unfortunately, only recently, strength degradation effects have drawn a small amount of attention [10,11]. Theoretically, in uMMCs where the fibre fracture stress is higher than that of the matrix, the monotonic strength of the material is dictated by

E.E. Gdoutos and Z.P. Marioli-Riga (eds), Recent Advances in Composite Materials, 251-261

the fibre fragmentation process. Yet, under conditions of cyclic loading, the fibre fragmentation process is dominated by the stress field ahead of the growing crack.

In this work, strength degradation is examined in conjunction with the onset of the above damage mechanisms. The approach allows us to predict crucial boundary conditions.

2. MODELLING THE STRENGTH OF uMMCs

There are several discrepancies between the residual strength under static and cyclic loading. Some of these are: a) the fragmentation process of the fibre is governed by both the global straining conditions (stress) and the length of the crack; b) interfacial debonding ahead of the crack tip will precede fibre failure; and c) the crack will produce areas of different residual strength depending on parameters like the amount of fibre bridging, the amount of broken fibres within the crack wake, the amount of interfacial debonding, etc.

2.1. The Static Strength of uMMCs

Kelly and Tyson [12] proposed that the ultimate strength of a uMMC in longitudinal tension is dictated by the failure of fibres of length equal to the critical (also known as ineffective). In this case, the load from the broken fibres is redistributed to he neighbouring fibres. The stress redistribution process triggers a "chain" reaction which can lead to additional fibre failures and matrix cracking. A simplistic representation is depicted in Figure 1. The onset and the rate of the above process depends on the interacting triangle between stress level - distribution of inherent manufacturing defects - material phases [8]. In general, an increasing stress level will trigger an increasing density of micro-cracks that emanate from manufacturing defects, i.e. fibre warts, voids in the matrix and at the interface, broken reaction layers, touching fibres, etc [13]. Consequently, the rate of fragmenting the fibres should be related to the rate that the density of micro-cracks reaches a critical state. In addition, the amount of residual stresses (which is varying with the consolidation technique) that is accommodated by either material phases should also be encountered as responsible for strength variations. The above complexities prohibit the development of a strength prediction model which is not utilising experimental and statistical measurements.

Using the assumption of the critical length and the global straining compatibility between the fibres and the matrix, in [7, 13] it was proposed that the ultimate tensile strength of a uMMC can be predicted by,

$$\sigma^{c}{}_{UTS} = \sigma_0 \left(\frac{meL_c}{L_o} \right)^{1/m} V_f + (1 - V_f)\sigma^{y}{}_{tm} \tag{1}$$

where σ_o and L_o are statistical reference values of fibre strength and length for a known failure probability m (Weibull Modulus), V_f is the volume fraction of the fibres and σ^{y}_{tm} is the true matrix yield stress (considering the pre-stressing caused by the thermal residual stresses).

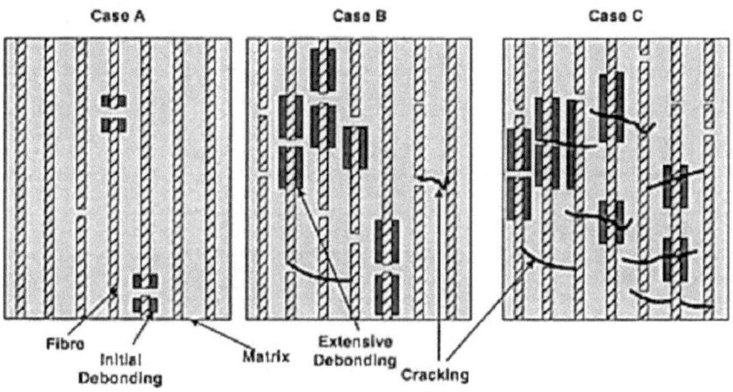

Figure 1. A schematic representation of the response of a medium strength interface uMMC under increasing tension loading. At low stress levels, case A, a limited number of fibres is anticipated to break. Limited debonding due to pre-existing stress concentrations (interfacial breaks, fibre voids, touching fibres, etc) is possible [8]. At medium stress levels, case B, the number of fibre breaks will increase and matrix cracking will commence. At high stress levels, case C, multiple cracking of the fibres will saturate to a value close to the critical fibre length [8]. At this stage, the majority of the fibre breaks will turn into matrix cracks.

The critical fibre length, L_c, according to the shear-lag analysis [14], is given by,

$$L_c = \frac{\sigma^{f}{}_{fr} d}{2\tau_f} \tag{2}$$

where $\sigma^{f}{}_{fr}$ is the fracture strength of the fibre, d is the fibre diameter and τ_f is the interfacial shear strength. The fracture strength of the fibre, for a given length, is predicted using the Masson and Bourgain equation [14],

$$\sigma^{f}{}_{fr} = \sigma_o \left(\frac{L_o}{L_c} \right)^{1/m} \tag{3}$$

Using an iterative solution between equations 1-3, the ultimate tensile strength can be predicted. A typical solution is shown in Figure 2.

TABLE 1A. Mechanical Properties of the SCS-6/Ti-15-3 [8]. The designation SCS-6 stands for the Textron Specialty Materials fibre.

E_f (GPa)	E_m (GPa)	σ^{y}_{m} (MPa)	σ_o (GPa)	L_o (mm)	m	d (μm)	τ_f (MPa)	τ_s (MPa)
400	90	750	3.6	25	9	140	124	81.5

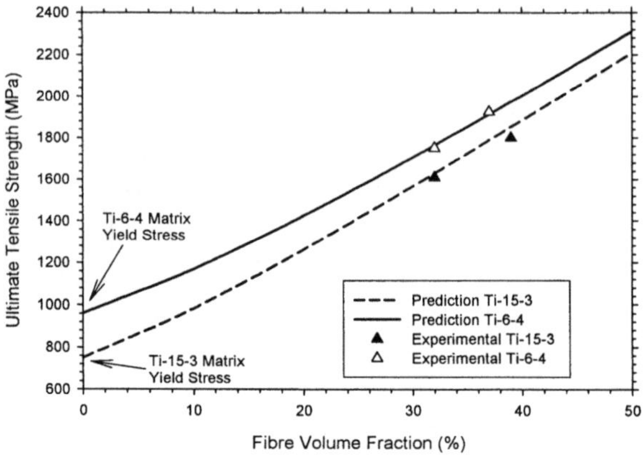

Figure 2. Comparison between the predicted and experimental values [8] of the ultimate tensile strength of a SCS-6/Ti-15-3 and SCS-6/Ti-6-4 for various fibre volume fractions. The mechanical properties used in the calculations can be found in Table.1.

TABLE 1B. Mechanical Properties of the SCS-6/Ti-6-4 [8]. The designation SCS-6 stands for the Textron Specialty Materials fibre.

E_f (GPa)	E_m (GPa)	σ^y_m (MPa)	σ_o (GPa)	L_o (mm)	m	d (μm)	τ_f (MPa)	τ_s (MPa)
400	110	960	3.6	25	9	140	138	81.5

2.2. *The Cyclic Strength of uMMCs*

Even though, the amount of strength degradation attributed to interfacial debonding is anticipated to be small (the debonded fibre will continue to support full load) [8], the issue is likely to require a statistical analysis. This is because extensive interfacial debonding could overload the fibre and thus increase the probability of a premature fibre failure. The density of interfacial debonding sites is strongly related to the cracking density, the volume fraction and the manufacturing process [13]. In general, uMMCs that have an amplified tendency to developed high values of secondary cracks are expected to increase the density of debonded sites. In [13] it was reported that SCS-6/Ti-15-3 exhibits a 300% higher density of secondary cracking than the SCS-6/Ti-6-4. In [15], it was suggested that the effectiveness of the constraint effect to control the spread of plasticity ahead of the crack tip (the high stiffness fibre is anticipated to relax part of the crack tip plastic strains) diminishes, when the rate of interfacial debonding exceeds the growth rate of the crack. Mathematically such rationale can be expressed by [16],

$$\sigma_{CD} = \frac{2\sqrt{2}}{\pi}\left(\sigma_{3d}\sqrt{\left(1-n_2{}^2\right)} - \sigma_2 n_1 + \sigma_2 n_2\right) \tag{4}$$

with $n_1 = \dfrac{a}{a+2FD}$, $n_2 = \dfrac{a+2FD-d}{a+2FD}$ and $\sigma_{3d} = \dfrac{4\tau_f L_D E_c}{dE_f}$.

The parameter σ_2 defines the flow resistance of the material given by [8],

$$\sigma_2 = \frac{\sigma_t{}^y{}_m}{E_m} E_f V_f + \sigma_t{}^y{}_m (1 - V_f) \tag{5}$$

where the parameter $\sigma_t{}^y{}_m$ (true matrix yield stress) is in turn given by,

$$\sigma_t{}^y{}_m = \sigma^y{}_m - \frac{V_f E_m E_f \Delta\alpha\Delta T}{E_c} \tag{6}$$

The parameters $\Delta\alpha$ and ΔT are the difference between the coefficients of thermal expansion between the matrix and fibre and the effective temperature change range, respectively. In the case of titanium/silicon carbide MMCs that have been produced by the foil-by-foil technique a temperature range of 555°C is usually considered [8].

In Eq(4) σ_{CD} represents the far-field stress that would cause constraint effect degradation, σ_{3d} is the critical value of the constraint effect, a is the cark length, d is the fibre diameter, L_D is the debonding length (taken as 4-5 times the fibre diameter to ensure statistical consistency of the interfacial shear strength [17]). The parameters E_c and E_f represent the elastic moduli of the composite (obtained by the Law of Mixtures) and fibre respectively. The parameter FD, is the distance between the beginnings of two successive fibres and for a square packing configuration is given by [18],

$$FD = d + \frac{\sqrt{\pi}\sqrt{V_f\left(\dfrac{d}{2V_f}\right)}}{2} \tag{7}$$

Examination of the boundary conditions for the parameters n_1 and n_2 reveals that the growth difference rationale is expressed in terms of – *the crack is able to debond two successive fibre rows without further propagation*. It is worth noting that Eq.(4) represents a fibre bridging (closure stress) free solution. Though, such approach provides conservative results. A typical solution of Eq.(4) is shown in Figure 3.

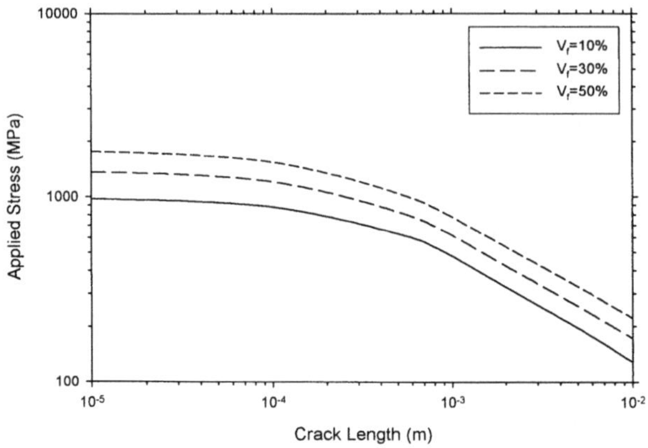

Figure.3: Constraint effect degradation for SCS-6/T-15-3 for different values of V_f.

Once the fibre constraint effect stops to provide the necessary strain relaxation, the crack tip plasticity is able to expand to a size that is controlled by the flow resistance of the matrix, $\sigma_t^y{}_m$ and the sliding friction provided by the debonded fibre as it slides against the matrix. If the conservative conditions of an identical local to global straining are imposed (strength changes ahead of the crack tip will trigger a widespread chain reaction), then the residual strength of the material can be also determined by Eq.(1) considering that the critical fibre length will be dominated by the sliding shear strength,

$$L_c = \frac{\sigma^f{}_{fr} d}{2\tau_s} \tag{8}$$

The corresponding residual strength due to the fibre constraint effect degradation is shown in Figure 4. From the analysis depicted in Figure 4, it is clear that strength degradation is promoted by the fibre volume fraction. However, due to the minimum strength degradation, such finding is unlikely to dismiss the use of uMMCs for stiffness oriented components.

After dismissal of the fibre constraint effect, the crack will propagate without causing any significant strength drop. Such status will terminate when the stress field ahead of the crack is high enough to cause fibre failure. In this case, the material will experience flow resistance degradation since the effective fibre volume fraction will decrease. Whether, the above decrease would be rapid or gradual, depends on many parameters the analysis of which is beyond the scope of this work. It is worth noting that fibre failure ahead of the crack tip would most certainly follow failure of the fibres within the crack wake. In [19], it was proposed that fibre failure at some distance D ahead of the crack tip would commence when the crack opening displacement is equal to a critical value. The above rationale is depicted in Figure 5.

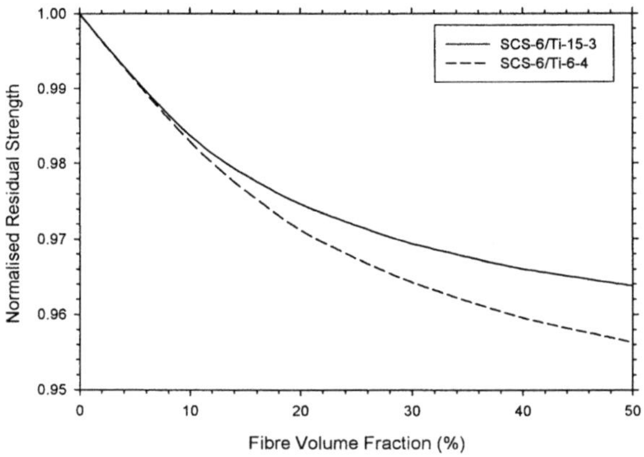

Figure 4. Normalised residual strength, due to constraint effect degradation, for SCS-6/Ti-15-3 and SCS-6/Ti-6-4 versus fibre volume fraction. Normalised residual strength represents the ratio between the strength after fibre constraint effect degradation and the strength for monotonic loading.

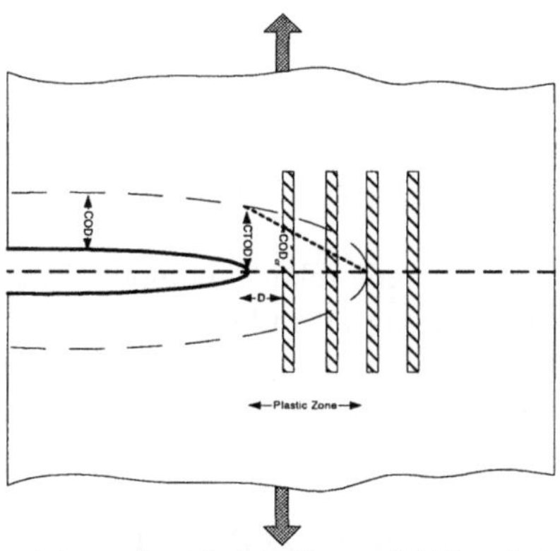

Figure 5. Conditions of flow resistance degradation. The straight line that connects the end of the plastic zone and the COD at the crack tip is for simplicity.

Figure 6 shows a plot of the distribution of COD_{cr} for the SCS-6 fibre embedded to Ti-6-4 and Ti-15-3 matrices as a function of V_f. The steps to acquire COD_{cr} can be found in [20].

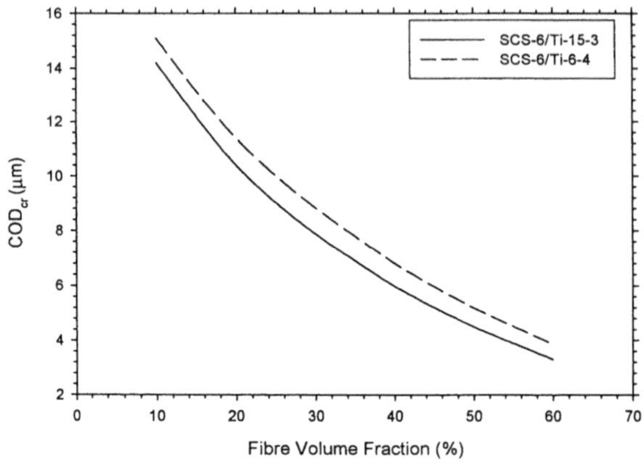

Figure 6. Critical COD for SCS-6 fibre embedded in Ti-6-4 and Ti-15-3 matrices. The predictions correspond to a Weibull modulus of 9 and a number of 8 fibres per row.

Considering the simplification of the COD as depicted in Figure 5, the boundary condition for flow resistance degradation can be written as,

$$\mathrm{CTOD} = \frac{\mathrm{PLASTIC\ ZONE\ x\ COD_{CR}}}{\mathrm{PLASTIC\ ZONE} - \mathrm{D}} \tag{9}$$

Due to the stress and crack length related nature of the crack tip plastic zone, solution of Eq.(7) is obtained using an in-house computer code. The flow chart of the code can be found in [4]. Considering that Eq.(9), for values of D<d, denotes failure of the first fibre row, the effective fibre volume fraction can be approximately given by,

$$\mathrm{V_f}^{\mathrm{eff}} = \frac{(\mathrm{PLASTIC\ ZONE - d) \ x\ V_f}}{\mathrm{PLASTIC\ ZONE}} \tag{10}$$

Thus, the corresponding strength can be calculated by,

$$\sigma^c{}_{UTS} = \sigma_o \left(\frac{me \dfrac{\sigma^f{}_{fr} d}{2\tau_s}}{L_o} \right)^{1/m} V^{eff}{}_f + \left(1 - V^{eff}{}_f\right) \sigma^y{}_{tm} \tag{11}$$

Solution of Eqs.(9-11) is depicted in Figure 7.

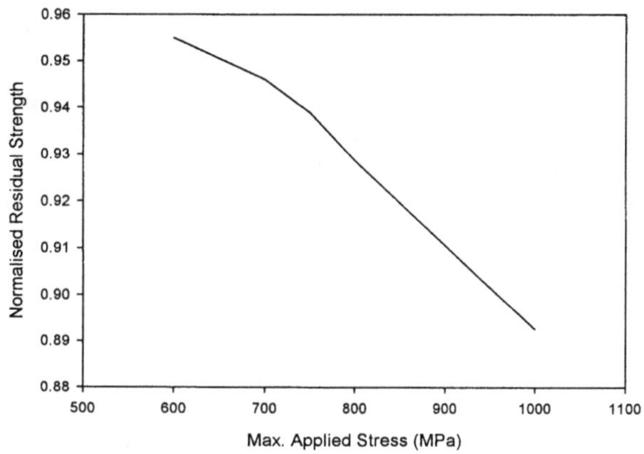

Figure 7. Normalised residual strength versus max. applied stress at the onset of flow resistance degradation for a 35%, SCS-6/Ti-15-3.

From Figure.7 it is worth noting that: A) for stress level 600 MPa the residual strength is approximately that predicted in Figure 4. This is due to the fact that at low stress levels (High Cycle Fatigue), flow resistance degradation can precede or match the degradation of the constraint effect [8] and B) the normalised strength distribution follows a steep decay with the applied stress. This is because at high stress levels, the onset of flow resistance degradation takes place on short plastic zone sizes. In such case, failure of the first fibre row can cause a significant reduction of $V^{eff}{}_f$.

Knowledge over the load redistribution, within the crack tip plastic zone, after fibre failure, can provide the necessary information to adjust Eq.(10) in order to account for a more severe strength degradation (choosing D>d

3. CONCLUSIONS – DISCUSSION

In this work, strength degradation under monotonic and cyclic loading conditions is discussed. In the case of monotonic loading, the strength of an unidirectional MMC can be obtained by using the Kelly's-Tyson's critical fibre concept into a modified version of the Law of Mixtures. Comparison between predicted and experimental results showed that such approach could provide accurate predictions.

On the other hand, strength degradation due to cycling loading is modelled through the strength changes caused by the onset of the two major cyclic mechanisms of damage: a) constraint effect degradation and b) flow resistance degradation. Constraint effect degradation, representing the condition where the ability of the high stiffness fibre to relax crack tip plastic strains diminishes, showed a maximum strength degradation of 4% for a 50% fibre volume fraction. Such finding is within the range of experimental data published by Robertson and Mall [21].

Strength degradation due to the onset of flow resistance degradation (failure of first fibre row), is modelled through the concept of an effective fibre volume fraction within the premises of the crack tip plastic zone. The approach allows the examination of the residual strength in terms of crack length or number of cycles. Knowledge of the redistribution of the fibre bearing load after failure can be also implemented to account for more severe cases.

4. DEDICATION

This work is dedicated to my son, Alexander Chris Rodopoulos born on the 11th of July, 2002.

5. REFERENCES

1. Johnson, W. S.: Fatigue Damage Accumulation in Various Metal Matrix Composites, *NASA Technical Memorandum 89116*, (1987), Langley.
2. Ritchie, R. O.: Mechanisms of fatigue crack propagation in metals, ceramics and composites: Role of crack tip shielding, *Mater. Sci. Engng.*, A103, (1988), 15-28.
3. Davidson, D. L., Chan, K. S., McMinn, A. and Leverant, G. R.: Micromechanics and fatigue crack growth in an alumina-fiber-reinforced magnesium alloy composite, *Metall. Trans.*, 20A, (1989), 2369-2378.
4. de los Rios, E. R., Rodopoulos C. A. and Yates J. R.: A model to predict the fatigue life of fibre-reinforced titanium matrix composites under constant amplitude loading, *Fatigue Fract. Engng. Mater. Struct.*, 19, 5, (1996), 539-550.
5. Walls, D. P., Bao, G. and Zok, F.: Mode I fatigue cracking in a fiber reinforced metal matrix composite, *Acta Metall. Mater.*, 41, (1993), 2061-2071.
6. de los Rios, E. R., Rodopoulos C. A. and Yates J. R.: A theoretical study of fatigue behaviour in unidirectional Ti-based metal matrix composites, *Inter. J. Fract.*, 101, (2000), 385-395.
7. Warrier S. G. and Majumdar, B. S.: Elastic shielding during fatigue-crack growth of titanium matrix composites, Metall. Mater. Trans., 30A, (1997), 277-286.
8. Gdoutos, E. E., Pilakoutas, K. and Rodopoulos, C. A.: Failure Analysis of Industrial Composite Materials, McGraw-Hill, (2000), NY.

9. de los Rios, E. R., Rodopoulos C. A. and Yates J. R.: Damage tolerant fatigue design in metal matrix composites, Int. Journal of Fatigue, 19, 5, (1997), 379-387.
10. Calcaterra, J. R. and Mall, S.: Strength degradation during fatigue of unidirectional and cross-ply SCS-6/Ti-15-3 composites, Inter. J. Fatigue, 21, (1999), 215-223.
11. Calcaterra, J. R., Mall, S. and Coghlan, S. C.: Degradation of residual strength in SCS-6/Ti-15-3 due to fully reversed fatigue, Metall. Mater. Trans., 30A, (1999), 307-313.
12. Kelly A. and Tyson, W. R.: Tensile properties of fibre reinforced metals II. Creep of fibre tungsten, J. Mech. Phys. Solids, 14, (1966), 177-186.
13. Greaves, I: The growth of naturally initiating fatigue cracks in titanium-silicon carbide MMCs, Ph.D Thesis, (1994), University of Sheffield.
14. Clyne T. W. and Withers P. J.: An Introduction to Metal Matrix Composites, Cambridge University Press, (1993), London.
15. de los Rios, E. R., Rodopoulos C. A. and Yates J. R.: Fatigue damage design principles of fibre-reinforced titanium matrix composites, ICAF'97, Edinbrough, R. Cook and P. Poole eds, (1997), 1051-1060.
16. Rodopoulos, C. A., Yates, J. R. and de los Rios, E. R.: Micro-mechanical modelling of fatigue damage in titanium metal matrix composites, Theor. Appl. Fract. Mech., 35, (2001), 59-67.
17. Yang, C. J., Jeng, S. M. and Yang, J.-M. Interfacial properties measurement for SiC fibre-reinforced titanium alloy composites, Scr. Metall. Mater., 24, (1990), 469-474.
18. Hull, D.: An Introduction to composite materials, 6th Edition, Cambridge University Press, (1992), NY.
19. de los Rios, E. R., Rodopoulos C. A. and Yates J. R.: A theoretical study of fatigue behaviour in unidirectional Ti-based metal matrix composites, Inter. J. Fract., 101, (2000) 385-395.
20. Rodopoulos, C. A.: Fatigue damage map for metal matrix composites - A useful tool for design against fatigue, 1st Hellenic Conference of Composite Materials, S. A. Paipetis, E. E. Gdoutos, eds, (1997), 545-569.
21. Robertson, D. D. and Mall, S.: Strength degradation in titanium matrix composites exposed at elevated temperature, 11th International Conference on Composite Materials, (1997), 365-373.

5. Structural Analysis

DAMAGE TOLERANT COMPOSITE JOINT DESIGN

C. C. CHAMIS* and L. MINNETYAN[+]
*Senior Aerospace Scientist
NASA Glenn Research Center, Cleveland, OH 44135, USA
[+]Professor
Clarkson University, Potsdam, New York 13699-5710, USA

Abstract Damage initiation and progressive fracture of adhesively bonded graphite/epoxy composite joints are investigated under various loading conditions. Load induced damage in both the adhesive bond and the adjoining laminate is considered. An integrated computer code is utilized for the simulation of composite structural degradation. The damage progression sequence and structural fracture resistance during different degradation stages are evaluated in detail. Results indicate that adhesive properties have a significant effect on structural durability even when damage initiation/progression is not in the adhesive bond. Design implications with regard to damage tolerance of adhesively bonded joints are examined. Influence of the type of loading and the choice of adhesive on damage initiation and progression for adhesively bonded composite joints are investigated.

1. INTRODUCTION

Graphite/epoxy composite structures have extensive uses in fulfilling many design requirements in aircraft structures, automotive, and marine applications. A cost effective method to fabricate composite structures is by adhesive bonding of structural elements. However, the durability and damage tolerance of adhesively bonded joints is difficult to evaluate due to the high level of interaction between the structural and adhesive response. The analysis and design of adhesively bonded joints has received significant attention in the past. The majority of previous investigations are based on stress analysis for the prediction of adhesive joint failures. The premise of this paper is that when adhesively bonded composite structural joints fail, there is often significant degradation of the composite laminates, as well as in the adhesive layer, prior to joint failure. Laminate configuration and properties affect adhesive stresses and joint durability. Responsively, adhesive properties and thickness influence damage progression in the laminate. Any method that attempts to fully evaluate the structural integrity of an adhesively bonded composite joint must take into account the combined adhesive/laminate structural degradation processes. Under tensile loading, internal damage in composites is often initiated as transverse cracking in one or more plies. At the presence of stress concentrations or defects, or under in plane compressive loading, initial damage may also include fiber fracture. Further degradation is in the

E.E. Gdoutos and Z.P. Marioli-Riga (eds), Recent Advances in Composite Materials,265-278

form of additional fiber fractures that usually lead to structural fracture by the coalescing of damage zones. Because of the numerous possibilities with material combinations, composite geometry, ply orientations, and loading conditions, it is essential to have an effective computational capability to predict the behavior of composite structures for any loading, geometry, material combinations, and boundary conditions. The predictions of damage initiation, growth, accumulation, and propagation to fracture are important in evaluating the load carrying capacity and reliability of composite structures. Quantification of the structural fracture resistance is also required to evaluate the durability/life of composite structures. The objective of this paper is to evaluate the effects of adhesive properties and thickness on the durability of composite structures via an integrated computational method.

The CODSTRAN (COmposite Durability STRuctural ANalysis) computer code [1] has been developed for this purpose. CODSTRAN is able to simulate damage initiation, damage growth, and fracture in composites under various loading, considering also the effects of residual stresses and environmental conditions. The simulation of progressive fracture by CODSTRAN has been verified to be in reasonable agreement with experimental data from tensile tests [2]. Further additions to CODSTRAN have enabled investigation of the effects of composite degradation on structural response [3], composite damage induced by dynamic loading [4], composite structures global fracture toughness [5], effect of the hygrothermal environment on durability [6], damage progression in composite shells subjected to internal pressure [7], the durability of stiffened composite shells under load combinations [8], design considerations for progressive fracture in composite shell structures and the simulation of damage propagation in thick composite shells under external pressure [9]. The purpose of this paper is to examine damage progression in composite joints for design evaluation. Due to the tendency of adhesively bonded structural elements to deform independently and to pull apart, composite joints present potential nucleation points for damage initiation and fracture. Damage propagation studies are conducted to evaluate adhesively bonded joint performance. In general, the choice of adhesive properties for structural durability depends on the geometry, composite structure, and laminate configurations of the joined elements.

For the purpose of the present discussion, the following terminology is used to describe the various stages of degradation in the composite structure: (1) *damage initiation* refers to the start of damage induced by applied loading; (2) *damage growth* is the progression of damage from the location of damage initiation to adjacent regions; (3) *damage propagation* is the rapid progression of damage to other regions of the structure. At any stage of damage progression, if there is a high level of structural resistance to damage progression, the structure is stable with regard to fracture. The corresponding state of structural damage is referred to as *stable damage.* On the other hand, if damage progression does not encounter significant structural resistance, it corresponds to an *unstable damage* state. Unstable damage progression is characterized by very large increases in the amount of damage due to very small increases in loading.

2. BRIEF DESCRIPTION OF CODSTRAN

CODSTRAN is an integrated, open-ended, stand alone computer code consisting of three modules: composite mechanics, finite element analysis, and damage progression modeling. The overall evaluation of composite structural durability is carried out in the damage progression module [1] that keeps track of composite degradation for the entire structure. The damage progression module relies on ICAN [10] for composite micromechanics, macromechanics and laminate analysis, and calls a finite element analysis module that uses anisotropic thick shell elements to model laminated composites [11].

Figure 1 shows a schematic of the computational simulation cycle in CODSTRAN. The ICAN composite mechanics module is called before and after each finite element analysis. Prior to each finite element analysis, the ICAN module computes the composite properties from the fiber and matrix constituent characteristics and the composite layup. The finite element analysis module accepts the composite properties that are computed by the ICAN module at each node and performs the analysis at each load increment. After an incremental finite element analysis, the computed generalized nodal force resultants and deformations are supplied to the ICAN module that evaluates the nature and amount of local damage, if any, in the plies of the composite laminate. Individual ply failure modes are determined by ICAN using failure criteria associated with the negative and positive limits of the six ply-stress components (σ_{L11}, σ_{L22}, σ_{L33}, τ_{L12}, τ_{L13}, τ_{L23}) a modified distortion energy (MDE) or combined stress failure criterion, and interply delamination due to relative rotation of the plies.

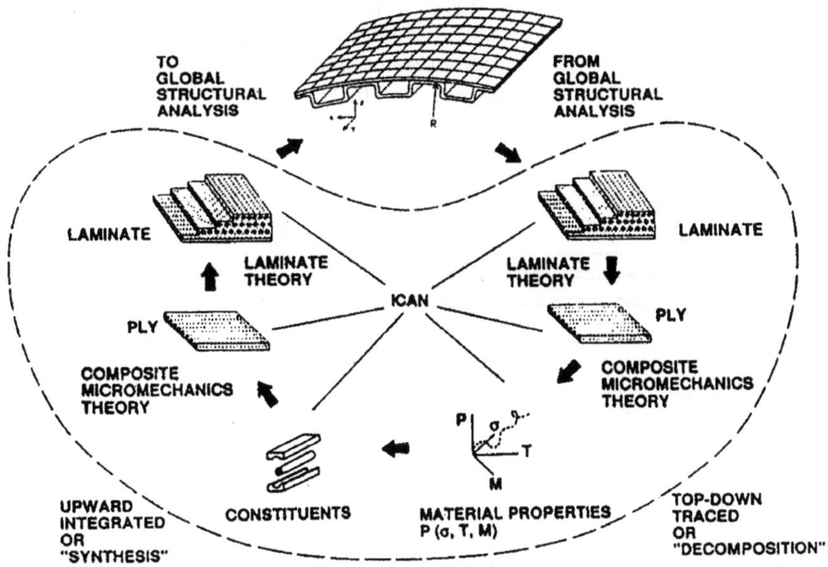

Figure 1. CODSTRAN Simulation Cycle

The generalized stress-strain relationships for each node are revised according to the composite damage evaluated by the ICAN module after each finite element analysis. The model is automatically updated with a new finite element mesh and properties, and the structure is reanalyzed for further deformation and damage. If ply failure criteria indicate damage during a load increment, CODSTRAN degrades the composite properties affected by the damage and reanalizes the structure under the same load. When there is no indication of further damage under a load, the structure is considered to be in equilibrium. Subsequently, another load increment is applied leading to possible damage growth, accumulation, or propagation.

3. DAMAGE PROGRESSION IN A DOUBLE LAP JOINT

A double lap joint of a graphite/epoxy laminate is considered as shown in Figure 2. The laminate consists of 48 plies that are configured as [0/±45/90]s6 with a total thickness of 6.35 mm (0.25 in). The 0° plies are oriented in the load direction and the 90° plies are oriented transverse to the load direction. The specimen has a width of 102 mm (4.0 in) and a length of 406 mm (16.0 in). The finite element model contains 260 nodes and 184 elements.

The composite system is made of AS-4 graphite fibers in a high-modulus, high strength (HMHS) epoxy matrix. The fiber volume ratio is 0.60 and the void volume ratio is 2 percent. The adhesively bonded double lap joint has been cured at a temperature of 177°C (350°F). The use temperature is 21°C (70°F). Two types of adhesives are considered. Type 1 adhesive has properties that are exactly the same as the HMHS matrix properties. Type 2 adhesive is a toughened epoxy resin adhesive (TERA) that is 42 percent stronger in tension compared to the HMHS adhesive but 43 percent weaker in compression and has an elastic modulus that is 44 percent lower than the elastic modulus of the HMHS adhesive.

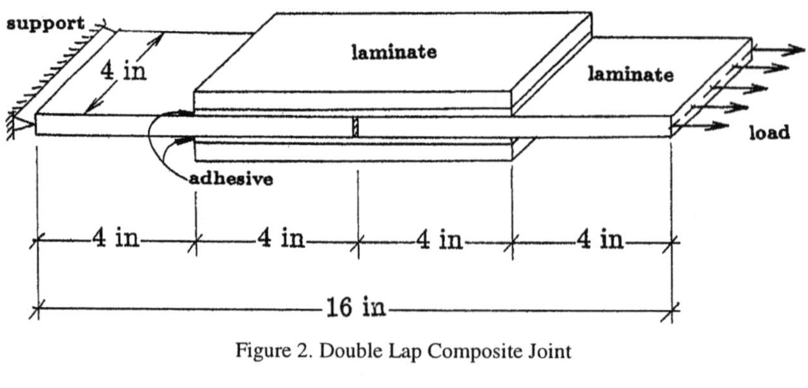

Figure 2. Double Lap Composite Joint

The double lap joint is first subjected to a gradually increasing tensile load. The specimen is loaded by restraining one end and applying a uniformly distributed tensile load to the other end. Damage progression is computationally simulated as the loading is increased.

In general, overall structural damage may include individual ply damage and also through-the-thickness fracture of the composite laminate. CODSTRAN is able to simulate varied and complex composite damage mechanisms via evaluation of the individual ply failure modes and associated degradation of laminate properties. The type of damage growth and the sequence of damage progression depend on the composite structure, loading, material properties, and hygrothermal conditions. A scalar damage variable, derived from the total volume of the composite material affected by the various damage mechanisms is also evaluated as an indicator of the level of overall damage induced by loading. This scalar damage variable is useful for assessing the overall degradation of a given structure under a prescribed loading condition. The rate of increase in the overall damage during composite degradation may be used as a measure of structural propensity for fracture. Computation of the overall damage variable has no interactive feedback on the detailed simulation of composite degradation. The procedure by which the overall damage variable is computed is given in reference [5]. In this paper, the composite structure is defined to be 100 percent damaged when all plies of all nodes develop some damage. Usually, structural fracture will occur near a stress concentration before the 100 percent damage level is reached.

Figure 3 shows the simulated damage progression and displacements with increasing tensile loading on the double lap joint. The progressive damage response of the double lap joint is evaluated for both HMHS and TERA adhesive properties. Damage initiation for the HMHS adhesive bonded joint is at a 16.0 KN (3.60 kip) load by adhesive failure at the beginning of the lap joint near the left end support (Figure 2). In case of the TERA bonded joint, damage initiation is at 18.4 KN (4.13 kips) by simultaneous adhesive failure at the beginning of the lap joint and by matrix failures in two 90° plies closest to the adhesive bond near the center of the lap joint.

Although the TERA bonded joint performs better than the HMHS bonded joint during damage initiation, in the subsequent damage growth and progression stages the opposite case holds true. In Figure 3, load versus damage performance of the HMHS bonded joint, as plotted by the solid line, is considerably better than the performance of the TERA bonded joint shown by the dashed line. Figure 3 shows also larger displacements the TERA bonded joint as compared to the displacements of the HMHS bonded joint under the same loading.

Damage progression in the double lap joint is also evaluated under compressive loading. Figure 4 shows damage levels and corresponding displacements under compressive loading. The overall damage progression response under compressive loading is similar to that of tensile loading. Under both tensile and compressive loads the TERA bonded joint performs better at damage initiation; whereas the HMHS bonded joint performs better during damage growth and progression. The overall volume of damage under compressive loading is lower than that under tensile loading since compressive loading does not cause extensive matrix cracking in the 90° plies of the laminate as is the case under tensile loading.

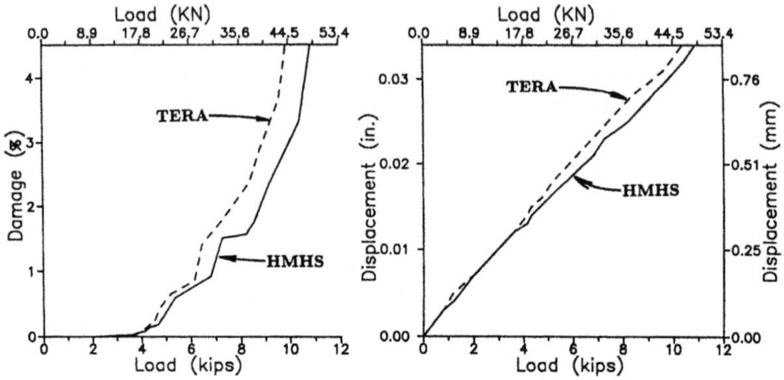

Figure 3. Damage and Displacement of Double Lap Composite Joint under Tension

The displacements of TERA bonded and HMHS bonded joints as shown in Figure 4 are the same prior to the damage initiation stage. However, after damage initiation, the TERA bonded joint behaves softer. Since the difference in displacements appears after damage initiation, it is a characteristic of the adhesive/laminate, combined damage progression response rather than an effect of the more compliant elastic deformation of the TERA bond.

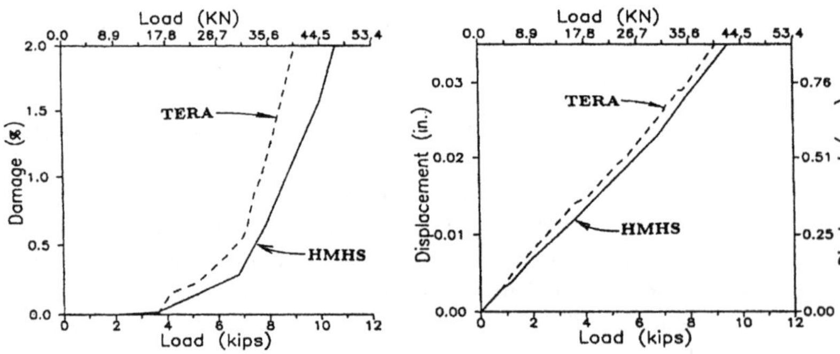

Figure 4. Damage and Displacement of Double Lap Joint under Compressive loading

The global Damage Energy Release Rate (DERR) is defined as the gradient of work done by external forces during structural degradation with respect to the produced damage. DERR can be used to evaluate structural resistance against damage propagation at different stages of loading. Figure 5 shows the DERR as a function of tensile and compressive loading for the HMHS and TERA bonded joint. The DERR for damage initiation is much higher under compressive loading, indicating greater resistance to damage initiation under compressive loading. However, after the damage initiation stage is completed DERR becomes much lower, indicating the lack of structural resistance to damage propagation. On the other hand, when the loading is tensile, as indicated by the solid lines in Figure 5, the initial DERR level is relatively low and damage growth immediately after damage initiation does not encounter significant structural resistance. However, there is a secondary peak in DERR during damage progression. The initially low DERR level is because of matrix cracking in the 90° plies near the adhesive bond due to the relatively low tensile strength of the matrix. The secondary peak in DERR preceeds the involvement of the ±45° plies and fiber fractures in the damage propagation process.

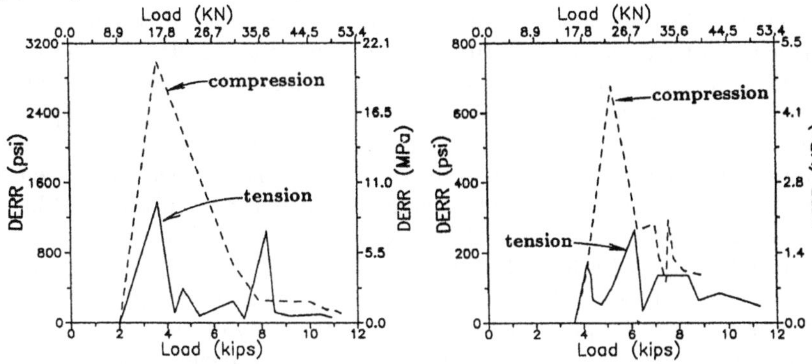

a) HMHS Adhesive b) TERA Adhesive

Figure 5. DERR Levels for Double Lap Composite Joints

Similar to the HMHS bonded joint, for the TERA bonded joint initially the DERR level is also higher under compressive loading. However, in the case of the TERA bonded joint, the secondary tensile DERR peak does not develop during damage progression. Also, comparing Figure 5(a) with Figure 5(b), it is seen that the overall DERR levels are considerably lower for the TERA bonded joint as compared to the HMHS bonded joint. Computational simulation results collectively indicate that the HMHS bond performs better than the TERA bond with regard to damage tolerance of the graphite/epoxy double lap joint.

4. STIFFENED COMPOSITE PANEL

The second investigated example is a composite panel stiffened by a hat type stringer that is adhesively bonded to the skin. The finite element model, shown in Figure 6a, uses stacked thick shell elements with tying equations between consecutive stacked layers. Figure 6b shows a cross section through the stiffened panel with physical dimensions. The composite system is made of the same high strength AS-4 graphite fibers in a high-modulus, high-strength epoxy matrix (AS-4/HMHS) similar to the double lap joint. The adhesive properties here are the same as the HMHS epoxy matrix. The influence of adhesive thickness on the global damage progression is investigated. Durability of the stiffened panel with adhesive thickness of 0.1323, 0.2647, and 0.5293 mm (0.00521, 0.01042, and 0.02084 in) is evaluated under tensile, compressive, and lateral pressure loads. The three adhesive bond thicknesses are designated as Type I, Type II, and Type III, respectively, in the order of increasing adhesive thickness.

The skin laminate consists of forty-eight 0.132 mm. (0.00521 in.) plies resulting in a composite thickness of 6.35 mm. (0.25 in.). The fiber volume ratio is 0.60. The laminate configuration is [0/±45/90]s6. The 0° plies are in the axial direction of the stiffener, along the x axis indicated in Figure 6. The width of the stiffened panel is 991 mm (39 in.) and it has a length of 559 mm (22 in.). Only half of the panel is represented in the computational model by defining a plane of structural symmetry at half-length of the panel. The stiffener hat sections are made from the same AS-4/HMHS composite structure as the skin laminate. The stiffener is adhesively bonded to the skin at all surfaces of contact. Composite constituent level progressive damage simulation in the adhesive layer as well as in the joined elements is integrated into the structural analysis of the stiffened panel. Axial tension or compression is applied by imposing a gradually increasing uniform axial loading at the clamped left edge of the panel. Damage initiation and progression are monitored as the panel is gradually loaded. Figure 7a shows the relationship between applied tensile loading and the produced damage during damage initiation and progression. Damage initiation is by ply failure in the top surface ply of the skin near the end of the stiffener. Figure 7a indicates that the stiffened panel experiences damage initiation sooner for a thicker adhesive bond. However, the effect of adhesive thickness on the damage initiation load is very small under tension. Figure 7b shows the DERR levels for the three adhesive thickness values investigated. Figure 7b indicates that although a thicker adhesive bond has a lower damage initiation load, structural resistance of the stiffened composite panel to damage progression under tension is considerably higher for the thicker adhesive bond.

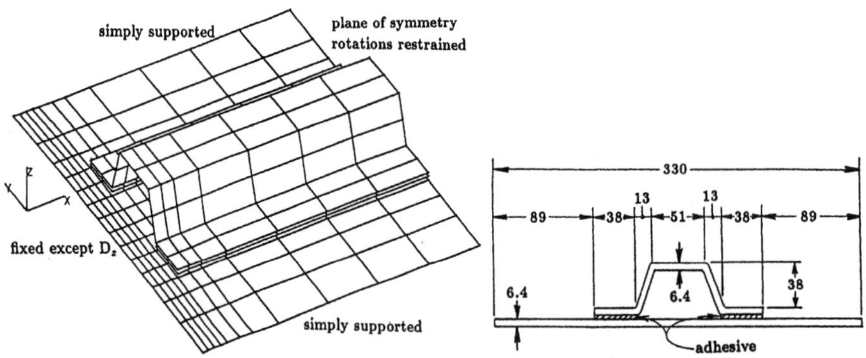

a) Finite Element Model

b) Cross Section (dimensions in mm)

Figure 6. Adhesively Bonded Stiffened Panel

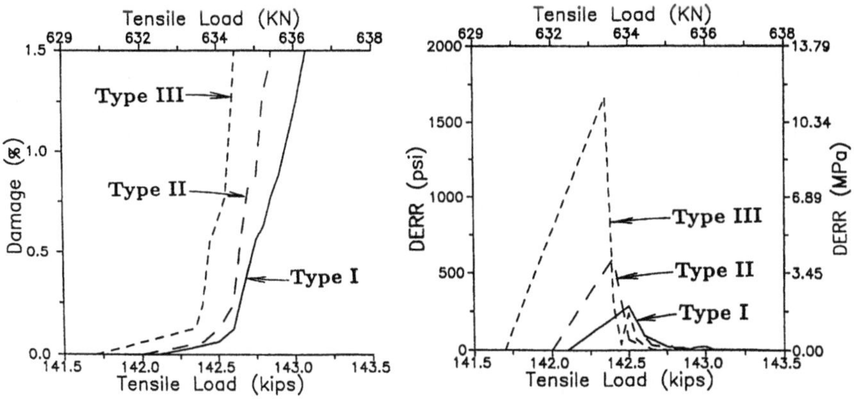

a) Damage Volume b) Damage Energy Release Rates

Figure 7. Damage Evolution Characteristics of Adhesively Bonded Stiffened Panel

Figure 8a shows damage initiation and progression under compressive loading. Similar to tensile loading, a thicker adhesive bond experiences damage initiation sooner. Figure 8b shows the DERR levels under compression. Unlike tensile loading, DERR levels do not monotonically increase with adhesive thickness. Type II bond with an adhesive thickness of two plies has the best DERR performance. However, if the thickness is increased to Type III, with an adhesive thickness of four plies, the DERR level is significantly degraded. Figure 8b indicates the existence of an optimal adhesive thickness under compressive loading.

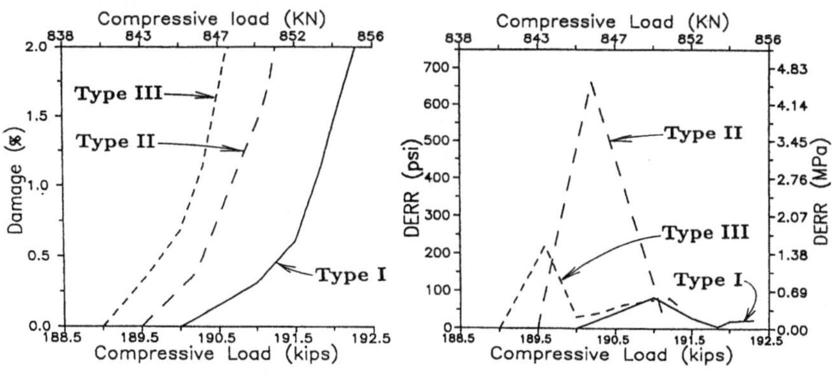

a) Percent Damage Volume b) Damage Energy Release Rates

Figure 8. Damage Evolution of Adhesively Bonded Panel Under Compressive Loading

The same stiffened panel is also investigated under lateral pressure applied to the skin. Figure 9a shows the damage progression under lateral pressure for Type I and

Type II adhesive bonds with one and two-ply adhesive thickness, respectively. The effect of adhesive thickness is not significant with regard to damage initiation and progression under normal pressure. Figure 9b shows the DERR levels under lateral pressure. The DERR levels for the thicker adhesive bond are higher, indicating better structural resistance against damage propagation for the thicker adhesive.

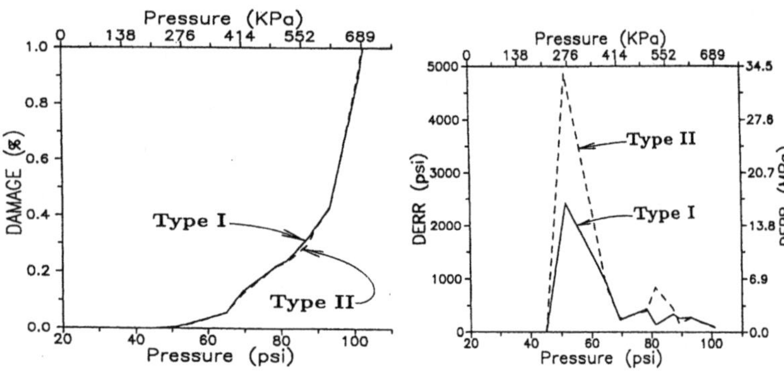

a) Percent Damage Volume b) Damage Energy Release Rates

Figure 9. Damage Evolution in Stiffened Panel under Lateral Pressure

5. COMPOSITE PIPE JOINT

A graphite/epoxy laminated thin composite pipe with a circumferential lap joint is considered next. The composite system is made of AS-4 graphite fibers in a high-modulus, high strength (HMHS) epoxy matrix. The HMHS matrix properties are representative of the 3501-6 resin. The fiber volume ratio is 0.60 and the void volume ratio is 1.0 percent. The cure temperature is 177°C (350°F) and the use temperature is 21°C (70°F). The adhesive bond at the joint is assumed to have the same properties as the HMHS epoxy matrix.

The specimen is built by joining two cylinders where one of the cylinders fits exactly inside the other cylinder concentrically at the joint. Each segment of the shell has a length of 305 mm (12 in.). The inside diameter of the inner pipe is 254 mm (10 in.). The length of axial overlap at the joint is 51 mm (2 in.). Figure 10a shows a schematic of the specimen considered. The specimen has an overall length of 559 mm (22 in.) as shown in Figure 10a. To evaluate possible local effects on structural degradation, edges of the joint are tapered axially as shown in Figure 10b. Two values of the slope *s* at the taper are considered in this investigation. At first a relatively steep taper slope of s=2 is assumed. Five closely spaced nodes are taken in the axial direction at the taper to enable accurate assessment of stress concentration effects. The number of circumferential nodes is 32. The overall finite element model of the specimen contains 704 nodes and 672 quadrilateral elements. Figure 10c shows projection of the finite element model into a plane containing the axis of the cylinder.

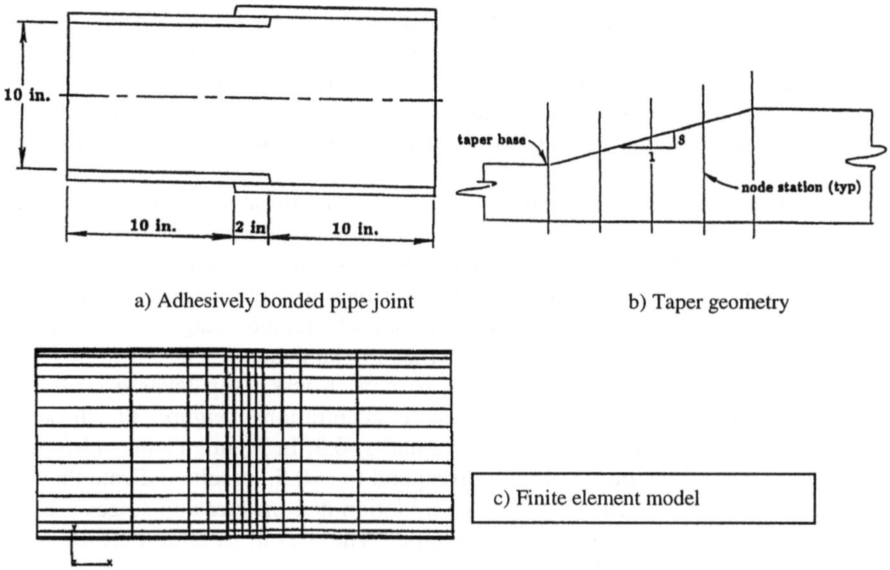

Figure 10. Concentric Pipe Joint

The jointed pipe specimen is first investigated under internal pressure. The composite pipe is simulated as a closed-end cylindrical pressure vessel by applying a uniformly distributed axial tension such that the generalized axial stresses in the pipe wall are half those developed in the hoop direction. The pipe is subjected to an internal pressure that is gradually increased through different stages of degradation. Damage progression is computationally simulated as the loading is increased.

For the first specimen the laminate consists of 12 plies that are configured as $[90/\pm20]_4$ with a total thickness of 1.68 mm (0.066 in). The ±20° angle plies are helically wound around the cylinder and their orientation is given with respect to the axial direction. The 90° plies are oriented in the circumferential direction.

Damage initiation occurs by matrix cracking of the innermost ±20° plies due to transverse tensile failures. The location of initial damage is within the smaller diameter pipe near the overlapping joint. The internal pressure at damage initiation is 1.32 MPa (192 psi). After damage initiation, damage growth is due to spreading of the transverse tensile failures to other ±20° plies on both sides of the shell joint. At a pressure of 2.03 MPa (295 psi) damage begins progression to the 90° or circumferential plies. Through-the thickness fracture occurs at a pressure of 3.94 MPa (571 psi).

To investigate the effects of combined pressurization and bending, the same specimen is subjected to a flexural load that increases in proportion to internal pressure. The ratio of bending load to pressure is taken to be 0.328 N-mm/Pa (20 in-lbs/psi). The results indicate the presence of flexural load to have a detrimental effect on durability and damage tolerance.

Solid lines in Figure 11 show the damage progressions with increasing internal pressure, with and without bending. When bending is present, the damage initiation stage moves to a very early loading stage and the fracture pressure is approximately one third of that for pressurization without bending. More detailed examination would indicate that bending induces local buckling of the thin shell adjacent to the lap joint.

To investigate the effect of laminate configuration on durability, a specimen with the same geometry but made of a cross-ply laminate is subjected to internal pressure. The ply layup is $[0/90]_4$. The overall damage progression is shown in Figure 11b in comparison with the $\pm 20^{\circ}$ angle ply laminate. For the cross ply laminate the damage initiation stage is delayed, yet the final structural fracture stage is quickened, indicating a lower damage tolerance for the cross-ply laminate.

Next, the same specimen is considered to have a less steep taper of s=0.5 to assess the possible effects of local stress concentrations. Figure 11 shows a comparison of results for taper slopes of s=2 and s=0.5. In Figure 11 the solid lines correspond to s=2 and the dashed lines correspond to s=0.5. Results indicate that the general characteristics of structural damage initiation and progression stages are not changed much by softening the slope at the taper. The overall structural effects have a much more significant influence on damage progression as compared to local stress concentrations.

A closer investigation of the stress state in the tapered end of the joint is of interest. Figure 12a shows the hoop stress variations of the inner and outer hoop plies at an internal pressure of 1.14 MPa (165 psi) prior to damage initiation. Distance along the taper is measured from the base reference toward the thicker part of the joint. The taper length is 3.353 mm (0.132 in.). Figure 12a indicates that hoop stresses are lower in the taper but high in the single thickness shell adjacent to the joint. The inner ply stresses are considerably higher compared to the outer ply stresses. Figure 12b shows the axial stresses that are the transverse stresses for the hoop plies. These stresses are raised within the taper. Another raised stress at the taper is the shear stress in the $\pm 20^{\circ}$ angle plies. However, the shear stress levels are too low to affect durability.

6. CONCLUSIONS

In the light of the durability investigation of the example jointed composite structures and from the general perspective of the available CODSTRAN (COmposite Durability STRuctural ANalysis) computer code, the following conclusions are drawn:

1. Computational simulation, with the use of established composite mechanics and finite element modules, can be used to predict the influence of local joint geometry as well as loading, on the safety and durability of composite joints.

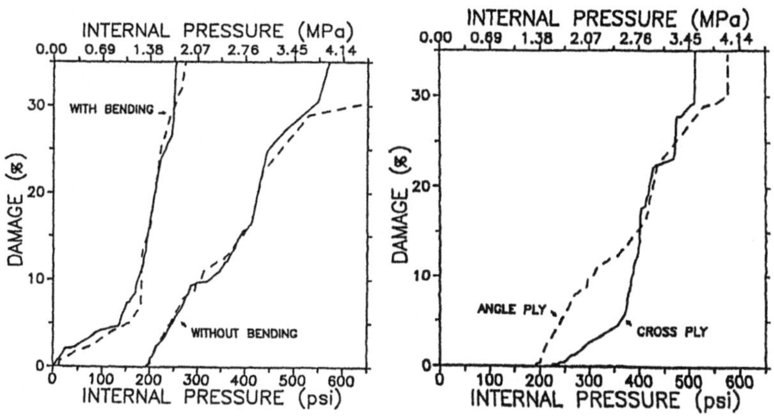

a) [90/±20]$_4$ concentric pipe joints b) comparison of [90/±20]$_4$ with [0/90]$_4$ pipe joint

Figure 11. Damage Progression in Concentric Pipe Joint

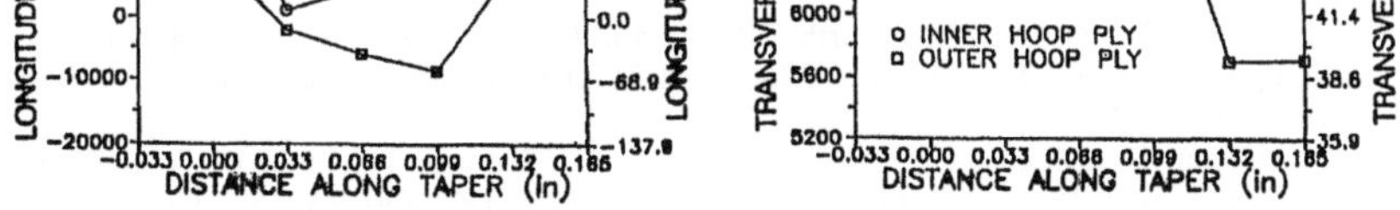

a) σ_{L11} hoop stresses b) σ_{L22} axial stresses

Figure 12. Graphite/Epoxy:[90/±20]$_4$ 90° Ply Stresses Due to Pressure Only

2. Adhesive properties and thickness affect the damage progression response for bonded joints. With regard to the structural durability of bonded graphite/epoxy laminates, a high modulus adhesive bond is preferable to a toughened but lower modulus adhesive bond.

3. Damage initiation, growth, and accumulation stages involve matrix cracking as well as fiber fractures in the fiber composite contiguous to the joint region.

4. Flexural loading effects are detrimental to damage tolerance characteristics of composite joints.

5. Changes in the local geometry such as the taper slope for concentric pipe joints have only a minor effect on the structural durability characteristics.

6 Joints of cross-ply laminates may have better performance with regard to damage initiation but are not damage tolerant when compared to angle ply laminates.
7. The demonstrated procedure is flexible and applicable to all types of constituent materials, structural geometry, and loading. Hybrid composites and homogeneous materials, as well as binary composites can be simulated.
8. Fracture toughness parameters such as the structural fracture load and damage progression characteristics are identifiable for any structure by the demonstrated method.
9. Computational simulation by CODSTRAN represents a new global approach to progressive damage and fracture assessment for any structure.

7. REFERENCES

1. Chamis, C.C. and Smith, G.T., "Composite Durability Structural Analysis," NASA TM-79070, 1978.
2. Irvine, T. B and Ginty, C. A., "Progressive Fracture of Fiber Composites," Journal of Composite Materials, Vol. 20, Mar. 1986, pp. 166-184.
3. Minnetyan, L., Chamis, C.C. and Murthy, P.L.N., "Structural Behavior of Composites with Progressive Fracture," Journal of Reinforced Plastics and Com- posites, Vol. 11, No.4, April 1992, pp. 413-442.
4. Minnetyan, L., Murthy, P.L.N. and Chamis, C.C., "Progression of Damage and Fracture in Composites under Dynamic Loading," NASA TM-I03II8, April 1990, 16 pp.
5. Minnetyan, L., Murthy, P.L.N. and Chamis, C.C., "Composite Structure Global Fracture Toughness via Computational Simulation," Computers & Structures, Vol. 37, No. 2, pp.175-180, 1990
6. Minnetyan, L., Murthy, P.L.N. and Chamis, C.C., "Progressive Fracture in Composites Subjected to Hygrothermal Environment," International Journal of Damage Mechanics, Vol. 1, No.1, January 1992, pp. 60-79
7. Minnetyan, L., Chamis, C.C. and Murthy, P.L.N., "Structural Durability of a Composite Pressure Vessel," Journal of Reinforced Plastics and Composites, Vol. 11, No. 11, November 1992, pp. 1251-1269
8. Minnetyan, L., Rivers, J.M., Murthy, P.L.N. and Chamis, C.C., "Structural Durability of Stiffened Composite Shells," Proceedings of the 33rd SDM Conference, Dallas, Texas, April 13-15, 1992, Vol. 5, pp. 2879-2886
9. Minnetyan, L., Chamis, C.C. and Murthy, P.L.N., "Damage Tolerance of Thick Composite Shells under External Pressure," Proceedings of the 10th DOD/NASA/FAA Conference on Fibrous Composites in Structural Design, Hilton Head, South Carolina, November 1-4, 1993, Session VII, pp. 85-95.
10. Murthy, P.L.N. and Chamis, C.C., Integrated Composite Analyzer (ICAN): Users and Programmers Manual, NASA Technical Paper 2515, March 1986.
11. Nakazawa, S., Dias, J.B. and Spiegel, M.S., MHOST Users' Manual, Prepared for NASA Lewis Research Center by MARC Analysis Research Corp., April 1987.

CORE FAILURE OF SANDWICH BEAMS

I.M. DANIEL[a], E.E. GDOUTOS[b], J.L. ABOT[a] and K.-A. WANG[a]
[a]*Robert R. McCormick School of Engineering and Applied Science, Northwestern University, USA Evanston, IL 60208, USA*
[b]*School of Engineering, Democritus University of Thrace, GR-671 00 Xanthi, Greece*

Abstract. Core failure modes were studied in composite sandwich beams under three-point bending and in cantilever beams under end loading. The beams consisted of unidirectional carbon/epoxy face sheets and a variety of core materials, including aluminum honeycomb, PVC closed-cell foams, polyurethane foam and balsa wood. The constituent materials were fully characterized and in the case of the core materials, failure envelopes were developed for biaxial states of stress. Deformation and failure mechanisms were studied experimentally by means of moiré gratings and birefringent coatings. Results were obtained for stress (strain) distributions in the linear and nonlinear/plastic range of the core, critical failure loads and their dependence on geometrical dimensions, material parameters and loading conditions.

1. INTRODUCTION

Sandwich construction is of particular interest and widely used, because the concept is very suitable and amenable to the development of lightweight structures with high in-plane and flexural stiffness. Sandwich panels consist typically of two thin face sheets (or facings, or skins) and a lightweight thicker core. Commonly used materials for facings are composite laminates and metals, while cores are made of metallic and non-metallic honeycombs, cellular foams, balsa wood or trusses. The facings carry almost all of the bending and in-plane loads and the core helps to stabilize the facings and defines the flexural stiffness and out-of-plane shear and compressive behavior.

The overall performance of sandwich structures depends not only on the properties of the facings, but also on those of the core and the adhesive bonding the core to the skins, as well as on geometrical dimensions. Sandwich beams under general bending, shear and in-plane loading display various failure modes. Their initiation, propagation and interaction depend on the constituent material properties, geometry and type of loading. Failure modes and their initiation can be predicted by conducting a thorough stress analysis and applying appropriate failure criteria in the critical regions of the beam. Possible failure modes include tensile or compressive failure of the facings, debonding at the core/facing interface, indentation failure under local-

E.E. Gdoutos and Z.P. Marioli-Riga (eds), Recent Advances in Composite Materials, 279-290

ized loads, core failure, wrinkling of the compression face and global buckling. Following initiation of a particular failure mode, this mode may trigger and interact with other modes and final failure may follow another failure path. A general review of failure modes in composite sandwich beams was given by Daniel et al. [1]. Wrinkling failures in sandwich columns and beams were studied by Gdoutos et al. [2].

In the present work core failure modes of composite sandwich beams under three-point bending and cantilever beams under end load were studied. A common failure of sandwich construction is the so-called "core shear failure," in which the core fails when the shear stress reaches its critical value. However, although the shear stress is usually the dominant stress, there are situations in which the normal stress in the core is of comparable magnitude or even higher than the shear stress. Such situations occur when the bending component at the point is high (long span beams) and in the vicinity of applied loads and reactions. Under such conditions a material element in the core is subjected to a multiaxial state of stress. Thus, it is important to characterize the core material under multiaxial states of stress.

A number of core materials, including aluminum honeycomb, various types of closed-cell PVC foams, a polyurethane foam, foam-filled honeycomb and balsa wood, were characterized under uniaxial and biaxial states of stress. Gdoutos et al. [3] characterized Divinycell H250 under multiaxial states of stress and showed that the failure surface of the material can be described by the Tsai-Wu failure criterion [4].

In the present work, core failure modes were investigated experimentally in composite sandwich beams under three-point bending and in end-loaded cantilever beams. Core failures were analyzed based on the state of stress and core material properties.

2. ANALYTICAL BACKGROUND

2.1 Stress Field in the Core of Sandwich Beams

Consider a sandwich beam of rectangular cross section with facings and core materials displaying linear elastic behavior, subjected to a bending moment, M, and shear force, V. The in-plane maximum normal stress, σ, and shear stress, τ, in the core, for a low stiffness core and thin facings are given by

$$\sigma = \frac{PL}{C_1 b d^2}\left(\frac{E_c}{E_f}\right)\frac{h_c}{h_f} \tag{1a}$$

$$\tau = \frac{P}{C_2 b h_c} \tag{1b}$$

where

$$M = \frac{PL}{C_1}, \quad V = \frac{P}{C_2} \tag{2}$$

P = applied concentrated load
L = length of beam
E_f, E_c = Young's modulus of facing and core material, respectively
h_f, h_c = thickness of facings and core, respectively
d = distance between centroids of the facings
b = beam width
C_1, C_2 = Constants depending on loading configuration
($C_1 = 4$, $C_2 = 2$ for three-point bending; $C_1 = C_2 = 1$ for cantilever beam)

The maximum normal stress, σ, for a beam under three-point bending occurs under the load, while for a cantilever beam under end loading it occurs at the support. The shear stress, τ, is constant along the beam span and through the core thickness, as verified experimentally [1,5].

2.2. Failure Mode Transition

When the normal stress in the core is small relative to the shear stress, it can be assumed that core failure occurs when the shear stress reaches a critical value. Furthermore, failure in the facings occurs when the normal stress reaches its critical value, usually the facing compressive strength. Under such circumstances we obtain from Eqs. (1a) and (1b) that failure mode transition from shear core failure to compressive facing failure occurs when

$$\frac{L}{h_f} = C\frac{F_f}{F_{cs}} \tag{3}$$

where

F_f = facing strength in compression or tension
F_{cs} = core shear strength
C = constant (C = 2 for a beam under three-point bending; C=1 for a cantilever beam under an end load)

When the left hand term of Eq. (3) is smaller than the right hand term, failure occurs by core shear, whereas in the reverse case failure occurs by facing tension or compression. Failure mode transition according to Eq. (3) is shown in Fig. 1 for a beam under three-point bending and a cantilever beam under end load. In the above discussion it is assumed that the beam is adequately reinforced locally at the points of load application and supports, to suppress premature failure due to indentation.

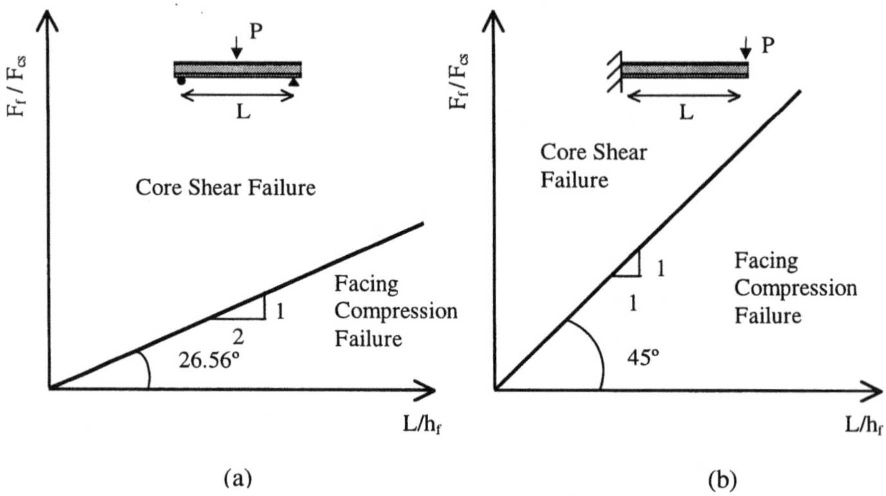

Figure 1. Failure envelopes for (a) a beam under three-point bending and (b) a cantilever beam under end loading

3. EXPERIMENTAL PROCEDURE AND RESULTS

3.1 Materials and Specimens

The sandwich beams were fabricated from 8-ply unidirectional carbon epoxy (AS4/3501-6) facings and various core materials. They include four types of a closed-cell PVC foam (Divinycell H80, H100, H160 and H250, with densities of 80, 100, 160 and 250 kg/m^3, respectively), an aluminum honeycomb (PAMG 8.1-3/16 001-P-5052, Plascore Co.), a polyurethane foam, a foam-filled honeycomb and balsa wood. All core materials were characterized in uniaxial tension and compression along the in-plane and through-the-thickness directions and under in-plane shear. Some of their characteristic properties are tabulated in Table I. The core materials (honeycomb or foam) were provided in the form of 25.4 mm (1 in.) thick plates. The honeycomb core was bonded to the top and bottom facings with FM73 M film adhesive and the assembly was cured under pressure in an oven following the recommended curing cycle for the adhesive. The foam cores were bonded to the facings using a commercially available epoxy adhesive (Hysol EA 9430) (Daniel and Abot [5]). Beam specimens 25.4 mm (1 in.) wide and of various lengths were cut from the sandwich plates.

3.2 Failure Envelopes

Two core materials, Divinycell H100 and H250 were fully characterized under multiaxial stress condition [3]. A series of biaxial tests were conducted including constrained strip specimens in tension and compression with the strip axis along the through-the-thickness and in-plane directions; constrained thin-wall ring specimens in compression and torsion; thin-wall tube specimens in tension and torsion; and thin-wall tube specimens under axial tension, torsion and internal pressure. From these tests and uniaxial results in tension, compression and shear, failure envelopes were constructed. It was shown that the failure envelopes were described well by the Tsai-Wu criterion [4].

For a beam loaded under combined bending and shear, the foam is subjected to longitudinal normal stress, σ_1, and in-plane shear stress, τ_5 (τ_{13}). The Tsai-Wu criterion for this case takes the form

$$f_1 \sigma_1 + f_{11} \sigma_1^2 = 1 - k^2 \tag{4}$$

where

$$f_1 = \frac{1}{F_{1t}} - \frac{1}{F_{1c}}, \quad f_{11} = \frac{1}{F_{1t}\, F_{1c}}, \quad f_{55} = \frac{1}{F_5^2}, \quad k = \frac{\tau_5}{F_5}$$

F_{1t}, F_{1c} = tensile and compressive strengths in the in-plane (1,2) direction.
F_5 = shear strength on the 1-3 plane.

It was assumed that the failure behavior of all core materials is described by the Tsai-Wu criterion. Failure envelopes of all core materials constructed from the values of F_{1t}, F_{1c} and F_5 are shown in Fig. 2. Note that the failure envelopes of all Divinycell foams are elongated along the σ_1-axis, which indicates that these materials are stronger under normal longitudinal stress than in-plane shear stress. Aluminum honeycomb and balsa wood show the opposite behavior. For all materials, the most critical cominations of shear and normal stress fall in the second and third qua d-rants (the failure envelopes are symmetrical with respect to the σ_1-axis).

4. SANDWICH BEAMS

The deformation and failure mechanisms in the core of sandwich beams in three-point bending and cantilever beams under end load were studied by means of moiré gratings and photoelastic coatings. Figure 3 shows moiré fringe patterns in the core of a sandwich beam under three-point bending for an applied load that produces stresses in the core within the linear elastic range [5]. The moiré fringe patterns corresponding to the u (horizontal) and w (through-the-thickness) displacements away from the applied load consist of nearly parallel and equidistant fringes from which it follows that

$$\varepsilon_x = \frac{\partial u}{\partial x} \cong 0, \quad \varepsilon_z = \frac{\partial w}{\partial z} \cong 0, \quad \gamma_{xz} = \frac{\partial u}{\partial z} + \frac{\partial w}{\partial x} = \text{constant} \tag{5}$$

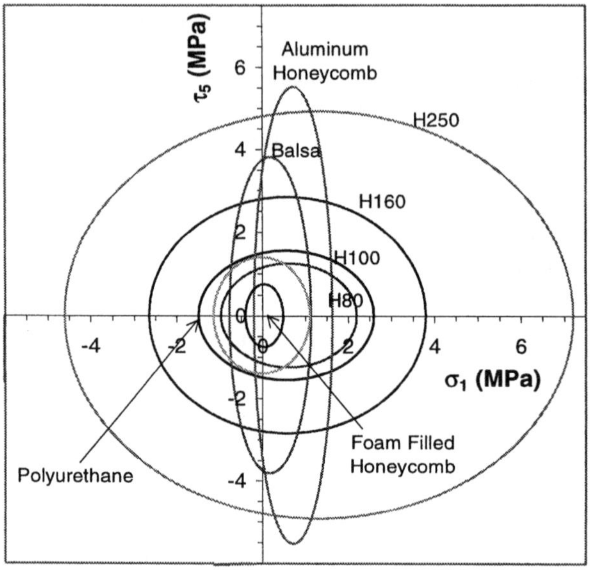

Figure 2. Failure envelopes for various core materials based on the Tsai-Wu failure criterion for interaction of normal and shear stress

Thus, the core is under nearly uniform shear stress, which is in agreement with the linear theory of sandwich beams.

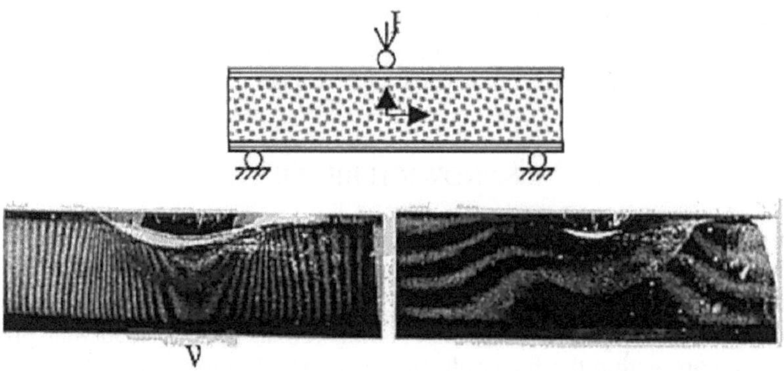

Figure 3. Moiré fringe patterns corresponding to horizontal and vertical displacements in sandwich beam under three-point bending (12 lines/mm, 300 lines/in; Divinycell H250 core)

Figure 4 shows isochromatic fringe patterns in the birefringent coating for a beam under three-point bending for various values of applied load P. The shear strain γ_{xz} is related to the fringe order by

$$\gamma_{xz} = f_e N \tag{6}$$

where f_e is the photoelastic constant of the coating material of thickness 1 mm (0.042) in.) and N is the fringe order. Note that the fringe patterns for applied loads less than 3.3 kN (740 lb) are almost uniform, indicating that the shear strain (stress) in the core is constant. For this load the average shear stress in the core is 2.55 MPa (370 psi), which is close to the proportional limit of the shear stress-strain curve of the core material (Divinycell H250). For higher loads the core begins to yield and the shear strain becomes highly nonuniform peaking at the center.

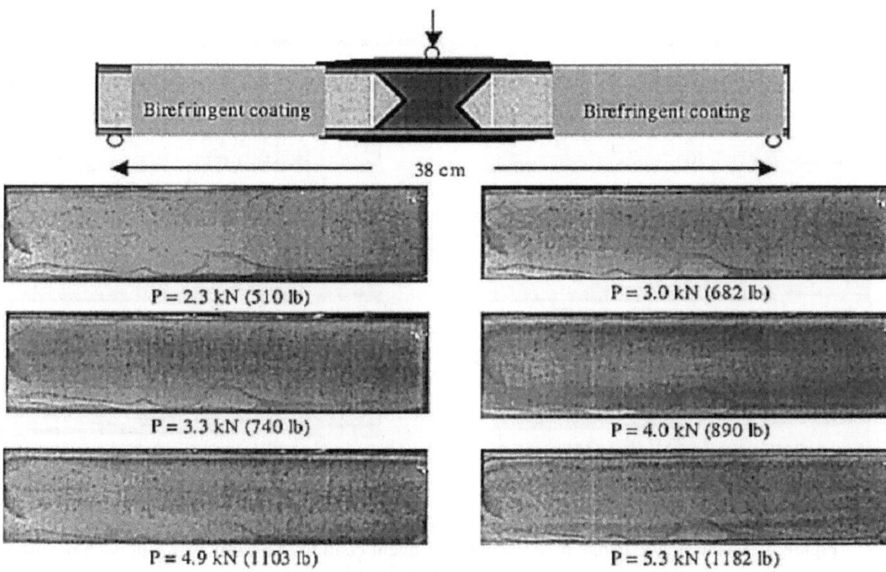

Figure 4. Isochromatic fringe patterns in birefringent coating of sandwich beam under three-point bending

From Figure 2 it is shown that the most critical situation for core failure is a combination of compressive and shear stresses. The criticality of the compression/shear stress biaxiality was tested with a cantilever sandwich beam loaded at the free end. The isochromatic fringe patterns in the birefringent coating in Figure 5 show how the peak birefringence which is proportional to the maximum in-plane shear strain moves towards the fixed end of the beam at the bottom where the critical combination of compressive normal stress and shear stress occurs.

Analysis of horizontal displacements by the moiré and birefringent coating methods shows that, for nonlinear deformation, the neutral axis is no longer located at the mid-plane of the core, but moves towards the tension side of the beam. This is due to nonlinear behavior of the core and facings. If most of the bending moment is taken up by the facings, the distance of the neutral axis from the tensile side of the beam is given by

$$x = \frac{h_c}{1+\frac{\varepsilon_c}{\varepsilon_t}} \tag{7}$$

where ε_c and ε_t represent the average compressive and tensile strains of the facings, respectively.

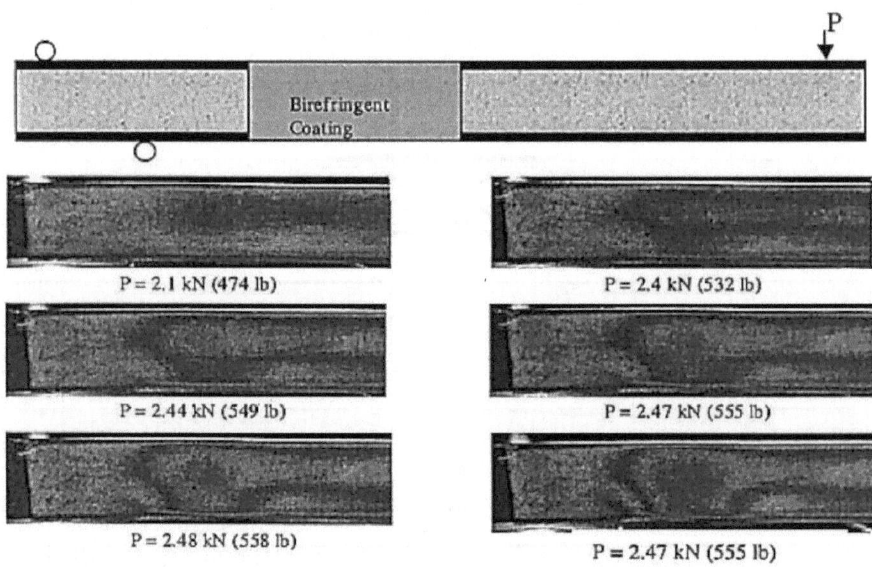

Figure 5. Isochromatic fringe patterns in birefringent coating of a cantilever sandwich beam under end load

5. FAILURE LOADS

From Eqs. (1) we obtain for the compression side

$$\frac{\tau}{\sigma} = -\frac{C_1 h_f E_f}{C_2 L E_c}\left(\frac{d}{h_c}\right)^2 \tag{8}$$

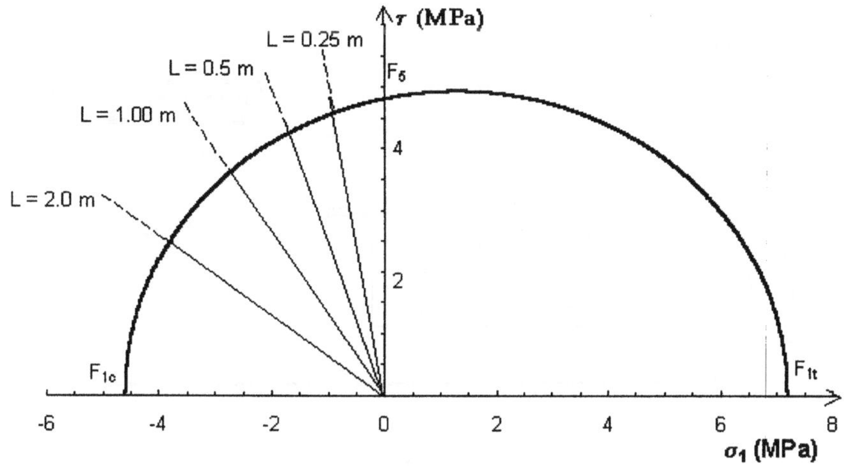

Figure 6. Determination of the critical values of the longitudinal normal and in-plane shear Stresses by the intersection of the failure envelope by the loading line

where σ is the maximum compressive stress. Equation (8) defines the slope of a straight line on the τ - σ plane. The intersection of this line with the failure envelope on the τ - σ plane defines the critical values of τ and σ for initiation of core failure (Fig. 6).

Equation (8), for $d \cong h_c$ is simplified as

$$\frac{\tau}{\sigma} \cong -\frac{C_1}{C_2}\frac{h_f}{L}\frac{E_f}{E_c} \tag{9}$$

From Eq. (9) we observe that a transition of core failure from shear failure to normal stress failure occurs as the beam length, L, and core modulus, E_c, increase and as facing thickness, h_f, and facing modulus, E_f, decrease. Furthermore, we observe that since for a beam under three-point bending $C_1/C_2=2$, and for a cantilever beam $C_1/C_2 = 1$ the cantilever beam is prone to fail by core failure in compression, whereas the beam under three-point bending is more prone to fail by shear core failure.

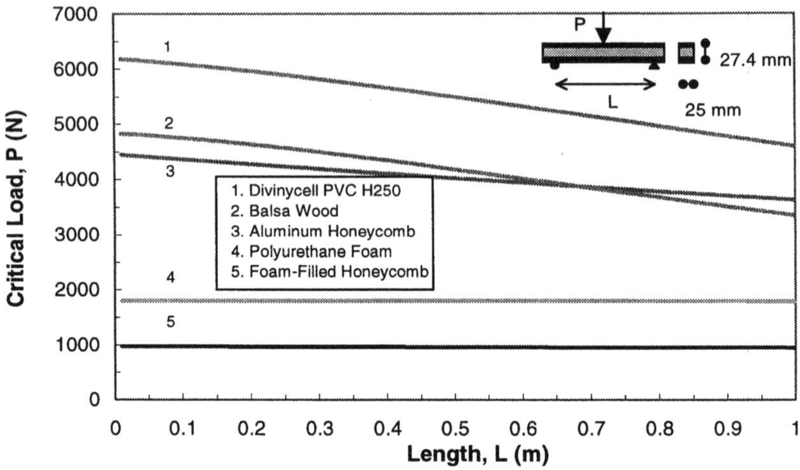

Figure 7. Critical load for core failure initiation versus length of a beam under three-point bending for a series of core materials

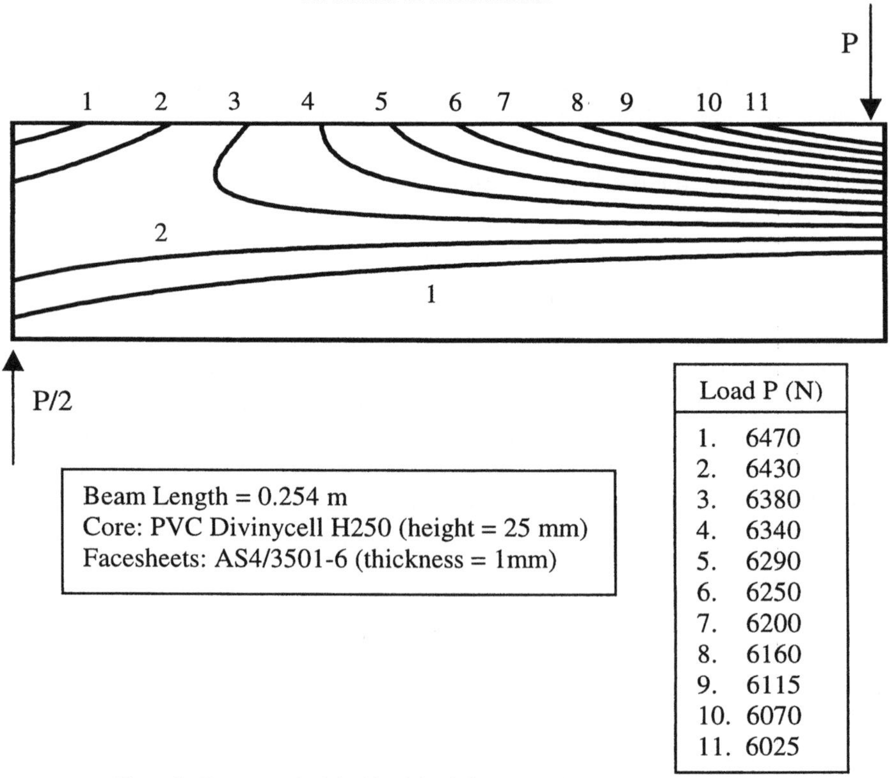

Figure 8 .Contours of critical load for failure initiation of half beam of length L=25.4 cm (10 in.) under three-point bending

Core failure initiation can be described by calculating the state of stress in the core and applying the Tsai-Wu failure criterion [4]. This yields a curve for critical load (at core failure initiation) versus span length. Figure 7 shows the variation of the critical load P_c with length L of a beam under three-point bending for various core materials. Note that for the same span length, the critical load is the highest for Divinycell H250 and lowest for foam-filled honeycomb. Figure 7 can be used to design a sandwich beam against core failure.

Finally, figure 8 presents the contours of the applied load for core failure of half sandwich beam of length 25.4 cm (10 in.) under three-point bending, according to the Tsai-Wu failure criterion. The contours give a picture of the distribution of the failure envelopes in the beam.

7. CONCLUSIONS

Core failure modes were investigated for composite sandwich beams under three-point bending and cantilever beams under end loading. A number of core materials were characterized and their failure behavior under multiaxial state of stress was studied. Deformation and failure mechanisms were studied experimentally by means of moiré gratings and birefringent coatings. The main conclusions of the present study are the following:

1. In the linear elastic range the core shear stress (strain) is uniform through the thickness according to the bending theory of sandwich beams and as confirmed experimentally.
2. In the nonlinear/plastic range the core begins to yield and the shear strain becomes nonuniform peaking at the center.
3. In the nonlinear region the neutral axis is no longer located at the mid-plane of the core, but moves towards the tension side of the beam. This affects the shear and bending stress distributions in the beam.
4. Core failure is accelerated when compressive and shear deformations are combined.
5. Core yielding and stiffness loss reduce core support for the facings, precipitating facing wrinkling.
6. The failure behavior under multiaxial states of stress of a series of lightweight core materials can be described well by the Tsai-Wu failure criterion.
7. The dependence of the critical load for core failure initiation on the geometrical configuration, core and facing material properties and loading conditions of the sandwich beam was analyzed.

8. REFERENCES

1. Daniel, I. M., Gdoutos, E. E., Wang, K.A. and Abot, J. L.: Failure Modes of Composite Sandwich Beams, *International Journal of Damage Mechanics*, to appear.

2. Gdoutos, E. E., Daniel, I. M. and Wang, K.A.: Compression Facing Wrinkling of Composite Sandwich Structures, *Mechanics of Materials*, to appear.
3. doutos, E. E., Daniel, I. M. and Wang, K.A.: Failure of Cellular Foams under Multiaxial Loading, *Composites Part A*, **33** (2002), 163-176.
4. Tsai, S. W. and Wu, E. M.: A General Theory of Strength for Anisotropic Materials, *Journal of Composite Materials*, **5** (1971), 58-80.
5. Daniel, I. M. and Abot, J. L.: Fabrication, Testing and Analysis of Composite Sandwich Beams, *Comp. Science and Technology*, **60** (2000), 2455-2463.

BUCKLING OF LONG SANDWICH CYLINDRICAL SHELLS UNDER EXTERNAL PRESSURE

G.A. KARDOMATEAS* and G.J. SIMITSES+
*Professor of Aerospace Engineering
+Professor Emeritus of Aerospace Engineering
Georgia Institute of Technology, Atlanta, Georgia, U.S.A.

Abstract. The paper deals with the theoretical prediction of buckling loads for sandwich long cylindrical shells with laminated facings and foam core. The loading is a uniform hydrostatic pressure, which means that the loading remains normal to the deflected surface during the buckling process. Several fiber materials are used in the laminated facings. The materials are: Boron/Epoxy, Graphite/Epoxy and Kevlar/Epoxy laminates with 0° orientation with respect to the hoop direction. These various materials are employed to provide comparative data that can be used in design. Shell theory results are generated with and without accounting for the transverse shear effect. Moreover, results based on three-dimensional elasticity are also generated for comparison purposes. The effect of the ratio of radius to thickness is assessed.

1. INTRODUCTION

Lightweight sandwich construction is of great interest in the design and manufacture of aircraft, spacecraft and marine vehicles because of high specific stiffness and strength. In addition, sandwich construction offers enhanced corrosion resistance, noise suppression and reduction in life-cycle costs. There are several issues and questions related to the use of sandwich construction that require attention and answers. One of the important issues is the prediction of buckling loads.

In this paper, theoretical predictions for buckling are generated by shell theory (very long cylindrical shell) and by three dimensional elasticity. The geometry, which is fully described in the next section, is a cylindrical configuration consisting of laminated facings and a soft core. The fibers in the facings have 0^0 orientation with respect to the hoop direction. The shell theory predictions are based on the formulae presented in [1,2] and specialized to an infinite length cylinder, whose behavior is similar to that of a sandwich ring. The three-dimensional elasticity is presented in a later section and is based on the approach presented in [3]. The effect of facing materials and of transverse shear are assessed for four values of the

E.E. Gdoutos and Z.P. Marioli-Riga (eds), Recent Advances in Composite Materials, 291-300

ratio of mean radius to total thickness. There are but few studies reported in the literature that deal with sandwich shell analyses, see [4,5,6].

TABLE 1. Material Properties

	Designation	E_{11} (kPa)	E_{22} (kPa)	υ_{12}	υ_{21}	G_{13} (kPa)
Boron/Ep	B(4)5505	2.21E+08	2.07E+07	0.23	0.0216	5.79E+06
Graph/Ep	T300 5208	1.81E+08	1.03E+07	0.28	0.0159	7.17E+06
Kevlar/Ep	49Epoxy	7.59E+07	5.52E+06	0.34	0.0247	2.28E+06
Alloy-Foam		4.59E+04	4.59E+04	0.33	0.33	1.72E+04

2. STRUCTURAL GEOMETRIES

Consider a sandwich ring of width B, mean radius R, core thickness c and facings thickness h. The laminated facings are made of Boron/Epoxy, Graphite/Epoxy and Kevlar/Epoxy, with properties given in Table 1. The core is Alloy-Foam and its properties are also given in Table 1. The values employed for the geometry are: c = 25.4 mm, h = 2.54 mm and B = 76.2 mm. This value for B was chosen in order to ensure that buckling is in the plane of the ring and not out of the plane. Moreover, R is chosen in such a manner that R/h_{tot} is 30, 60, 120 and 240. Note that the sandwich is symmetric about its mid-surface and $h_{tot} = c + 2h$.

3. SHELL THEORY PROCEDURE

Since the shell is considered to be very long, the buckling analysis reduces to that for a ring [1]. This was verified by Birman [7], who derived the buckling equations for a cylindrical sandwich shell under hydrostatic pressure using Sander's shell theory [8], and let the length of the shell approach infinity. If the transverse shear effect is neglected, the expression for the pressure becomes (classical)

$$p_{cl} = 3\frac{(EI)_{eq}}{BR^3} \tag{1}$$

where p_{cl} is the thin ring critical pressure and (EI)eq is the equivalent bending stiffness, given by

$$(EI)_{eq} = B\left[E_f \frac{h^3}{6} + 2E_f h\left(\frac{h}{2}+\frac{c}{2}\right)^2 + E_c \frac{c^3}{12}\right] \tag{2}$$

This expression was taken from [9].

If the transverse shear effect is accounted for then

$$p_{cr} = 3\frac{(EI)_{eq}}{BR^3}/(1+4k_s) \tag{3}$$

where k_s accounts for the transverse shear and is given by:

$$k_s = \frac{(EI)_{eq}}{CR^2} \tag{4}$$

where

$$C = \int_A KGdA$$

K being a shear correction factor taken as equal to one, and G is the transverse shear modulus of the sandwich cross section.

Two different expressions for C are employed herein. In the first case, it is assumed that only the core provides shear resistance. In this case,

$$C = BcG_c \tag{5}$$

and

$$k_{s_1} = \frac{(EI)_{eq}}{BcG_c R^2} \tag{6}$$

where G_c is the shear modulus of the core (see Table 1). For this case,

$$p_{cr_1} = 3\frac{(EI)_{eq}}{BR^3}/(1+4k_{s_1}) \tag{7}$$

In the second case, an effective shear modulus for the sandwich, $\overline{G}$, which includes the contribution of the facings, is derived based on the compliances of the constituent phases [10]. The expression for $\overline{G}$ is given by

$$\frac{(2h+c)}{\overline{G}} = \frac{2h}{G_f} + \frac{c}{G_c} \tag{8}$$

which leads to

$$\overline{G} = (2h+c) \Big/ \left(\frac{2h}{G_f} + \frac{c}{G_c} \right)$$

where G_f is the shear modulus, G_{13}, of the facing material (see Table 1). Therefore, in this case, the expression for p_{cr}, becomes:

$$p_{cr_2} = 3\frac{(EI)_{eq}}{BR^3} / \left(1 + 4k_{s_2}\right) \tag{9}$$

where

$$k_{s_2} = \frac{(EI)_{eq}}{B(c+2h)\overline{G}R^2} \tag{10}$$

4. THREE-DIMENSIONAL ELASTICITY PROCEDURE

By considering the equations of equilibrium in terms of the second Piola-Kirchhoff stress tensor, subtracting these at the perturbed and initial conditions, and making order of magnitude assumptions on the products of stresses and strains/rotations, based on the fact that a characteristic feature of stability problems is the shift from positions with small rotations to positions with rotations substantially exceeding the strains, Kardomateas [3] obtained buckling equations of the form:

$$\begin{aligned}&\frac{\partial}{\partial r}\left(\sigma_{rr}-\tau_{r\theta}^{0}\omega_{z}+\tau_{rz}^{0}\omega_{\theta}\right)+\frac{1}{r}\frac{\partial}{\partial\theta}\left(\tau_{r\theta}-\sigma_{\theta\theta}^{0}\omega_{z}+\tau_{\theta z}^{0}\omega_{\theta}\right)\\&+\frac{\partial}{\partial z}\left(\tau_{rz}-\tau_{\theta z}^{0}\omega_{z}+\sigma_{zz}^{0}\omega_{\theta}\right)+\frac{1}{r}\left(\sigma_{rr}-\sigma_{\theta\theta}+\tau_{rz}^{0}\omega_{\theta}+\tau_{\theta z}^{0}\omega_{r}-2\tau_{r\theta}^{0}\omega_{z}\right)=0\end{aligned}\tag{11}$$

In the previous equations, σ_{ij}^{0}, τ_{ij}^{0} and ω_j^{0} are the values of stresses and rotations, respectively, at the initial equilibrium position (pre-buckling state), and σ_{ij}, τ_{ij} and ω_j are the corresponding values at the perturbed position (buckled state).
The boundary conditions associated with Equation (11) were obtained from the traction (stress resultant) relationships in terms of the second Piola-Kirchhoff stress tensor, and by further considering the fact that because of the hydrostatic pressure loading, the magnitude of the surface load remains invariant under deformation, but its direction changes (since hydrostatic pressure is always directed along the normal to the surface on which it acts). By writing these equations for the initial and the perturbed equilibrium position and then subtracting them and using the previous arguments on the relative magnitudes of the rotations Kardomateas [3] obtained the following boundary conditions:

$$\begin{aligned}&\left(\sigma_{rr}-\tau_{r\theta}^{0}\omega_{z}+\tau_{rz}^{0}\omega_{\theta}\right)\hat{l}+\left(\tau_{r\theta}-\sigma_{\theta\theta}^{0}\omega_{z}+\tau_{\theta z}^{0}\omega_{\theta}\right)\hat{m}\\&+\left(\tau_{rz}-\tau_{\theta z}^{0}\omega_{z}+\sigma_{zz}^{0}\omega_{\theta}\right)\hat{n}=p\left(\omega_{z}\hat{m}-\omega_{\theta}\hat{n}\right)\end{aligned}\tag{12}$$

4.1. Pre-buckling State

The problem under consideration is that of a sandwich hollow cylinder deformed by uniformly distributed external pressure, p and of infinite length (generalized plane deformation assumption). Then, not only the stresses, but also the displacements do not depend on the axial coordinate. Alternatively, this is the assumption we would make if the cylinder were securely fixed at the ends. An elasticity solution to this problem was provided by Kardomateas [10]. The solution is an extension of the classical one by Lekhnitskii [11] for a homogeneous, orthotropic shell and was provided in closed form. All three phases, i.e., the two face-sheets and the core were assumed to be orthotropic. Moreover, there were no restrictions as far as the individual thicknesses of the face-sheets and the sandwich construction could be asymmetric.

In this configuration, the axially symmetric distribution of external forces produces stresses identical at all cross sections and dependent only on the radial coordinate r. We take the axis of the body as the z- axis of the cylindrical coordinate system and we denote by R_1 and R_2 the inner and outer radii. Let us also denote each phase by i where i=f_2 for the outer face-sheet, i=c for the core and i= f_1 for the inner face-sheet.

Then, the pre-buckling stresses in each of the phases, i.e. for $i = f_1$, c, f_2, are:

$$\sigma_{rr}^{0(i)}(r) = p\left(C_1^{(i)} r^{k_i-1} + C_2^{(i)} r^{-k_i-1}\right) \tag{13a}$$

$$\sigma_{\theta\theta}^{0(i)}(r) = p\left(C_1^{(i)} k_i r^{k_i-1} - C_2^{(i)} k_i r^{-k_i-1}\right) \tag{13b}$$

$$\sigma_{zz}^{0(i)}(r) = p\left(-C_1^{(i)} \frac{\left(a_{13}^i + a_{23}^i k_i\right)}{a_{33}^i} r^{k_i-1} - C_2^{(i)} \frac{\left(a_{13}^i - a_{23}^i k_i\right)}{a_{33}^i} r^{-k_i-1}\right) \tag{13c}$$

$$u^{0(i)}(r) = p\left(C_1^{(i)} \frac{\left(\beta_{11}^i + \beta_{12}^i k_i\right)}{k_i} r^{k_i} - C_2^{(i)} \frac{\left(\beta_{11}^i - \beta_{12}^i k_i\right)}{k_i} r^{-k_i}\right) \tag{13d}$$

Furthermore, the pre-buckling radial displacement is found to be:

$$\tau_{\theta z}^{0(i)}(r) = \tau_{rz}^{0(i)}(r) = \tau_{r\theta}^{0(i)}(r) = 0 \tag{13e}$$

The constants $C_1^{(i)}$, $C_2^{(i)}$, are found from the conditions on the cylindrical lateral surfaces (traction-free) and the conditions at the interfaces between the phases of the sandwich structure [10].

4.2. Perturbed State

In the petrurbed position we seek plane equilibrium modes as follows:

$$\begin{aligned} u_i(r,\theta) &= U_i(r)\cos n\theta \\ v_i(r,\theta) &= V_i(r)\sin n\theta \\ w_i(r,\theta) &= 0 \end{aligned} \tag{14}$$

Substituting these in the strain vs displacement and rotation vs displacement relations and then using the stress -strain relations in terms of the stiffness constants, c_{ij}^{i}, the buckling Equations (1) result in two linear homogeneous ordinary differential equations of the second order for $U_i(r)$, $V_i(r)$. For example, the first equation is:

$$\begin{aligned} & c_{11}^{(i)} U_i'' + \frac{c_{11}^{(i)}}{r} U_i' - \left[\left(c_{66}^{(i)} + \frac{\sigma_{\theta\theta}^{0(i)}}{2}\right) n^2 + c_{22}^{(i)}\right] \frac{U_i}{r^2} \\ & + \left(c_{12}^{(i)} + c_{66}^{(i)} - \frac{\sigma_{\theta\theta}^{0(i)}}{2}\right) \frac{nV_i'}{r} - \left(c_{22}^{(i)} + c_{66}^{(i)} + \frac{\sigma_{\theta\theta}^{0(i)}}{2}\right) \frac{nV_i}{r^2} = 0 \end{aligned} \tag{15}$$

The associated boundary conditions are as follows:
(a) At the inner and outer bounding surfaces, we have two traction conditions at each of the surfaces
(b) At the face-sheet/core interfaces, we have four conditions at each of the interfaces:
(1) two Traction Continuity:
(2) two Displacement Continuity:

Detailed forms of these boundary conditions for each interface or bounding surface can be found in Kardomateas and Simitses [12].

4.3. Solution of the eigen-boundary-value problem

We have formulated in the foregoing an eigenvalue problem for differential equations, with p the parameter (two point boundary value problem). An important point is that $\sigma_{rr}^{0(i)}(r)$, $\sigma_{\theta\theta}^{0(i)}(r)$, $\sigma_{rr}^{0(i)}(r)'$, depend *linearly* on the external pressure, p (the parameter) through expressions in the form of Equations (13) and this makes possible the direct application of standard solution techniques.

With respect to the method used there is a difference between the present problem and the one for the homogeneous orthotropic body solved by Kardomateas [3]. The complication in the present problem is due to the fact that the displacement field is continuous but has a slope discontinuity at the face-sheet/core interfaces. This is the reason that the displacement field was not defined as one function but as three distinct functions for $i=f_1$, c and f_2, i.e. the two face sheets and the core. Our formulation of the problem employs, hence, "internal" boundary conditions at the face-sheet/core interfaces, as outlined above. Due to this complication, the shooting method [13] was deemed to be the best way to solve this eigen-boundary-value problem for differential equations. A special version of the shooting method was formulated and programmed for this problem. Details of the shooting method can be found in Kardomateas and Simitses [12].

5. RESULTS AND DISCUSSION

Table 2 gives the critical pressure from the Elasticity formulation for a range of mean radius over total thickness ratios, in comparison with the classical shell and the two shear deformable shell formulas. In all cases, n =2 was used in the buckling modes, Equations (14) (as is the case with external pressure). It is seen that the classical (no shear) formula can yield results highly non-conservative, in excess of three times the elasticity value for the lower ratio of R/h_{tot} and Boron/Epoxy case. Both shear correction formulas yield reasonable results with the shear correction formula based on the core only being in general conservative as opposed to the shear correction formula based on an "effective shear modulus", $\overline{G}$, which is non-conservative.

In the results presented in Table 2, the face sheets were quite thin and the shear correction formula based on the core only, Equation (7), seemed to be more accurate. In order to further examine this premise, the critical load was calculated for a construction in which the total thickness remains the same but the face sheet thickness is increased at the expense of the core. The results, listed in Table 3, show that the shear correction formula based on an effective modulus (which includes the core), Equation (9), is now more accurate.

TABLE 2. Critical Pressure in N/m^2-Sandwich w/ Alloy-Foam Core

Face Sheet	R/h	3D Elasticity	Classical Shell (% vs 3D Elast)	Shell with Shear Based on Gc (% vs 3D Elast)	Shell with Shear Based on $\overline{G}$ (% vs 3D Elast)
Boron/ Epoxy	30	277,305	862,343 (+310.9%)	253,721 (-8.5%)	323,361 (+16.6%)
	60	70,416	107,793 (+53.0%)	67,383 (-4.3%)	76,087 (+8.0%)
	120	11,817	13,474 (+14.0%)	11,717 (-0.85%)	12,203 (+3.3%)
Graphite/ Epoxy	30	258,549	706,307 (+273.2%)	238,236 (-7.9%)	298,643 (+15.5%)
	60	61,528	88,288 (+43.5%)	59,207 (-3.8%)	65,825 (+7.0%)
	20	9,918	11,036 (11.3%)	9,829 (-0.9%)	10, 168 (+2.5%)
Kevlar/ Epoxy	30	171,351	296,324 (+72.9%)	162,433 (-5.2%)	188,347 (+9.9%)
	60	31,418	37,040 (+17.9%)	30,712 (-2.2%)	32,397 (+3.1%)
	120	4,476	4,630 (+3.4%)	4,403 (-1.6%)	4,470 (-0.13%)

TABLE 3. Critical Pressure in N/m^2 - Effect of increased face sheet thickness: f = 0.3 in, c = 0.6 in and B = 3 in. -Sandwich w/ Alloy-Foam Core

Face sheet	R/h	3D Elasticity	Classical Shell (% vs 3D Elast)	Shell with Shear Based on Gc (% vs 3D Elast)	Shell with Shear Based on $\overline{G}$ (% vs 3D Elast)
Graphite/ Epoxy	30	393,573	1,466,490 (+372.6%)	188,038 (-52.2%)	542,378 (+37.8%)
	60	105,699	183,311 (+73.4%)	67,900 (-35.8%)	128,553 (+21.6%)
	120	19,297	22,914 (+18.7%)	16,081 (-16.7%)	20,709 (+7.3%)

TABLE 4. Critical Pressure in N/m^2 - Homogenous (no sandwich)

	R/h	3D Elasticity	Classical Shell (% vs 3D Elast)	Shell with Shear Based on Gc (% vs 3D Elast)	Shell with Shear Based on $\overline{G}$ (% vs 3D Elast)
Graphite/ Epoxy	30	1,641,360	1,675,930 (+2.1%)	1,657,330 (+0.97%)	1,660,400 (+1.2%)
	60	208,228	209,491 (+0.61%)	208,905 (+0.33%)	209,002 (+0.37%)
	120	26,180	26,186 (+0.03%)	26,168 (-0.05%)	26,171 (-0.03%)

In order to compare with the homogeneous, monolithic, Table 4 gives the critical pressure for a construction made of graphite/epoxy homogeneous, i.e. no sandwich. It is seen that the differences from the elasticity values are very small, even with the classical shell formula. This illustrates the nature of sandwich construction, in which buckling is a more demanding issue.

6. ACKNOWLEDGEMENT

The financial support of the Office of Naval Research, Ship Structures and Systems, S & T Division, Grants N00014-90-J-1995 and N00014-0010323, and the interest and encouragement of the Grant Monitor, Dr. Y.D.S. Rajapakse, are both gratefully acknowledged.

7. REFERENCES

1. Smith, Jr., C.V. and Simitses, G.J.: Effect of Shear and Load Behavior on Ring Stability, *ASCE, Journal of the EM Division* **95** (1969), 559-569.
2. Simitses, G.J. and Aswani, M.: Buckling of Cylinders Under Uniform Lateral Loading, *ASME Journal of Applied Mechanics* **41** (1974), 827-829.
3. Kardomateas, G.A.: Buckling of Thick Orthotropic Cylindrical Shells Under External Pressure, *ASME Journal of Applied Mechanics* **60** (1993), 195-202.
4. Birman, V. and Simitses, G.J.: Theory of Cylindrical Sandwich Shells with Dissimilar Facings Subjected to Thermo-mechanical Loads, *AIAA Journal* **37** (2000), 362-367.
5. Birman, V., Simitses G.J. and Shen, L.: Stability of Short Sandwich Cylindrical Shells with Rib-Reinforced Facings, in *Recent Advances in Applied Mechanics*, J.T. Katsikadelis, D.E. Beskos and E.E. Gdoutos (editors), National Technical University of Athens, Greece, (2000), 11-21.
6. Birman, V. and Simitses, G.J.: Stability of Long Cylindrical Sandwich Shells with Dissimilar Facings subjected to Lateral Pressure, in *Advances in Aerospace Materials and Structures*, G. Newaz (editor), ASME AD –vol. 58, (1999), 41-51.
7. Birman, V.: Private Communication, April 2002
8. Sanders, Jr., J.L.: Nonlinear Theory for Thin Shells, *Quarterly of Applied Math* **21** (1963), 21-36.
9. Huang, H. and Kardomateas, G.A.: Buckling and Initial Postbuckling Behavior of Sandwich Beams Including Transverse Shear, in press, *AIAA Journal* (2002).
10. Kardomateas, G.A.: Elasticity Solutions for a Sandwich Orthotropic Cylindrical Shell Under External Pressure, Internal Pressure and Axial Force, *AIAA Journal* **39** (2001), 713-719.
11. Lekhnitskii, S.G.: *Theory of Elasticity of an Anisotropic Elastic Body*, Holden Day, San Francisco, 1963, also Mir Publishers, Moscow, 1981.
12. Kardomateas, G.A. and Simitses, G.J.: Elasticity Solution for the Bucking of Long Sandwich Cylindrical Shells Under External Pressure, submitted to the *ASME Journal of Applied Mechanics* (2002).
13. Press, W.H, Flannery, B.P., Teukolsky, S.A and Vetterling, W.T.: *Numerical Recipes*, Cambridge University Press, Cambridge, 1989.

6. Applications

DESIGN AND MANUFACTURING OF HIGH PERFORMANCE STRUCTURAL COMPONENTS BY RESIN TRANSFER MOULDING

Z. P. MARIOLI-RIGA
Research & Development Department
Hellenic Aerospace Industry S.A.
GR-320 09 Shimatari, Greece

ABSTRACT *The ability to reduce the manufacturing and operating cost of high performance structural components and vehicles governs the competitiveness in the market. The use of RTM technologies in Aerospace applications gives added potential in recent days due to the combined benefit of weight – cost reduction. Complicated components could be easily manufactured by one injection , provided that the mould has been well designed and the positioning of gate ports accurately defined .by flow simulation. In the present work a sequenced methodology was established for the development of an aircraft structural stiffened skin. A mould with inserted heating elements and preformed reinforcements were used for the manufacturing of component .The experimental Investigation lead to various techniques for eliminating lack of wetting and improving the quality of final product.*

1. INTRODUCTION

As the demands for the composite materials increase many processing methods of composite materials have been developed and refined in order to meet the requirements of quality and productivity. Recently RTM is extensively studied and employed to produce components with high performance by reducing the cost. Mainly in the aircraft manufacturing industry the competitiveness of new aircraft is governed by the ability to reduce the cost and save further weight

The use of RTM technologies in aerospace applications could result efficiently to the cases: a)where structural components have been designed and manufactured using prepregs by reducing the cost and b) by replacing metallic components with the benefit of maintenance and manufacturing cost reduction and simultaneously the improvement of fatigue and corrosion performance.[1],[2]

Especially the possibility of the RTM manufacturing technology to integrate composite parts is expected to considerably reduce the cost of composite structural components. When large aircraft components are considered RTM offers unique advantages compared to other composite manufacturing techniques due to its ability to incorporate Spars, Ribs, Stiffeners, and Skins in one production [3][4]

E.E. Gdoutos and Z.P. Marioli-Riga (eds), Recent Advances in Composite Materials, 303-312

In many ways the development of an aerospace or, more broadly, advanced composites industry was permitted by the development of preimpregnated reinforcements (prepreg)

A variety of individual processes, such as autoclave moulding, vacuum bag moulding, compression moulding, were developed that utilized the new form of semi-finished material known as prepreg. Despite the similarities between the various processes it is unusual to refer to prepreg mouldings. Each process has its own literature and the commonalities between them are sometimes lost. By contrast, RTM has as many different processes under the RTM umbrella as there are prepreg processes; but it is more or less commonplace to speak of RTM, rather than, for example, rigid tool RTM with semi-rigid preforms.[5][6][7]

All the process variants have common features. Unresinated fibres are held within a tool cavity and a differential pressure is applied to a supply of resin such that the resin flows into the reinforcement completely wetting it out. The tool may be essentially rigid, semi-rigid, or contain flexible elements. Any consolidation pressure required to give the required reinforcement volume fraction (Vf%) may be applied by mechanical clamps, from a tooling frame or press, or by the use of an internal vacuum or external applied pressure in non-rigid tooling. Reinforcements may be of any fibre, and the use of all forms has been reported, from unidirectional (UD) through woven or knitted cloths to needled and random mats and fully three-dimensional reinforcement preforms. Volume fractions from below 20% to above 60% have been reported. The reinforcement may be laid onto the mould by hand, formed to shape by the mould closure, assembled by a wide variety of preforming techniques or may utilize specially woven or braided constructions. [8][9][10]

In the present work a standard methodology was established for developing a real aircraft subcomponent by RTM. The specific component is a portion of a large structure. It was decided to establish the methodology and specific parameters on the subcomponent first and then to incorporate it to the development of the large , fin box structure. The design analysis was made considered the loading history applied on the fin box of an existing fighter aircraft Based on the design of the skin stiffened panel and the flow simulation analysis a mould was developed and manufactured. Specific arrangements were made for the injection moulding and after series of experimental trials quality product was obtaine with no considerable evidence of porosity.

2. METHODOLOGY SEQUENCE FOR COMPONENT DEVELOPMENT

- FEM ANALYSIS & DESIGN
- MATERIALS SELECTION
- MOULD DEVELOPMENT
- FLOW SIMULATION / OPTIMIZATION OF MOULD AND PROCESS
- COST ANALYSIS
- MANUFACTURING SET- UP
- PREFORM PREPARATION IN THE MOULD
- INJECTION AND CURING PROCESS

3. COMPONENT & MATERIALS SELECTION

The selected aircraft component was a part from the fin box of a fighter aircraft a skin stiffened panel that was studied and investigated for RTM manufacturing.

The main goal of this investigation was to manufacture an aircraft skin plate carrying a stiffener via the RTM method in one injection. The specific stiffened plate is considered to be the upper left region of a fin box and it is subjected to shear and compression. The material system to be used for the manufacturing of the subcomponent is TENAX HTA fibres and RTM6 resin. The plate is rectangular (200mm in width, 250mm in height), it is 4mm thick and carries a stiffener 25mm in height. Its geometry is shown in detail in figure 1. The plate is part of the enclosed region shown in figure 2 and figure 3.

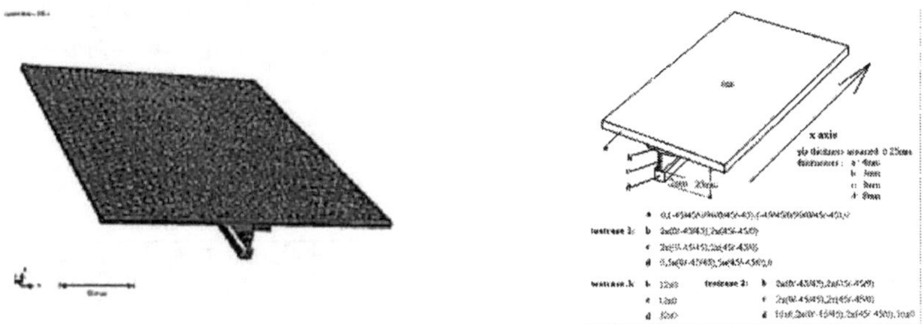

Figure 1: Geometry of stiffened skin

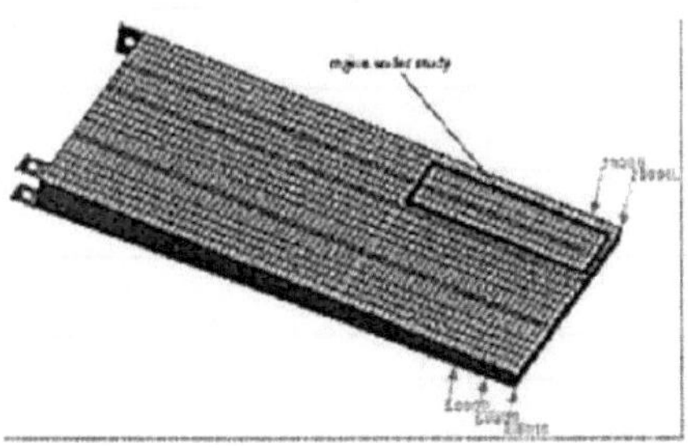

Figure 2: Finite element model of the fin box

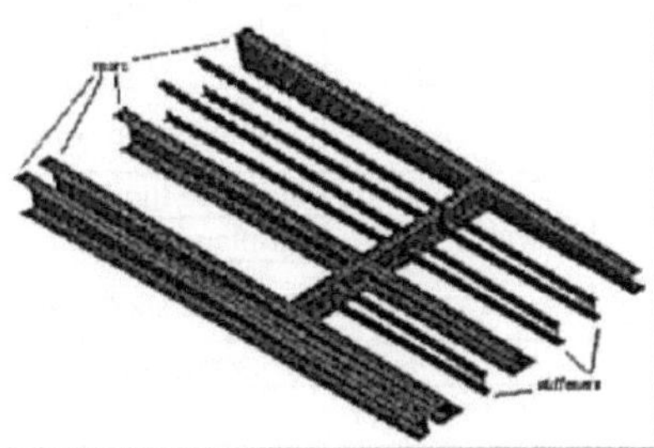

Figure 3: Finite element model of the fin box (the skin is not shown)

4. FINITE ELEMENT ANALYSIS

In order to get an idea of the load carrying capability of the skin stiffened plate a finite element analysis was performed using the Patran/Nastran software. The plate was subjected to a linear static analysis and to a buckling analysis (eigenbuckling). The analysis in total is referred within the subject of another paper work. Table I shows the mechanical properties that were assumed for a single ply of this material. When a real structure is simulated via the finite element method the most important parameters, that define how close the model is to the real structure, are the assumptions that are made about the boundary conditions and the applied loads. In the present case two different methodologies were adopted depending on the type of analysis. a) Linear static analysis b) Buckling analysis

In order to estimate the margin of safety of the plate the Tsai-Wu stress failure criterion was used taking into account different strengths in tension and compression. Different stacking sequences of the preform were investigated in order to conclude to the reinforcement design considering the applied loads conditions (figure 1):

Table I-mechanical properties for one pl

	Property		Design Value
Longitudinal	Tension modulus	E_{11t}	129 Gpa
	Compression modulus	E_{11c}	129 Gpa
	Buckling modulus	E_{11b}	112 Gpa
	Tension strength σ_{11t}		1330 Mpa
	Compression strength σ_{11c}		1250 Mpa
Transversal	Tension modulus	E_{22t}	8.8 Gpa
	Compression modulus	E_{22c}	8.8 Gpa
	Buckling modulus	E_{22b}	7.6 Gpa
	Tension strength σ_{22t}		50 Mpa
	Compression strength σ_{22c}		230 Mpa
Shear	In-plane modulus	G_{12}	4.4 Gpa
	In-plane buckling modulus	G_{12b}	4.0 Gpa
	In-plane shear strength σ_{12}		67 Mpa
	Poisson's ratio	v_{12}	0.3

test case 1:

region a: 0°,2x(-45°/45°/0°/90°/0°/45°/-45°),0° (16 plies)

region b: 2x(0°/-45°/45°),2x(45°/-45°/0°) (12 plies)

region c: 2x(0°/-45°/45°),2x(45°/-45°/0°) (12 plies)

region d: 0°, 5x(0°/-45°/45°),5x(45°/-45°/0°),0° (32 plies)

test case 2:
region a: 0°,2x(-45°/45°/0°/90°/0°/45°/-45°),0° (16 plies)
region b: 2x(0°/-45°/45°),2x(45°/-45°/0°) (12 plies)
region c: 2x(0°/-45°/45°),2x(45°/-45°/0°) (12 plies)
region d: 10x0°, 2x(0°/-45°/45°),2x(45°/-45°/0°),10x0° (32 plies)

test case 3:
region a: 0°,2x(-45°/45°/0°/90°/0°/45°/-45°),0° (16 plies)
region b: 12x0° (12 plies)
region c: 12x0° (12 plies)
region d: 32x0° (32 plies)

The conclusions obtained from the relevant study ,were as follows :

a) Buckling will occur before static failure of the skin for all test cases.
b) As far as the static strength of the skin is concerned it seems that the orientation of the fibres in the stiffener does not seem to play any significant role (failure index is equal for all test cases and well below unity).
c) As far as buckling behaviour is concerned it seems that configuration number 2 seems to give the highest buckling load and should be candidate for manufacturing.

From the above it is evident that the selected region of the skin can withstand the applied loads.

5. FLOW SIMULATION

For establishing the manufacturing concept and decide for the mould development a flow simulation analysis was carried out.

For the flow simulation of the resin in the RTM process, the software RTM-Worx by Polyworx was used. The trials, of the stiffened panel, were simulated with the Rtm-worx program. The initial purpose was to determine the permeability of the reinforcements taking into consideration the experimental filling time and comparing it with the filling time of the simulation.

For a selected fiber volume fraction of about 55-60%, the value of permeability found to be approximately 4e-11m^2. The filling time of the simulation was 580 sec. The simulation results were compared with the actual of the experimental skin stiffened panel which had better quality compared to some previous unsuccessful trials and the experimental filling time was 10 minutes.

From the simulation models was found that the permeability varies between the values 1e-11m2 and 1e-12m2 depending on the fiber volume fraction. The higher the value of the fibre volume fraction, the less the value of the permeability gets. Based on that fact we can estimate the value of the permeability in accordance to the fiber volume fraction and focus on the simulation of the resin flow.

In figure 4 the first trials are presented showing the flow simulation when one filling gate is located in the center of the stiffener and the four venting ports are located at the corners of the skin plate .By this arrangement of ports and gates it was revealed that the filling process is delayed due to the high fiber content in the stiffener which blocked the flow of the resin from the stiffener gate.

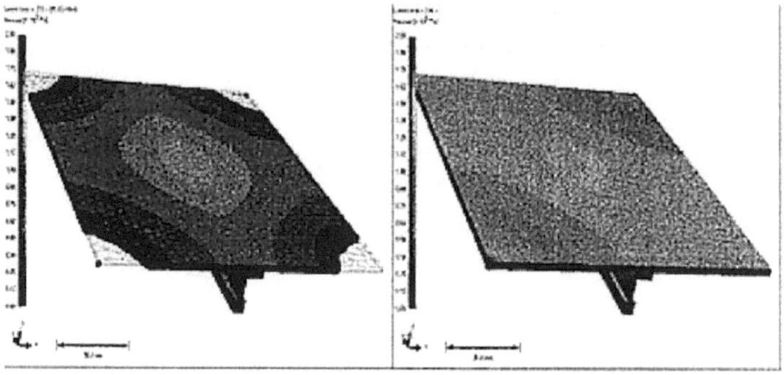

Figure 4. Simulation of resin flow with one gate on the stiffener

For optimizing the above filling process a simulation model was developed and improvements were incorporated in order to apply modifications related to gates and venting ports, on the RTM mould . Resin runners were added around the plate of the stiffened panel in order to evaluate if the runners facilitate the flow, reduce the filling time and avoid the creation of dry spots. The final results are presented in the following pictures.

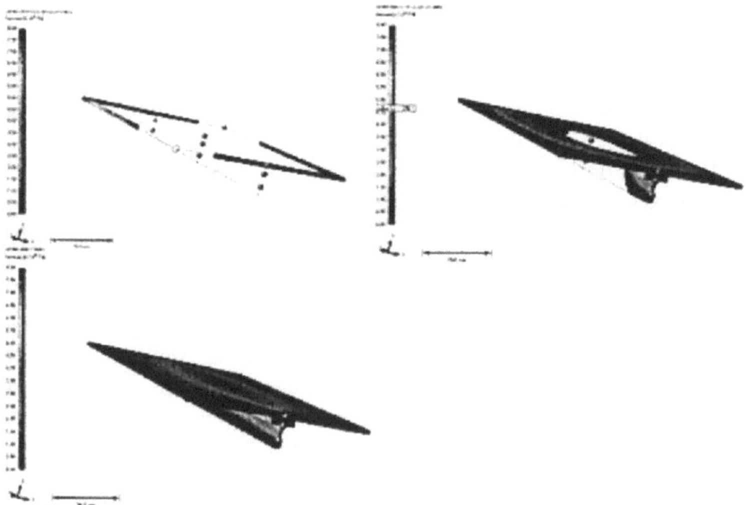

Figures 5 Simulation of flow with incorporated resin runners around.

The existence of runners improves the permeability and resin flow , reduces the possibility of dry spots formation and produces better composite material quality . From the simulation models was found that having the resin fed by two injection gates at the sides of skin panel the filling process is accelerated and improves the quality of the product.

6. MOULD DEVELOPMENT

The mould was designed according to the results of flow simulation. Two injection gates were manufactured on opposite sides of the skin plate of the stiffened panel and three venting ports were located along the stiffener . For improving the wetting process of reinforcements by resin, runners of a diameter 4-5mm around the plate of the stiffened panel were created, as well as across the stiffener where the venting ports are located.

Figures 6 Mould and injection arrangement for the stiffened skin

7. EXPERIMENTAL SET-UP & TESTING TRIALS

An Isojet RTM injection machine was used for the resin transfer process
The mould was cleaned and a released agent was applied. Triaxial fabric layers (±45°,0°) SAERTEX, were connected by powder binder together in a sequenced given by the FEM analysis in a manner to produce flexible preform The perform was positioned in the cavity of the mould as shown in the following figure . The mould was sealed and closed. Resin was degassed. Resin was injected under controlled parameters of: a) Resin Tank Temperature: 80°C b) Feeding Tube Temperature: 80°C c) Injection Pressure: 3 bar d) Injection Volume: 3cc/min-20cc/min e) Constant Mould Temperature: 120°C

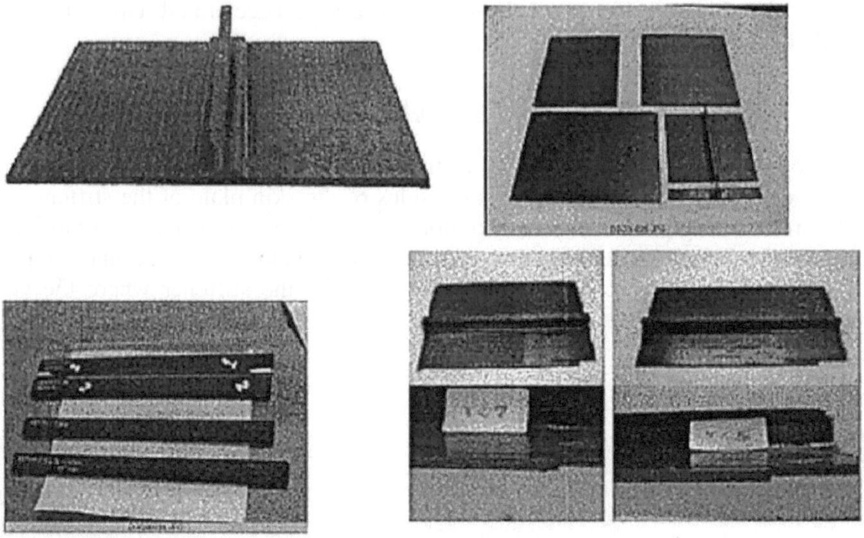

Fig 8 Specimens cut from stiffened panels

When fully filled, surplus resin was left to run out of the vent tubes until no bubbles were observed. Then the vent lines were blocked and the pressure inside the mould was kept at 3 bar The cure and post cure cycles were applied.
When the mould was at about room temperature, the mould was dismounted and the product was taken out.

Specimens were cut from the three series of stiffened panels that presented the best
quality, stiffened panels series 5, 6 and 7, from the flat region along the stiffener. They were tested in tension. Their length was 250 mm and their thickness was 5 mm. The width of the specimens is presented in the table below.
During the tensile testing the jaws of the machine had a constant speed of 3 mm/min.

For the determination of the fibre content of the stiffened panel the ASTM-D-3171-76 standard was applied: (procedure A).Specimens, cut from the manufactured stiffened panels had a fiber volume fraction about *50%-60%*

Table 2 Comparison of theoretical and actual strength of RTM product

Stiffened panel trial a/a	Fracture Load (KN)	Theoretical Fracture Load (KN) *	Cross Section (mm2)	Width (mm)	Actual Fracture Load/Theoretical Fracture Load(%)
1st series	30.1	36.893	72.7	14.85	81.7
2nd series	33	38.756	78	15.6	85%
3nd series	36.96	39.321	76	14.9	94.2

*Theoretical values were calculated based on "Classical Laminate Theory"

8. DISCUSSION AND CONCLUSIONS

The development of high quality products by RTM requires a sequenced methodology and accurate process with close control of RTM parameters .A remarkable cost for RTM technologies , is the design and production of quality tooling mould. To enable efficient design of mould the process of injection and resin flow inside the mould should be modeled and the filling gates should be carefully located especially when large structures are considered.

The quality evolution in Stiffened skin panels of present investigation , was observed after inspection and mechanical testing carried out on 3 series of manufactured components , each one characterized by different RTM parameters.

It was found that by increasing the pressure the resin lean areas which were revealed in the first trials were disappeared

The location of injection ports at the sides as well as the development of resin runners around the preforms of skin improved dramatically the quality of product. The preforms were manufactured about 7 mm shorter than the internal mould dimensions, so that free resin runners were created, in order to facilitate the resin flow and, consequently, the impregnation of the preform.

The "compaction pressure" of the mould should be considered of great importance parameter for the quality of wetting results. It was found that the higher compaction time leads to less porosity results.

Besides , internal resin pressure should be higher than 3 bars and be kept for longer time before disassembling the injection line, in order to allow for the impregnation of those resin lean areas, as well as the flow rate should be a less than 45÷55 cm3/min around 35÷45 cm3/min, for the purpose of better wetting results.

The specimens cut from the final products reached finally the 95% of the theoretical strength fact which proves a very efficient process development

The products made with fiber volume fraction around 50% revealed very good quality and strength. Additionally optical inspection on cut sections was made and no considerable porosity, was appeared. For this type of defects it was decided that the inspection by ultrasonic C-scan was not the appropriate method

9. ACKNOWLEDGEMENTS

The author would like to acknowledge the contribution of Ms. A.Pasparaki and Mr. Ch. Vrettos to the work presented in this paper.

10. REFERENCES

1. Kevin Potter *Resin Transfer Moulding* Chapman & Hall Edition london 1997
2. Douglas S. Cairns, Dell R. Humbert, John F. Mandell, *Modeling of resin transfer molding of composite materials with oriented unidirectional plies.* Composites Applied science & manufacturing 1997
3. E.F.Bruhn *Analysis & Design of flight vehicle structures*
4. MIL-HBK-17
5. Fisher, K., *RTM and core materials offer product advances. in: High Performance Composites,* Sept/Oct. 1995.
6. Ni, J., Li, S., Sun, X., Lee, J.L., *Mould filling analysis in vacuum-assisted resin transfer moulding.* Part II: SCRIMP Based on Grooves. in: Polymer composites, Vol 19, No 6, 1998.
7. Han, K., Ni,J., Toth, J., Lee, J.L., *Analysis of an Injection/Compression Liquid Composite Moulding Process. in: Polymer composites,* Vol. 19, No. 4, 1998.
8. D. Abraham, S. Matthews and R. Mcllhagger, *A comparison of physical properties of glass fibre epoxy composites produced by wet lay-up with autoclave consolidation and esin transfer moulding.1998* elsevier Composites part 795-801
9. N. R. L. Pearce, F. J. Guild and J. Summerscales, *An investigation into the effects of fabric architecture on the processing and properties of fibre reinforced composites produced by resin transfer moulding.* 1998 elsevier Composites part 19-27
10. Young Goo Kim, Dai Gil Lee and Park Kyoun Oh, *Manufacturing of the composite screw rotors by resin transfer molding*

HIGH FRICTION, LOW WEAR COMPOSITES BASED ON FIBRE REINFORCED CERAMICS

R. KOCHENDÖRFER
DLR-Institute of Structures and Design,
German Aerospace Center,
Pfaffenwaldring 38-40
70569 Stuttgart, Germany

Abstract. Carbon fibre reinforced ceramic matrix composites manufactured via the liquid silicon infiltration (LSI) route have a high potential for tribological applications, such as brake disks and pads. Manufacturing costs can be reduced by using simplified manufacturing techniques, such as hot pressing and short fibres without deleterious effects on tribological behaviour. Graphite modifications and novel ceramic coating techniques developed by DLR have been proven to increase the thermal conductivity of these C/C-SiC materials, so that a considerable improvement in wear and friction performance under high energy conditions can be achieved. These developments mean that a series manufactured brake disk which lasts the lifetime of a vehicle is feasible in the near future.

1. INTRODUCTION

New C/C-SiC materials, developed for the thermal protection of spacecraft at the DLR, Stuttgart, have demonstrated their high potential for the application in brake disks [1]. The aim of this work was to investigate the tribological behaviour of the C/C-SiC composite material and its individual main components, carbon fibre reinforced carbon (C/C) and silicon carbide (SiC), and the influence of the materials thermal conductivity, modified by graphite components. This led to significantly improved, new C/C-SiC friction materials with high wear resistance and a high coefficient of friction (COF) even under high energy test conditions.

Extreme light and damage tolerant ceramic matrix composites (CMC) represent a class of new material for the use not only as a braking material in transportation technology, but also for high performance brake systems in fields of mechanical engineering. Various designs and prototypes in original size were developed and tested under real conditions. A new ceramic coating has been developed which shows extraordinary low wear rates and high thermal stability. The light, quasi wear-free ceramic brake disks promise technical benefits for special brake applications in ultralight and high performance cars.

In comparison to grey cast iron, which represents the state of the art material for brakes, produced in million fold numbers of disks, C/C-SiC materials are still expensive due to the lack of any serial production and due to the higher material

E.E. Gdoutos and Z.P. Marioli-Riga (eds), Recent Advances in Composite Materials, 313-323

costs. However, as the density of C/C-SiC lies in the range of 2 g/cm³, brakes equipped with these lightweight materials lead to drastic savings in the unsprung mass of chassis components. Additionally, C/C-SiC composites show a much higher thermal stability, a negligible thermal distortion, a high thermal shock resistance and they can be manufactured via a liquid phase process which allows a high degree of freedom in the design and the shape of new lightweight disks.

The currently used tribo-system in automotive vehicles, consisting of grey cast iron disks and organic pads is restricted to a maximum operating temperature of 600-800 °C. With increasing temperature, grey cast iron looses its strength and thermal shock resistance rapidly (Fig. 1). In comparison, C/C-SiC composites are thermally stable up to 1400 °C and retain their mechanical properties. Real tests with two-dimensional reinforced C/C-SiC brake disks and pads pressed against each other showed surface temperatures which considerably exceed 1000 °C, demonstrating the high thermal stability of the ceramic material and its principal feasibility for brake disks.

However, high temperatures on the friction surfaces generally lead to a non-stable coefficient of friction, i.e. the COF decreases during braking and an over-heating of the brake's periphery may occur due to high heat radiation. Also, only pads made of C/C-SiC are able to withstand the high temperatures.

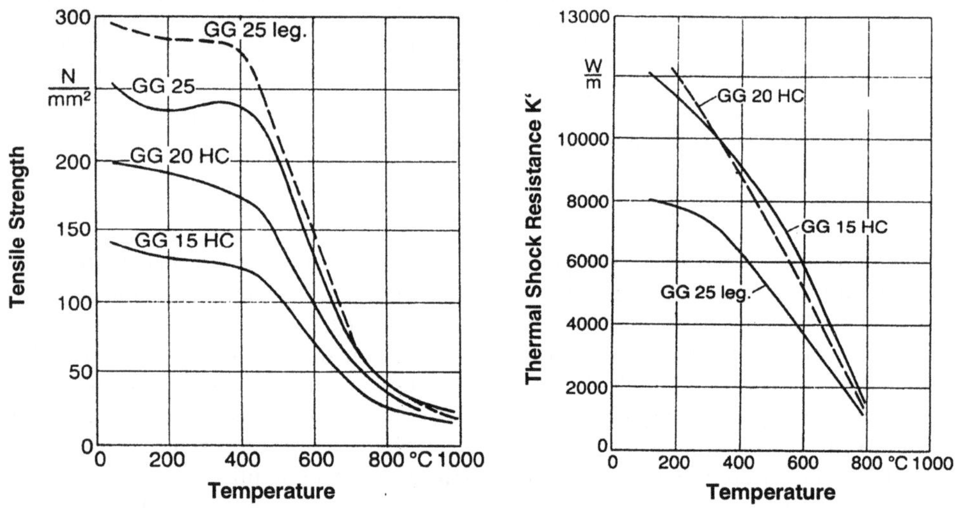

Figure 1. Thermal stability of different alloys of grey cast iron [2]

2. C/C-SIC MANUFACTURING ROUTE

The carbon fibre reinforced silicon carbide composites (C/C-SiC) are fabricated by the Liquid Silicon Infiltration process (LSI) developed by DLR. This three-step process is based on the carbonization of conventional CFRP green bodies at temperatures of about 900 °C in nitrogen atmosphere and the subsequent infiltration of the arising porous Carbon/Carbon with molten silicon under vacuum at 1650 °C. At the contact surfaces silicon reacts with the carbon to silicon carbide and forms the ceramic matrix. All process steps are carried out only once to shorten the manufacturing times. The material can be modified with the design of the CFRP green bodies, the process parameters and additives, so that different qualities of C/C-SiC can be achieved. Due to the microstructure and the phase composition, the mechanical and thermophysical properties can be tailored. Compared to traditional C/C, C/C-SiC is a dense material with an open porosity of about 3 % [3].

3. TESTING SPACE MATERIAL QUALITY IN FRICTION AND WEAR

The standard C/C-SiC material DLR uses for space applications is based on orthotropic woven fabrics of high tenacity (HT) carbon fibres and on a high yield carbonaceous precursor (XP 60). The C/C-SiC material consists of about three quarters of carbon, respectively carbon reinforced carbon (C/C), and about one quarter of silicon carbide (SiC) by weight. The content of metallic silicon (Si) was about 4 % by weight (Fig. 2).

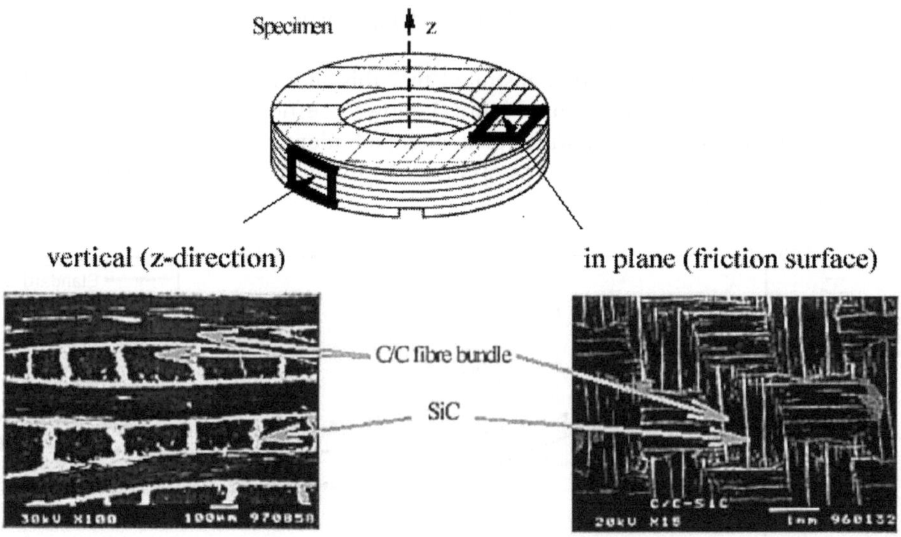

Figure 2. Microstructure of standard C/C-SiC, reinforced with laminated, orthotropic fabrics of carbon fibres

During reentry of space vehicles, as much of the evolved heat as possible should be re-radiated into space again. For this purpose, a high coefficient of emission as well as a high allowable surface temperature is favourable because of the T^4-rule.

Standard C/C-SiC is a suitable space material due to the low values of thermal conductivity perpendicular to the surface and high in-plane values to equalize temperature distribution (Table 1).

Table 1. Thermomechanical properties of basic C/C-SiC material and monolithic components [4]

Properties	C/C-SiC	SiC(1)	Graphite(2)
Thermal conductivity [W/mK]			
λ_{II} (200 °C / 1200 °C)	18.5 / 17	120 / 20	100 / 43
$\lambda_{\perp} = \lambda_z$ (200 °C / 1200 °C)	9 / 7.5		
CTE [10^{-6} K^{-1}]			
α_{II} (100 °C / 1500 °C)	-1 / 2.5	6	4.3
$\alpha_{\perp}$ (100 °C / 1500 °C)	2.5 / 6.5		

(1) β-SiC (CVD-coating) (2) V 1633; Ringsdorff, Bonn

However, brake tests demonstrated that this thermal behaviour is not ideal for friction and wear applications. Figure 3 shows coefficient of friction (COF) curves for different material qualities (disk and pad identical material if not indicated) during an high energy (150 kJ) stop brake from an average circumferential velocity of 15 ms^{-1} down to zero.

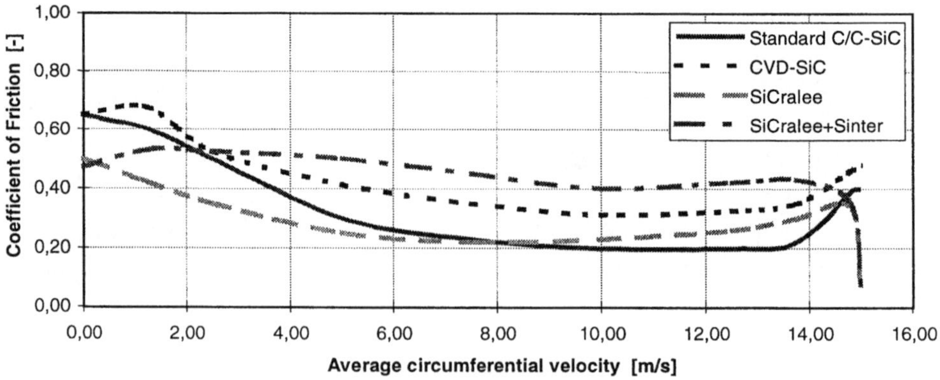

Figure 3. COF behaviour at a braking energy of 150 kJ for different braking materials [5]

Optimum is a constant and high COF value independent from the braking velocity. Obviously, standard C/C-SiC does not match this requirement, mainly because of the too low, in depth conductivity inducing so called "hot spots". If this local overheating occurs, the friction coefficient is reduced. This effect can be slightly receeded by coating the standard C/C-SiC material with a CVD SiC layer. Silicon carbide has a high thermal conductivity and its wear rate is much lower compared to standard C/C-SiC (Fig. 4), however, the CVD layers are thin – to an order of 150 μm – too thin for friction applications in brakes.

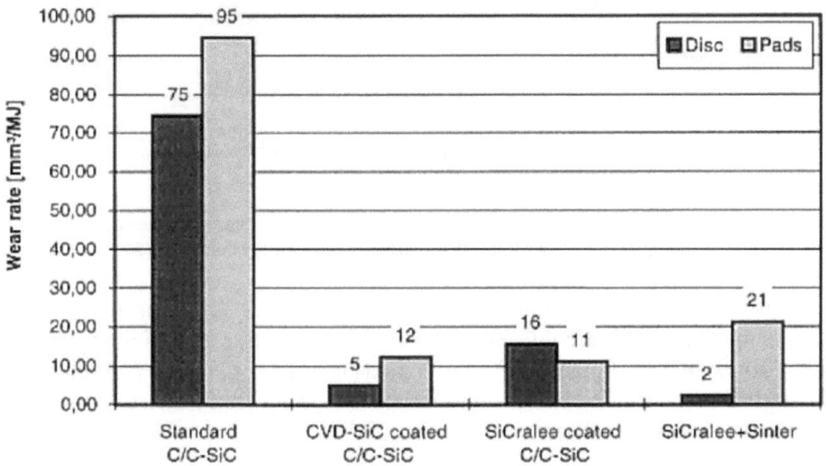

Figure 4. Wear rates of different braking materials at high energy conditions (150 kJ) [5]

4. MEASURES TO INCREASE IN-DEPTH THERMAL CONDUCTIVITY

The result of efforts to increase the thermal conductivity perpendicular to the friction surface is summarized in Fig. 5. Standard C/C-SiC has the lowest, 100 % graphite the highest COF. *C/C-SiC + 45 % graphite* represents a standard material, where 45 % of the friction surface was replaced by graphite bolt inserts.

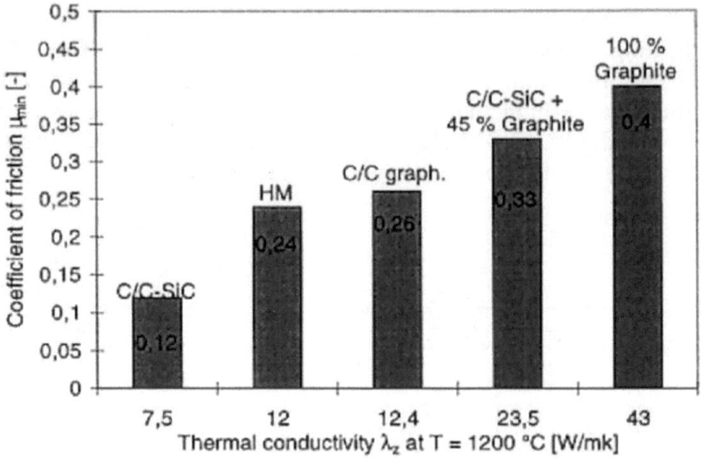

Figure 5. Thermal conductivity λ_z at 1200 °C and minimum COF of tested specimens under high braking energy (v_o = 16 ms^{-1}; p = 0.35 MPa;a = 0.36 ms^{-2} = constant) [5]

C/C graph represents a standard C/C-SiC which was graphitized at 2700 °C in the C/C stage prior to siliconisation and *HM* represents a material where the high tenacity fibres of the standard C/C-SiC material were replaced by graphitized high modular fibres. Fig. 5 clearly demonstrates that a high thermal conductivity correlates with a high coefficient of friction.

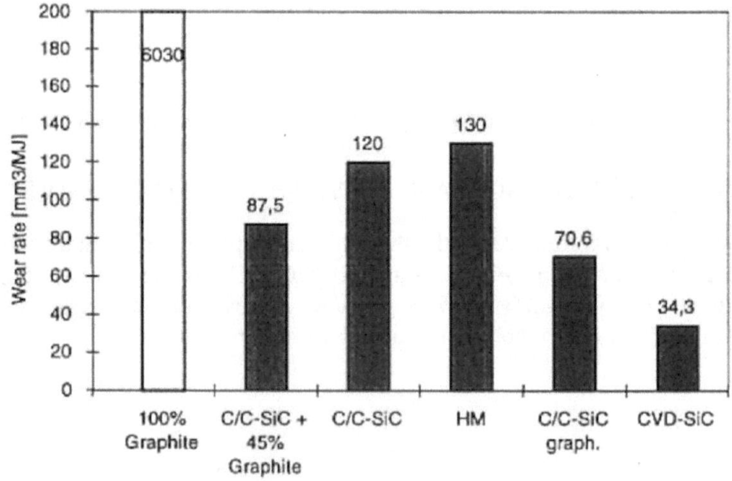

Figure 6. Wear rates of tested specimens under high braking energy (v_o = 16 ms^{-1}; p = 0.35 MPa;a = 0.36 ms^{-2} = constant)

The message of the corresponding wear rate values (Fig. 6) is not as clear. However, it seems that a combination of a wear resistant surface on top of a core material with a high thermal conductivity would be a favourable approach.

5. EXTERNAL COATING OF THE FRICTION SURFACES

Various technologies are available to coat fibre reinforced ceramics with ceramic layers e.g. plasma spraying or chemical vapour deposition (CVD). The use on a friction surface requires an intimate and stable join between coating and substrate which has to work at room temperature as well as at incandescent conditions. As all existing coating techniques need an additional manufacturing step, resulting in higher component costs and each are limited in terms of thickness, material composition etc., DLR developed its own method for coating the brake disk's surface.

During siliconization, the last processing step of the C/C-SiC composite manufacture, additional silicon as well as carbon are added to the surfaces. By varying the type of carbon material, the amount of shares of the two components and the processing conditions, a layer containing only silicon and silicon carbide, permanently and strongly fixed with the C/C-SiC substrate has been achieved by the chemical reaction of carbon and silicon to SiC. The thickness of this SiSiC-coating can be adapted to the requirements of the brake and amounts usually between 0,2 to 2 mm. To improve the surface finish and in order to achieve the brake disk's end contour, the ceramic coating is ground with diamond tools in a concluding step.

The coefficient of thermal expansion (CTE) for the ceramic coating is in the range of 3 to $4*10^{-6}$ 1/K while the fibre ceramic substrates show normally lower values. Depending on this CTE mismatch, a more or less microcracked surface occurs as a result of the higher contraction of the SiSiC coating during cooling after processing. The most pronounced microcrack pattern in the coating can be observed for two-dimensional reinforced substrates whereas isotropic reinforcements lead to nearly crack-free surfaces. Although microcracked, these so-called SiCralee-coatings show an extremely good adhesion on the substrate's surface even under thermal shock conditions.

The SiCralee curve in Fig. 3 is representative for SiCralee coated brake disks and pads with a coating thickness of 1 mm. The COF is more constant than for standard C/C-SiC uncoated or CVD-SiC coated but at the same low value as standard C/C-SiC. Replacing SiCralee pads by sinter metallic pads increases the COF and maintains it at a near constant level above 0.4.

This combination shows the lowest wear rate for the disk (Fig. 4), e.g., a brake disk which need never be changed during the lifetime of an automobile seems to be feasible.

6. MEASURES TO REDUCE MANUFACTURING COSTS

The CMC materials shown up until now in this paper are based on orthotropically layered carbon fabrics as in-plane reinforcement. Especially for automotive brake disks, the costs of the carbon fabrics and of common processing techniques for the CFRP preform, like resin transfer moulding or autoclave technique, are too high for a serial production. The most promising way to reduce the costs for the brake disks and to simplify the CFRP manufacture is using a short fibre reinforcement and applying hot pressing techniques.

Compared to fabric reinforced C/C-SiC, the main advantage of short fibre reinforced CMC materials are the lower costs of the fibre material (chopped fibre: 20 to 40 €/kg; fabric: 40 to 90 €/kg) combined with a minimum of waste during the complete manufacture of C/C-SiC brake disks (chopped fibre: about 5 %; fabric: 20 to 40 %).

Due to the nearly isotropic fibre orientation of C/C-SiC SF (short fibre), the thermal conductivity transverse to the friction surface of the brake disks is generally higher compared to the orthotropic material based on laminated fabrics (Fig. 7). This leads to lower surface temperatures on the brake disks, resulting in higher and more constant coefficients of friction and lower wear rates.

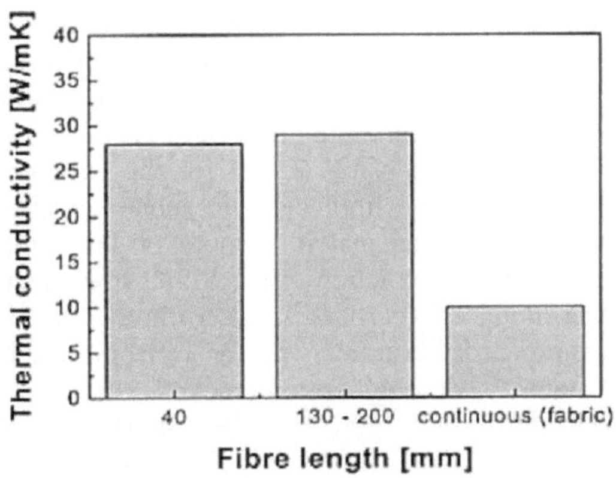

Figure 7. Influence of the fibre length on the transverse thermal conductivity of different C/C-SiC brake disks.

The actual manufacturing process of C/C-SiC SF consists of three main steps (Fig. 8). In the first step, two half shells are made via hot pressing as CFRP green bodies. Commercially available high tenacity carbon fibres are chopped to lengths between 10 and 200 mm and mixed together with carbon rich precursors on the basis of phenolic resins to homogenous compounds, which are filled in the press mould [6]. For densifying and curing a pressure between 5 and 10 MPa and a maximum temperature of 250 °C is applied. Actually, a relatively long process time of 2.5 hours is typical for a single part manufacture in laboratory scale, including the heating up of the hydraulic press, the mould and the compound. Significantly shorter pressing times are possible by preheating the filled mould as well as by using more reactive precursors. The fibre volume content of the CFRP preform ranges from 30 to 60 % in dependence on the required material properties. In the second step, the CFRP green body is pyrolysed at a maximum temperature above 900 °C in inert gas atmosphere. Due to the nearly isotropic fibre reinforcement of the CFRP green body, the risk of delamination during pyrolysis is much lower compared to materials based on laminated fabrics. Therefore, the total process time for pyrolysis could be significantly reduced from 160 to 90 hours. Both C/C half shells are then joined together by a carbonaceous paste and subsequently infiltrated with molten silicon at $T_{max} > 1420$ °C under vacuum within a total process time of 50 hours, including the necessary heating-up and cooling-down cycles of the furnaces.

The reaction of silicon with the carbon matrix forms the silicon carbide matrix of the C/C-SiC composite and leads to a very stable joint of the two parts. The friction surface and the load introduction areas are finally ground with diamond tools. This near net shape manufacture has to prove in the near future whether C/C-SiC brake disks can fit not only the technical and tribological but also the economical requirements of a series product.

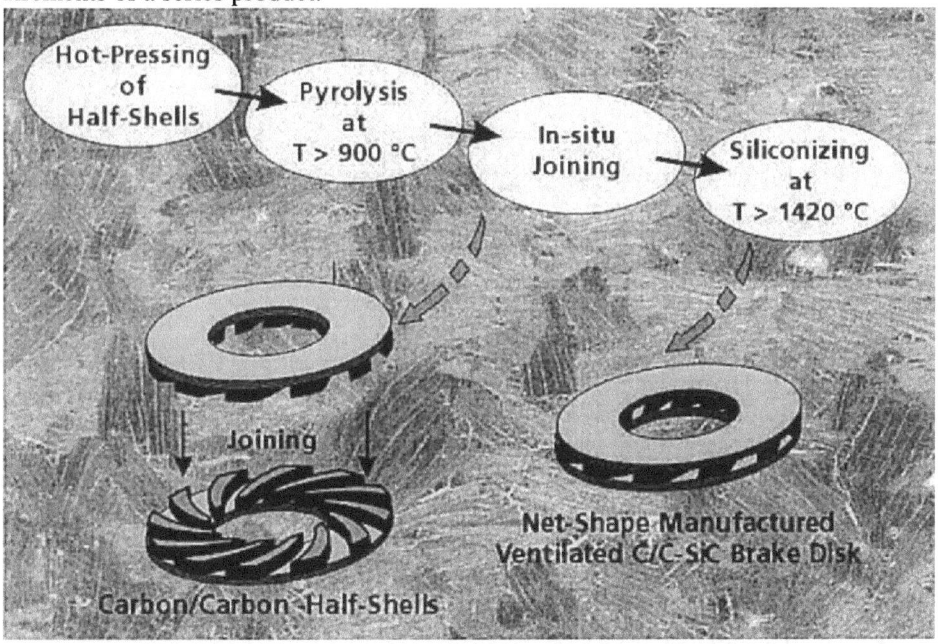

Figure 8: Patented net shape manufacture of ventilated, short fibre reinforced C/C-SiC brake disks by using hot pressing techniques and in-situ joining methods [7].

A drawback of short fibre reinforced CMC materials is their lower mechanical properties in combination with a bigger scatter compared to continuously reinforced materials.

Figure 9 indicates that a mix of fibre length between 130 and 200 mm within the material increases flexural strength above 130 MPa (standard deviation 20 MPa), thus offering a low cost brake disk material with sufficiently high mechanical and thermal properties. Due to the fact that this material quality shows a coefficient of thermal elongation which is almost identical to that of the SiCralee coating, it is well suited to be combined with this wear resistant surface.

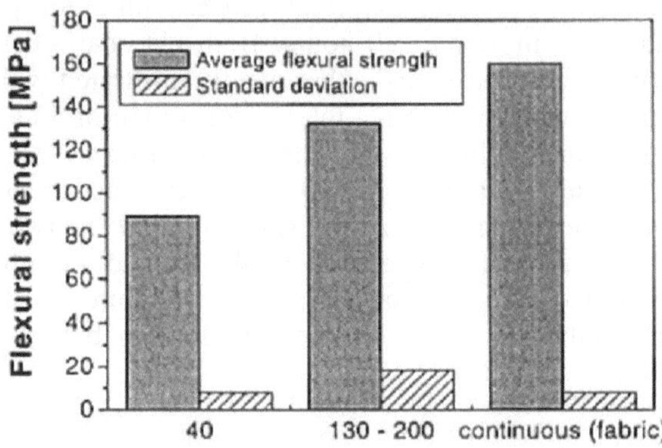

Figure 9: Influence of the fibre length on the flexural strength (four point bending) of different C/C-SiC brake disks.

7. CONCLUSIONS

The main drawback of C/C-SiC brake disks lies in their current costs which are at least one order of magnitude higher than for grey cast iron disks. In order to reach the optimistic goals of the car manufacturers to introduce CMC brake disks in the near future, cost efficient manufacturing routes have to be developed. DLR has been investigating different designs for solid as well as internally ventilated disks taking benefit of the main advantages of the LSI process:

- Applying press techniques for the manufacture of the CFRP body
- Pre-shaping in the green or carbon material's stage
- In situ joining of elements in the carbon/carbon stage by reaction bonding
- Implementation of SiCralee-coatings into the siliconization step

8. REFERENCES

1. Krenkel, W., CMC Materials for High Performance Brakes, ISATA Conference on Supercars, Aachen, Germany, 1994

2. Burckhardt, M., Bremsdynamik und PKW-Bremsanlagen, Fahrwerkstechnik, Editor Reimpell, J., Vogel-Verlag, Würzburg, Germany, 1991
3. Renz, R., Krenkel, W., C/C-SiC Composites for High Performance Emergency brake systems, ECCM-9, 9th European Conference on Composite Materials, Design and Applications, Brighton, England, June 4-7, 2000
4. Heidenreich, B., Krenkel, W., Development of C/C-SiC Materials for Friction Applications, EUROMAT 99, Munich, Germany, September 27-30, 1999
5. Krenkel, W., Renz, R., Henke, T., Ultralight and Wear Resistant Ceramic Brakes, EUROMAT 99, Munich, Germany, September 27-30, 1999
6. Heidenreich, B., Renz, R., Krenkel, W.: Short Fibre Reinforced CMC Materials for High Performance Brakes, 4th International Conference on High Temperature Ceramic Matrix Composites (HT-CMC 4), Munich, Germany, October 1-3 2001, DGM/DLR, High Temperature Ceramic Matrix Composites, S. 809-815, Wiley-VCH, (2001)
7. Krenkel, W., Renz, R., CMC Brake Disks: Design, Manufacture and Quality Assurance Aspects, Materials Week 2000, International Congress on Advanced Materials, their Processes and Applications, Munich, Germany, September 25-28, 2000

DESIGN AND EXPERIMENTAL ANALYSIS OF MODULAR PEDESTRIAN BRIDGE WITH CONCRETE DECK AND IPC TRUSS GIRDER

G. GIANNOPOULOS[a*], J. VANTOMME[a],
J. WASTIELS[b] and L. TAERWE[c]
([a])Royal Military Academy, Civil Engineering Department, Avenue de la Renaissance 30, B-1000 Brussels, Belgium
([b])Vrije Universiteit Brussel, Department of Mechanics of Materials and Constructions, Pleinlaan 2, B-1050 Brussels, Belgium
([c]) University of Ghent, Department of Structural Engineering, Magnel Laboratory for Concrete Research, Technologiepark-Zwijnaarde 9, B-9052 Ghent, Belgium

Abstract. In the present work the design and experimental analysis of a pedestrian bridge is presented. The actual bridge consists of a supporting truss system built with IPC sandwich panels and a horizontal concrete deck. The IPC is a kind of ceramic which is reinforced with glass fibers. The connection between the sandwich panels the truss system and the concrete deck is done using specially designed steel elements. The aim of this work is the examination of the behavior of the prototype that is constructed and evaluated via static and dynamic tests. The results of the experimental analysis are necessary for the final evaluation of the structure before the construction of bigger bridges and in order to have an overview of the critical areas of the structure that have to be reconsidered during the design phase of a different structure.

1. INTRODUCTION

As it is well known the owner of a steel bridge has to deal with the steady increase of the cost for preserving its structural integrity. The principal cause for the cost increase is the corrosion that takes place due to environmental conditions. Furthermore the replacement of certain parts of such bridge after 20 or more years contributes to the increase of the maintenance [1]. Recently the application of FRP like the ones used in aerospace industry has been extended to this field also. FRP seem to be very attractive for civil engineering structures, due to the high stiffness to weight ratio and due to their corrosion resistance. The application of such materials can lead to a reduced amount of in situ work required for the setup of such structures [2]. However, organic matrices, such as polyesters or epoxies are relatively expensive, inflammable and not resistant to high temperatures. These disadvantages limit their use in construction applications. Due to the above mentioned disadvantages, there is demand for a composite material with different type of

E.E. Gdoutos and Z.P. Marioli-Riga (eds), Recent Advances in Composite Materials, 325-335

matrix. After more than ten years of research in the Vrije Universiteit Brussel, a new non alkaline mineral polymer, a kind of structural ceramic, named IPC (inorganic posphate cement) was developed. This ceramic can be reinforced by glass fibers in order to increase the toughness, the tensile strength and the stiffness of the resulting composite material. This new material combines, in addition to the well know features of fiber reinforced plastics, many advantages: low
manufacturing cost based on reactive mineral powders, a non-inflammable behavior, an environmental friendly composition and a chemical resistant behavior. The present research project wants to investigate whether the combination of glass fiber reinforced IPC with concrete may be a feasible way of producing cost effective, corrosion and fire resistant bridges. The glass fiber reinforced IPC material is combined with traditional materials such as concrete, resulting in a new bridge concept.

In this paper the work on a 13 meter pedestrian bridge is presented. This bridge is composed from a concrete deck and three truss girders, composed from IPC sandwich panels. The faces of the sandwich panels are reinforced by mat and unidirectional glass fibers. For the connection of the panels, special steel connections have been designed. With this design, it is possible to develop a rapidly assembled system.

2. MATERIALS

1.1 IPC

Application of concrete materials in structures that may experience tensile load is usually avoided. This is due to the lack of strength of such materials in tensile loading, which leads to failure for very small loading values. A possible solution to this problem could be the use of glass fibers. However, ordinary concrete material attacks glass fibers due to its alkaline nature. The Inorganic Phosphate Cement developed at the Vrije Universiteit Brussel is acidic (pH=1) in fresh state, but neutral (pH=7) after hardening. Because of the non-alkaline conditions, during and after the fabrication process, glass fibers can be used as reinforcement. The IPC is a two-component system, with a calsiumcilicate mineral in powder form, and a phosphoric acid based aqueous solution of inorganic metal oxides. The viscosity of the mixture allows to obtain typically 10 to 15% volume fraction of short or continuous fibers in mat or woven shape, using the hand lay-up or spray-up fabrication [3].

The tensile behavior of the composite IPC-glass fibers is non-linear since the tensile failure strain of the matrix is very limited, much smaller than the tensile failure strain of the glass [4] [5]. The behavior can be approximated as bilinear, with an initial stiffness before matrix cracking and a reduced stiffness after matrix cracking.

The IPC-glass laminates are manufactured by hand lay-up with either quasi-unidirectional (89%, 11%) woven glass fiber or mat glass fibers having a surface density of 300 g/m^2 (IPC/mat). The fiber volume fraction is 15% for IPC reinforced by quasi-unidirectional glass fibers and 11% for IPC reinforced by mat. The density of the laminate is taken 2000 kg/m^3.

After numerous tests that have been performed with different materials, it was decided that the core material of the sandwich panels should be polyurethane. The density for this material is about 20 kg/m^3.

The skins of the sandwich panels are manufactured by hand lay-up on the polyurethane foam, without using any adhesive material. The sandwich panels are room temperature cured for 24 hours, and post cured at 60 Celsius for another 24 hours.

3. DESCRIPTION OF THE STRUCTURE

For the realization of the structure, a truss from IPC sandwich panels was designed [6], [7]. In figure 1 the actual bridge is presented.

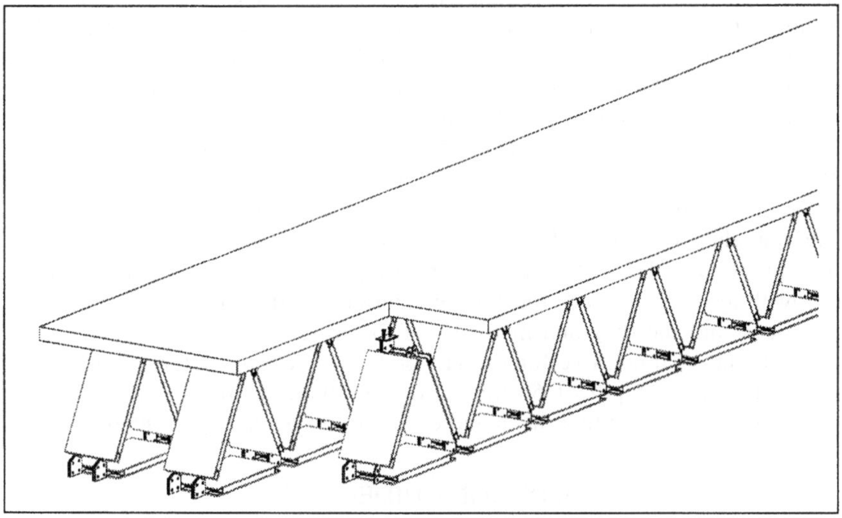

Figure 1. Representation of the actual bridge.

However, the full IPC truss solution is not feasible for many reasons. The buckling of the upper chord panels, the low bending stiffness of the bridge structure and the vibration problems due to the low weight are the reasons that restrict the application of a full IPC truss system. In view of these problems for the top chord of the bridge, a reinforced concrete slab is chosen [8]. This material is able to take up the compressive stresses and to reduce the vibration problems due to the increase of the own weight of the structure. In addition the bending stiffness of the whole structure is increased and the stresses due to concentrated loads can be distributed all over the structure [9, 10].

The structure is composed of 3 truss girders. Each girder has a width of 1 meter and a length of 13.344 m. The concrete deck is put on the three girders; its thickness is 120 mm. The truss system of each girder consists of horizontal and diagonal sandwich panels, each with the same dimensions in order to reduce the amount of different panel configurations. As a consequence a faster assembly procedure can be achieved with evident results on the logistics of such a structure. In order to minimize the amount of panels, 1.05 m long and 0.5 m wide panels are used. However the panels that are used are not identical as to their composition. The faces of the diagonal panels are constructed from 10 layers of IPC reinforced with mat glass fibers, leading to a 10 mm thickness, while for the bottom chord panels 29 layers of IPC reinforced with unidirectional glass fibers are implemented, leading to a face thickness of 12mm. The difference between the diagonal and the horizontal panels is explained by the presence of tensile forces in the bottom chord of the bridge. The core thickness is the same for all panels: 50 mm; this dimension was chosen in order to increase the critical buckling load of the panels, due to the compressive forces that are applied to some of the diagonal panels.

The connection system of the panels is made of steel. This connection system is realized using steel inserts for the sandwich panels, pins and steel connection plates. The joining of the bridge in-situ is thus reduced to the assembly of the panels. The steel insert consists of a 50 x 80 x 5 mm U profile which is placed between the skins of the panels. For the anchoring of the inserts in the sandwich panels, 10 mm diameter holes are realized and steel bolts of 8.8 quality with an interspace of 42 mm are used. For the connection of the panels, steel plates made of S355 steel are used; their thickness is 8 mm. This steel has a yield stress of 355 MPa and a failure stress of 510 MPa. For the connection between the panels, pins of 20 mm diameter are used having also a steel quality of 8.8. In order to assure the feasibility of the assembly procedure, the holes in the connection plates are 1 mm larger in diameter.

A dowel system is used for the realization of the connection between the concrete deck and the diagonal members of the truss. The dowels are subjected to shear, tensile and compressive forces. The dowels have a diameter of 16 mm and an effective embedment depth of 60 mm. 2 dowels with an interspace of 90 mm are used per plate [11]. In figure 2 the actual connection system is presented.

4. DESIGN GUIDELINES

The design of the IPC-concrete bridge is based on the requirements for pedestrian bridges. The design has been performed according to the requirements of the Belgian standard NBN B 03-101 [12], [13], [14]. The requirements for the design are the ability to carry a distributed load of $5kN/m^2$ and the ability to carry a concentrated load of 10 kN. This is the Serviceability Limit State loading case. In order to overcome uncertainties for the IPC material, a safety factor of 3 is used for the ultimate strength of this material; long term phenomena like creep and fatigue are not well studied yet. In the ultimate limit state, the own weight of the structure is multiplied by 1.35, while the pressure and the concentrated load are multiplied by a factor of 1.5. The admissible deflection for such a structure is between 1/200 and 1/300 of its length. For the evaluation of the overall behavior of the structure a finite element analysis has been performed which is presented in another work [15].

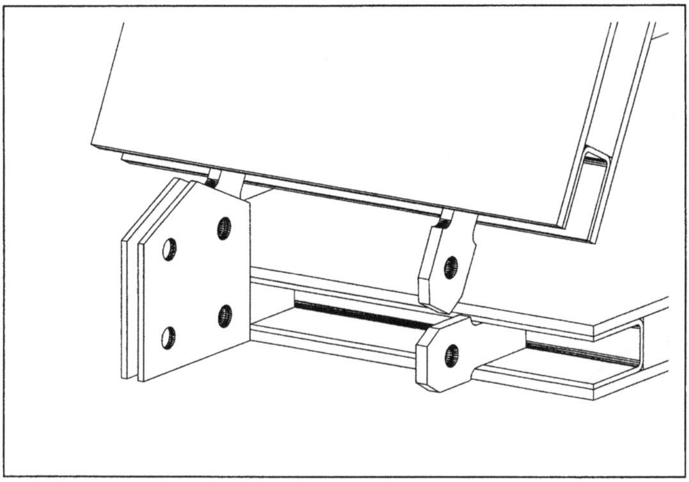

Figure 2: Representation of the connection system.

5. EXPERIMENTAL ANALYSIS OF THE STRUCTURE

1.2 Experimental setup.

Special attention was given to the analysis of the deformation of the bridge. According to design guidelines, the maximum deformation of the bridge should be within 1/200 and 1/300 of the length of the structure. As a consequence a deformation higher than 1/200 of the length is unacceptable, while in the other hand a deformation smaller than 1/300 of the length is of course acceptable.

Due to the fact that in the experimental analysis a distributed load could not be applied, the loading procedure as well as the loads applied had to be modified in order to be compliant with the real conditions. Due to the limitations of the experimental devices the bridge could only be loaded in four points with the same force for all points. The simulation of the guidelines loading is performed on the basis of having the same maximum bending moment in the middle of the structure. In figure 3 the bending moment for both loading conditions is presented.

As it is shown in figure 3 the bending moment in the middle of the structure remains constant for a bigger length in the case of the four point loading, while in the case of the distributed loading the maximum value exists only in one point of the structure.

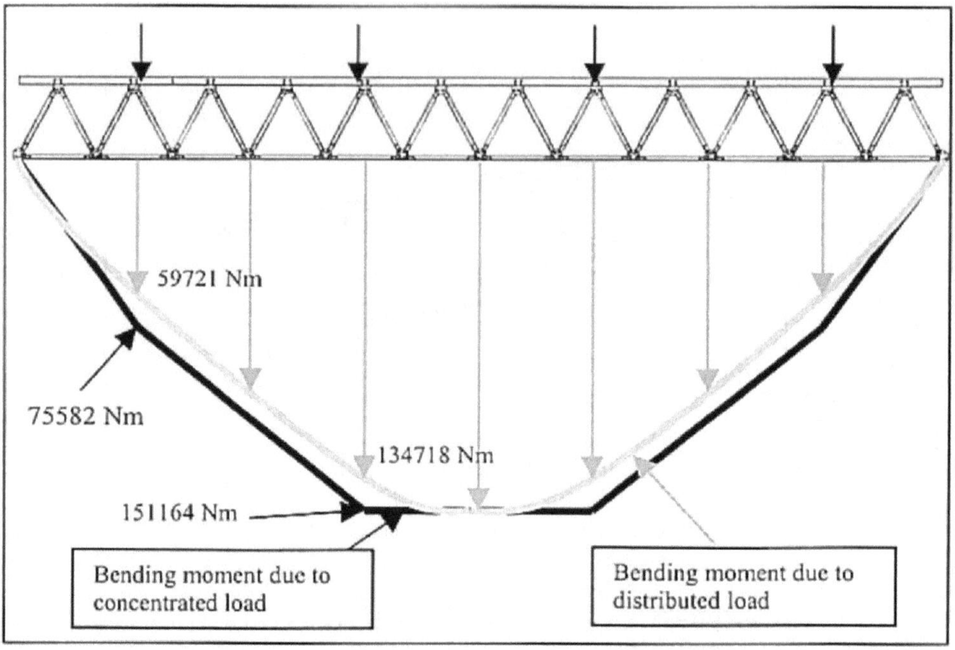

Figure 3: Diagram of the bending moment for the distributed and concentrated loading.

The loading procedure was split in 8 load steps. It is important to mention here that for the loading of the bridge a special structure with beams was applied on the top of the concrete deck, which increased the own weight of the structure. The mass of this structure is 1200 Kg and this weight is distributed on the four points that the load is applied. In table 1 the force at each load step is presented.

In figure 4 the experimental setup for the bridge is presented; it shows a special structure, which was constructed in order to distribute in four points the same force using only 2 load jacks. In addition the beams shown in figure 4 were also used for the support of the LVDTs in order to have a stable base for these devices.

Table 1: Load steps used for the static analysis of the structure (The load corresponds to each one of the four loading points and not to the load applied by each load jack).

LOAD STEP	LOAD (N)
1	2500
2	5000
3	7500
4	10000
5	12500
6	15000
7	17500
8	19000

Figure 4: Experimental setup

One of the most important parameters during the experimental evaluation of the structure was the measurement of the deformation all over the structure. In order to perform this measurement a number of 10 LVDTs were installed on the concrete deck. In addition 2 strain gages were applied to each panel of the truss system in order to investigate the deformations on these panels. A number of 42 strain gages were applied. Two of them were applied on the concrete deck in the middle of its span and 2.28 m from its end. It is important to mention here that due to the symmetry of the structure, only the half bridge was instrumented. In figure 5 the instrumentation of the panels is presented. However, all the four end diagonal panels were instrumented in order to control any possible failure at these areas since these panels were the most loaded diagonal panels. The end diagonal panels are forced in compression while the next two diagonal panels are forced in tension. It is obvious that due to compressive loading the two end diagonal panels are possible to fail in buckling, where the next two panels is possible to fail due to tensile loading.

As it has been already mentioned, the length of the bridge is approximately 13.5 m. This means that for the cabling of the strain gages the length of the cables has to be taken into account. For this reason special cables were used that in order to compensate this problem. It is very important to mention here that apart from the special cables used, the acquisition system is capable of taking into consideration the resistance of the connecting cable and thus it changes automatically its internal resistance in order to have for all channels the same loses and thus the best possible measurements. Finally we should also mention here that the acquisition system used for the tests is capable of measuring simultaneously 60 channels.

In figure 6 the instrumentation of the concrete deck in order to measure its displacement is presented.

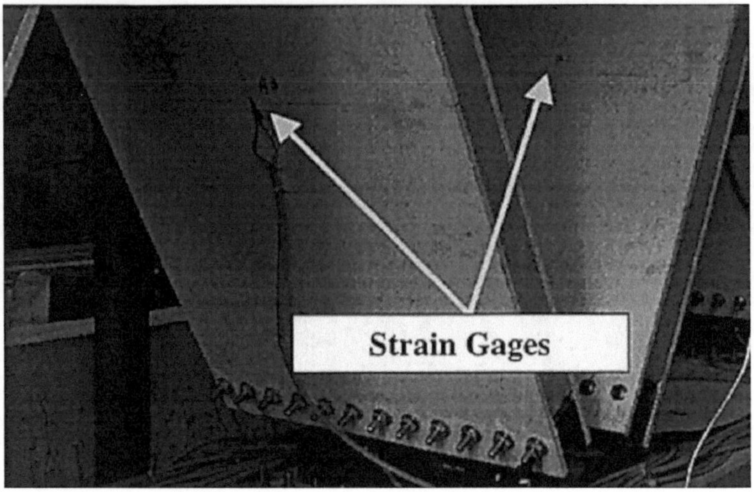

Figure 5: Strain gages on the diagonal panels.

1.3 Static analysis results.

As it has been already presented the loading of the bridge took place in 8 different load steps until the serviceability limit state. Since the design of the structure has been performed on the basis of the deformations and not on the basis of resistance the most important parameter is the overall deformation in function of the load. In figure 7 the force/deformation curve is presented.

In addition to the measurement of the displacement of the bridge, measurements on the panels have been also performed. In the curve in figure 8 the deformation of the most loaded horizontal panel is presented.

6. CONCLUSIONS

IPC sandwich panels can be used for the construction of bridges. The paper deals with a prototype bridge from IPC sandwich panels and a concrete deck. The aim of this work is the design, realization and testing of an actual IPC pedestrian bridge. The tests performed on this bridge showed that the use of IPC can lead to relatively light constructions which can be used in the real world. In addition the deformations measured on different parts of the structure are within the limits of the design guidelines. The deformation to length ratio in the service load is about 1/320 and thus the criterion of the design guidelines is fully accomplished.

However, there are still some issues to be addressed concerning this kind of structures. As it has been shown in this work, there are some permanent deformations on the structure which are due to the non-linear behavior of the IPC and the connection system. As a future work the design of a new connection system has to be performed. Furthermore the concrete deck should be substituted by another material which is lighter and can take bending moments. With the above mentioned alternative solutions the structure can be widely used in a number of applications.

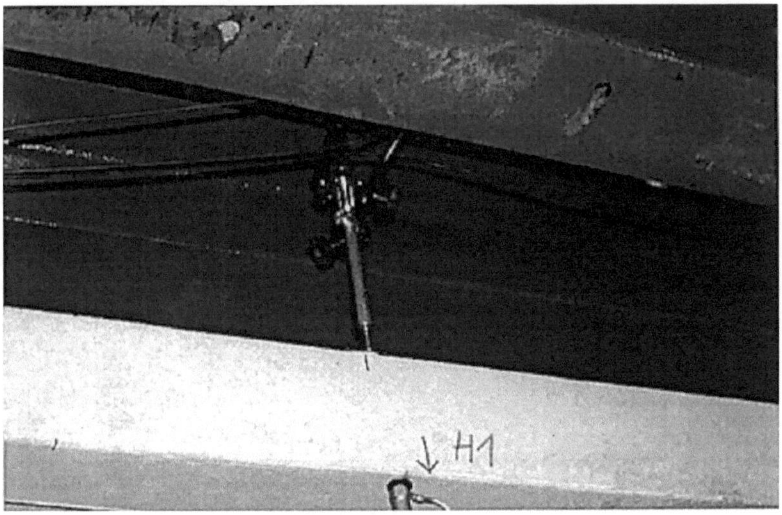

Figure 6: LVDT for the deformation measurement of the concrete deck.

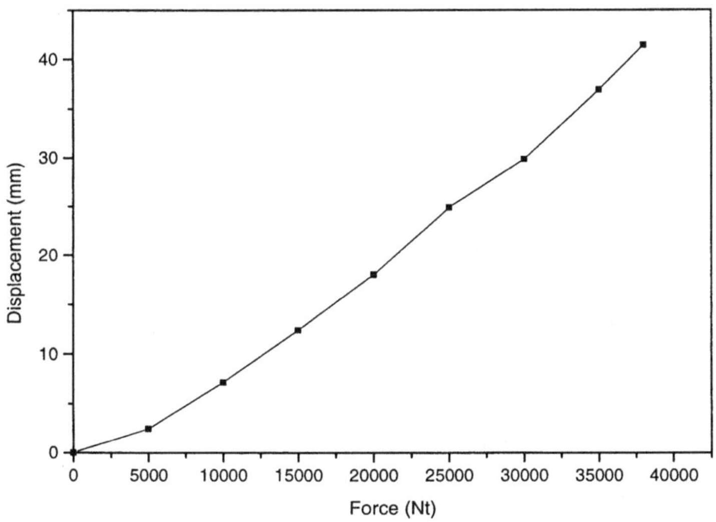

Figure 7: Force displacement curve until serviceability limit state loading.

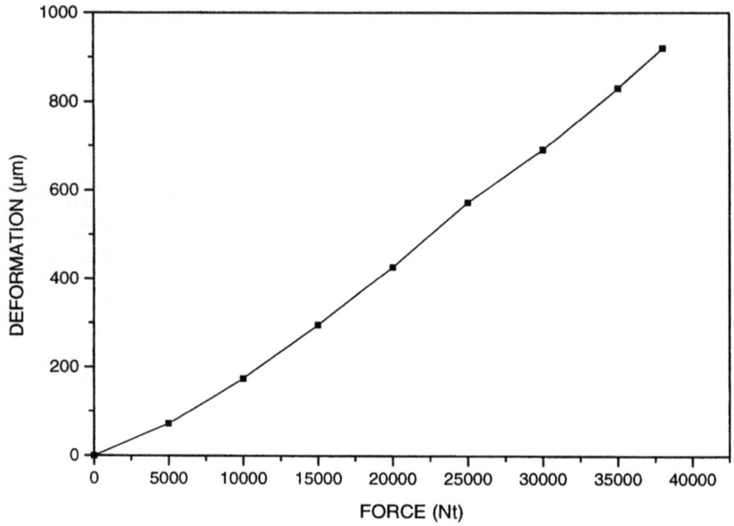

Figure 8: Deformation of the most loaded horizontal panel (middle panel)

7. ACKNOWLEDGEMENTS

Funding by the Flemish Fund for Scientific Research (FWO) under the contact G.0191.98 is greatly acknowledged.

8. REFERENCES

1. Noordzij R, Design and manufacturing of a full composite pedestrian bridge, Second international conference on composites in infrastructure, Ed H Saadatmanesh and MR Ehsami, Tucson, AZ, 1998: 443-455.
2. Scrimping saves money and time, Composites technology, September/October 1999: 21-28.
3. Wastiels J, Sandwich panels in construction with HPFRCC faces: new possibilities and adequate modelling, High Performance fibre Reinforced Cement Composites, Ed HW Reinhardt, AE Naaman, 1999: 143-151.
4. Aveston J, Cooper GA, Kelly A, Single and multiple fracture, The properties of fibre composites, IPC Science and Technology Press, London (UK), 1971: 15-24.
5. J. Gu, Wu X, Cuypers H, Wastiels J, Modeling of the tensile behaviour of an E-glass fibre reinforced phosphate cement, Computer methods in composite materials VI, Ed SV Hoa, WP De Wilde, WR Blain, Computational Mechanics Publications, Southampton (UK), 1998: 589-598.
6. DeRoover C, Vantomme J, Wastiels J, Croes K, H Cuypers, Taerwe L, Blontrock H, Modelling of a modular pedestrian bridge composed of a concrete deck and a truss girder with IPC sandwich panels, Fifth international conference on computational structures technology - CST2000, Leuven (Belgium), 2000: 327-336.
7. DeRoover C, Vantomme J, Wastiels J, Croes K, Taerwe L, Blontrock H, A new non-alkaline cement material reinforced by glass fibres (IPC) for the construction of bridges, Third international conference on advanced composite materials in bridge and structures - ACMBS III, Ottawa (Canada), 2000: 429-436.
8. Triantafillou TC, Meier U, Innovative design of FRP combined with concrete, Advanced composite materials in bridges and structures (ACMBS III), Neale KW, Labossiere P, Milan, 1994: 491-500.

9. Genders C, Eerste composietbrug in Nederland, Cement en beton, 6 June 1997: 16-19
10. Composite cooperation, Bridge design and engineering, May 1997: 76.
11. Blontrock H, Taerwe L, Nurchi A, Vantomme J, De Roover C, Wastiels J, Croes K, Testing of a dowel connection for a bridge with a concrete deck and a sandwich panel truss structure, Connections between concrete and steel, Rilem symposium, Stuttgart, September 2001.
12. Eurocode 1: Basis of design and actions on structures, Institut belge de normalisation (IBN), Brussels.
13. Eurocode 2: Design of concrete structure, Institut belge de normalisation (IBN), Brussels.
14. Eurocode 3: Design of steel structures, Institut belge de normalisation (IBN), Brussels.
15. G. Giannopoulos, J. Vantomme, J. Wastiels, L. Taerwe, 3D Numerical analysis of modular pedestrian bridge with concrete deck and IPC truss girder, ECCM –10, Brugges 3-6 June 2002, Belgium.

STRENGTHENING OF HISTORIC STRUCTURES WITH ADVANCED COMPOSITES: A REVIEW

T.C. TRIANTAFILLOU
Dept. of Civil Engineering
University of Patras
GR-265 00 Patras, Greece

Abstract. The unique advantages offered by FRP materials have made them ideal candidates for increasing the strength, stiffness and ductility characteristics of historic structures, especially in cases where traditional strengthening materials and techniques have proven inadequate and/or inapplicable. In this paper, which serves as an introduction to the papers for the special theme session on "Strengthening of Historic Structures with FRP", the author presents a short overview of FRP applications to historic structures made of masonry, timber, metals (cast iron, steel) and concrete, with emphasis on concepts and key case studies.

1. INTRODUCTION

Historic structures constitute a significant portion of the building stock throughout the world. Many of these structures have suffered from the accumulated effects of inadequate construction techniques and materials, seismic and wind loads, foundation settlements and environmental deterioration, and are structurally deficient or marginal for current use. In addition, changed usage and more stringent seismic design requirements have resulted in many historic structures being designated in need of upgrading through strengthening.

Traditional strengthening materials and techniques have proven inadequate and/or inapplicable in many cases of historic structures in need for upgrading. The unique advantages offered by fibre reinforced polymers (FRP), namely low weight, corrosion resistance, availability in practically unlimited lengths, easy handling etc., have made these materials ideal candidates for increasing the strength, stiffness and ductility characteristics of historic structures.

Most of the applications worldwide of the FRP strengthening technique to historic structures are found in old masonry buildings. However, structures made of other materials too, such as timber and cast iron, or even old concrete, have received upgrading through the use of FRP. Despite the potential of the technique and the success of applications so far, it is believed that the use of FRPs in combination with old constructions, of historic nature, is still to be developed properly, due to the following reasons: the concerns of incompatibility of FRPs with traditional materials; the lack of knowledge and education on this particular technique; the lack

E.E. Gdoutos and Z.P. Marioli-Riga (eds), Recent Advances in Composite Materials, 337-344

of simple, yet reliable analytical models; the lack of standards related to materials and design.

In this paper the author presents a short overview of FRP applications to historic structures made of masonry, timber, metals (cast iron, steel) and concrete, with emphasis on concepts and key case studies.

2. MASONRY

One of the early studies on the use of FRP reinforcement for strengthening historic structures was conducted by Triantafillou & Fardis (1993, 1997). In this pioneering work the authors present detailed concepts, models and analytical results on the applicability and effectiveness of unbonded FRP tendons used to apply circumferential prestressing to the external walls of historic masonry buildings (Figure 1). Their results demonstrate that the transverse confining stresses applied by the FRP tendons increase the strength (under both gravity and horizontal loads) of masonry substantially.

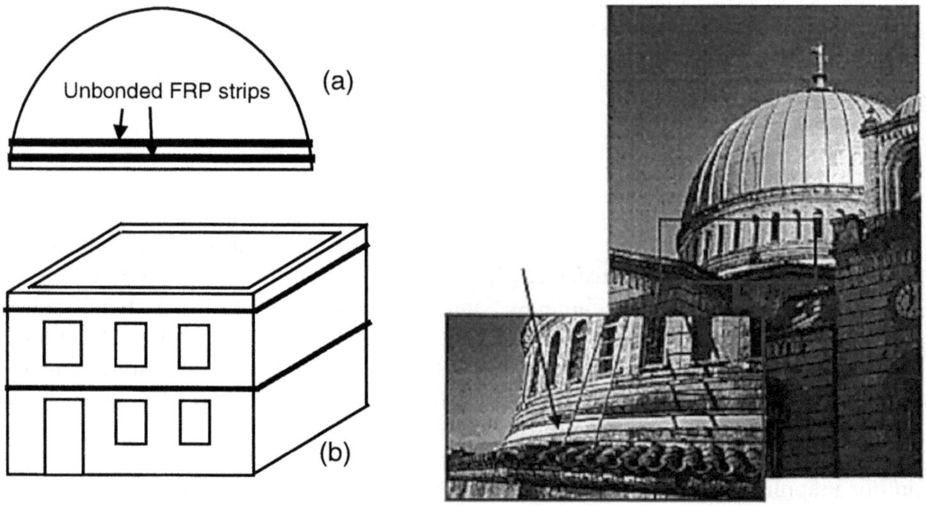

Figure 1. Circumferentially applied FRP tendons (a) at the base of domes, (b) in the form of belts around buildings; (c) application in the Church of Panaghia Faneromeni, Egion, Greece.

The use of FRP in the form of epoxy-bonded strips to strengthen masonry structures under out-of-plane or in-plane loading was first proposed by Schwegler (1994), see Figure 2a,b. He demonstrated the effectiveness of the technique through full-scale testing of one-story masonry walls and developed a model for the analysis of FRP-strengthened masonry based on the stress field theory. In a more recent study, Triantafillou (1998) developed and verified experimentally simple design models for the dimensioning of the epoxy-bonded FRP reinforcement in masonry walls under in-plane bending and shear or out-of-plane bending, all combined with axial load. Key advantage of the FRP strengthening technique in this case is that it may be applied in cases where standard techniques (e.g. reinforced concrete jacketing) are either not optimum (e.g. due to increased dimensions of cross sections, added weight to structures, disruption of occupancy) or not applicable (e.g. in case of building façades with arches and columns that cannot carry the RC jackets down to the foundation levels for architectural reasons, see Figure 2c).

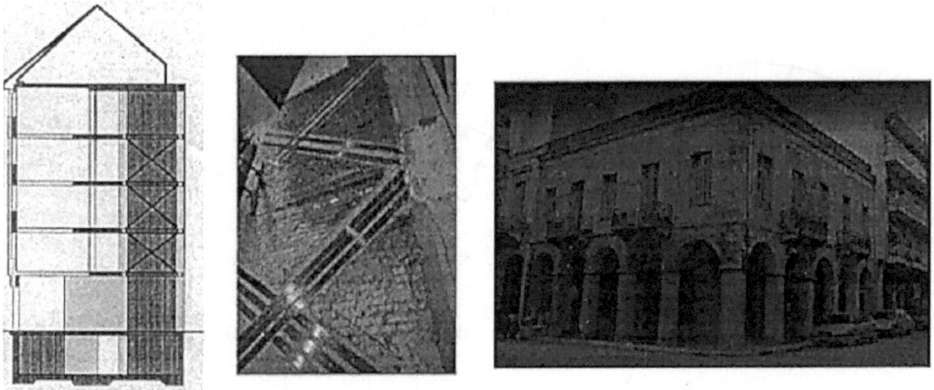

(a) (b) (c)

Figure 2. (a) and (b) Strengthening of masonry walls with epoxy-bonded CFRP strips (Schwegler & Kelterborn 1996). (c) Old masonry building where RC jacketing is inapplicable in the façade.

(a) (b) (c)

Figure 3. (a) and (b) Arrangement of CFRP strips applied to the extrados of vaults. (b) Application of transparent GFRP to the intrados of vaults.

Research has also been conducted on the use of FRP strips or sheets as strengthening materials applied to the extrados and intrados of arches and vaults (barrel or cross type) in historic masonry structures (Figure 3). Test results as well as simple analytical models indicate that this technique provides substantial strength increase of such structures by preventing the formation of hinges (e.g. Blasi et al. 2000, Faccio et al. 2000), see Figure 4, or by modifying the collapse mechanism. One important advantage of FRPs in such applications is their flexibility, which allows the material to follow curved surfaces easily. An interesting variation of the technique involves bonding of transparent sheets in the intrados of vaults, which makes them macroscopically invisible (Figure 3c).

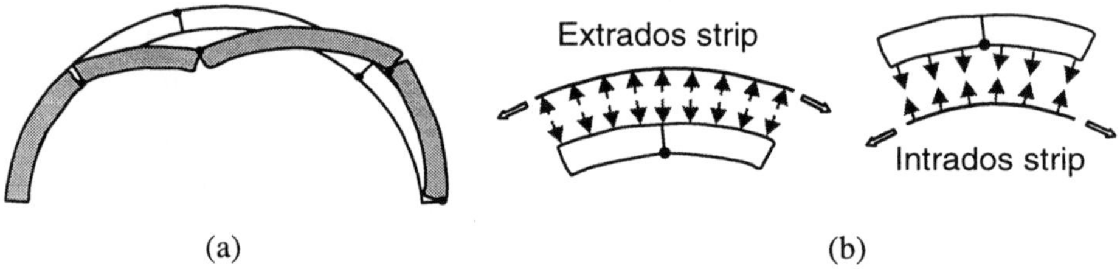

Figure 4. (a) Collapse mechanism of an arch with unsymmetrical load. (b) Interface stresses.

3. TIMBER

Using advanced composites in combination with timber has been very tempting. FRP materials have been used in the form of carbon/epoxy strips or tendons. When bonded on the outer faces of timber elements, the strips increase their strength, stiffness and ductility (Plevris & Triantafillou 1992). Similar results may also be obtained with externally placed tendons.

The first major application of CFRP on historic timber took place in Switzerland in 1992, on the covered wooden bridge in Sins, built in 1807. To sustain higher traffic loads, the cross-beams were strengthened with 1 mm thick strips epoxy-bonded to the upper and lower sides, at a total width ranging from 200 – 300 mm (Meier 1995), Figure 5a.

Another project was performed in 1998 on a similar structure, the 100 m long covered road bridge in Murgenthal-Fulenbach (Switzerland), built originally in 1864. In this truss structure some areas of the lower tension chords were damaged by insects and the load bearing capacity was considered insufficient to carry the loads. The affected bottom chords of the truss were trengthened with 50 mm wide CFRP strips epoxy-bonded to their tension face, Figure 5b (Kropf & Meierhofer 2000).

Wingspread, built in 1938 in Racine, Wisconsin, USA, is another example of CFRP-timber strengthening. This house, the largest ever designed by Frank Lloyd Wright (over 4,000 m^2 floor area), was a brilliant conception, but contained some inherent structural deficiencies. These did not manifest themselves significantly for

STRENGTHENING OF HISTORIC STRUCTURES WITH ADVANCED COMPOSITES: A REVIEW

T.C. TRIANTAFILLOU
Dept. of Civil Engineering
University of Patras
GR-265 00 Patras, Greece

Abstract. The unique advantages offered by FRP materials have made them ideal candidates for increasing the strength, stiffness and ductility characteristics of historic structures, especially in cases where traditional strengthening materials and techniques have proven inadequate and/or inapplicable. In this paper, which serves as an introduction to the papers for the special theme session on "Strengthening of Historic Structures with FRP", the author presents a short overview of FRP applications to historic structures made of masonry, timber, metals (cast iron, steel) and concrete, with emphasis on concepts and key case studies.

1. INTRODUCTION

Historic structures constitute a significant portion of the building stock throughout the world. Many of these structures have suffered from the accumulated effects of inadequate construction techniques and materials, seismic and wind loads, foundation settlements and environmental deterioration, and are structurally deficient or marginal for current use. In addition, changed usage and more stringent seismic design requirements have resulted in many historic structures being designated in need of upgrading through strengthening.

Traditional strengthening materials and techniques have proven inadequate and/or inapplicable in many cases of historic structures in need for upgrading. The unique advantages offered by fibre reinforced polymers (FRP), namely low weight, corrosion resistance, availability in practically unlimited lengths, easy handling etc., have made these materials ideal candidates for increasing the strength, stiffness and ductility characteristics of historic structures.

Most of the applications worldwide of the FRP strengthening technique to historic structures are found in old masonry buildings. However, structures made of other materials too, such as timber and cast iron, or even old concrete, have received upgrading through the use of FRP. Despite the potential of the technique and the success of applications so far, it is believed that the use of FRPs in combination with old constructions, of historic nature, is still to be developed properly, due to the following reasons: the concerns of incompatibility of FRPs with traditional materials; the lack of knowledge and education on this particular technique; the lack

E.E. Gdoutos and Z.P. Marioli-Riga (eds), Recent Advances in Composite Materials, 337-344

4. METALS

The use of either unstressed or stressed composites for strengthening of historic metallic (cast iron, steel) bridges is gradually gaining momentum, due to ease of installation, minimal disruption, minimal change made to the overall appearance of the structures, corrosion resistance and excellent mechanical properties of FRPs. The feasibility of strengthening metallic beam-type elements with FRP has been demonstrated in a number of studies, which have focused not only on mechanical performance aspects, but also on those related to durability as well as accelerated curing of the adhesive through induction heating. Of critical importance in the case of bonding carbon fibres on steel or iron is the protection of the metal from galvanic corrosion, as carbon is an excellent conductor of electricity. This may be achieved by eliminating the direct contact between carbon and metal through the use of an insulating material, e.g. polyester fabric, epoxy resin etc.

An example of FRP application to strengthen cast iron is the Tickford Bridge in Buckinghamshire, built in 1810 (Figure 7a). The bridge was strengthened with 14 layers of carbon fibre fabric, applied to the three largest spandrel rings and the lower main chord in a very short time period. This reinforcement was still no more than 10 mm thick, and after the bridge had been repainted even the sharpest eyes would be hard pushed to detect a change in appearance. A second example is the Hythe Bridge in Oxford, constructed in 1874. This structure was upgraded in flexure with the objective of raising its capacity from 7.5 tones to 40 tones; this was achieved through the use of prestressed CFRP strips, attached to the bottom face of the inverted T cast iron beams (Fig. 7b, Luke 2001).

(a)

(b)

Figure 7. (a) The Tickford Bridge. (b) Strengthening of the inverted T beams in the Hythe Bridge with four prestressed CFRP plates.

5. CONCRETE

Concrete structures are, in general, not considered of historic nature, with very few exceptions. One of them is the Osaka Castle in Japan (Figure 8), constructed in 1931 with reinforced concrete and steel elements. CFRP wrapping of the columns in this impressive structure was considered as the optimum solution for increasing its resistance against earthquakes.

Figure 8. The Osaka Castle is an example of a historic structure made with modern materials (concrete and steel) that received CFRP strengthening.

6. CONCLUSIONS

Strengthening of historic structures with unbonded or bonded FRP materials appears to be an attractive alternative to traditional upgrading techniques, especially in cases where ilementa tion of these techniques is impractical. The paper presented a short overview of FRP applications to historic structures made of masonry, timber, metals (cast iron, steel) and concrete, with emphasis on concepts and key case studies. It is believed that the combination of advanced composites with traditional materials in structures of historic nature, first proposed a little less than 10 years ago, is a growing field. Key issues, e.g. materials incompatibility, lack of materials and design standards, education and training, need to be addressed better before composites find their proper position as strengthening materials of the architectural heritage building stock.

7. REFERENCES

1. Blasi C., Carfagni S., Bottardi M. and Gottardo R. (2000). The Use of Carbon Fibres in Restoring the Stables of the Medici villa of Poggio a Caiano – Florence. In: Humar J. and Razaqpur A.G. (eds) *Advanced Composite Materials in Bridges and Structures,* Montreal, The Canadian Society for Civil Engineering, 553-560.
2. Faccio P., Foraboschi P. and Siviero E. (2000). Collapse of Masonry Arch Bridges Srengthened using FRP Laminates. In: Humar J. and Razaqpur A.G. (eds) Advanced Composite Materials in Bridges and Structures, Montreal, The Canadian Society for Civil Engineering, 505-512.
3. Kropf, F.W. and Meierhofer, U. (2000). Strengthening, Retrofitting and Upgrading of Timber Structures with High-Strength Fibres. Structural Engineering International, IABSE, 3, 178-181.

4. Luke, S. (2001). The Use of Carbon Fibre Plates for the Strengthening of Two Metallic Bridges of an Historic Nature in the UK. Proceedings of the International Conference on FRP Composites in Civil Engineering (CICE 2001), Hong Kong.
5. Meier, U. (1995). Strengthening of Structures Using Carbon Fibre/Epoxy Composites. Construction and Building Materials, 9, 341-351.
6. Meier, U. (2001). Strengthening and Stiffening of Historic Wooden Structures with CFRP. Proceedings of the International Conference on FRP Composites in Civil Engineering (CICE 2001), Hong Kong.
7. Katsumata, H., Kimura, K. and Murahashi, H. (2001). Experience of FRP Strengthening for Japanese Historical Structures. Proceedings of the International Conference on FRP Composites in Civil Engineering (CICE 2001), Hong Kong.
8. Plevris, N. and Triantafillou, T.C. (1992). FRP-Reinforced Wood as Structural Material. Journal of Materials in Civil Engineering, ASCE, 4:3, 300-317.
9. Silman, R. (1997). Major Repairs to Frank Lloyd Wright's Largest House: Wingspread, Racine, WI, USA. Proceedings of the IABSE Conference on Composite Construction – Conventional and Innovative, Innsbruck, 922-923.
10. Schwegler, G. (1994). Masonry Construction Strengthened with Fiber Composites in Seismically Endangered Zones. Proceedings of the 10th European Conference on Earthquake Engineering, Acapulco, Mexico.
11. Schwegler, G. and Kelterborn, P. (1996). Earthquake Resistance of Masonry Structures Strengthened with Fiber Composites. Proceedings of the 11th World Conference on Earthquake Engineering, A.A. Balkema, Rotterdam, 454-458.
12. Triantafillou, T.C. and Fardis, M.N. (1993). Advanced Composites for Strengthening Historic Structures. IABSE Symposium on Structural Preservation of the Architectural Heritage, Rome, 541-548.
13. Triantafillou, T.C. and Fardis, M.N. (1997). Strengthening of Historic Masonry Structures with Composite Materials. Materials and Structures, RILEM, 30, 486-496.
14. Triantafillou, T.C. (1998). Strengthening of Masonry Structures Using Epoxy-Bonded FRP Laminates. ASCE Journal of Composites for Construction, 2:2, 96-104.

EFFECTIVE STRENGTHENING OF REINFORCED CONCRETE BEAMS WITH ANCHORED FRPs

C. B. DEMAKOS[1] and D. P. KOUTSOUKOS [2]
Assoc. Professor. [1], Reinforced Concrete Laboratory
Dept. of Civil Engineering
Technological Educational Institution of Piraeus
Piraeus, Greece
Structural Engineer[2], P.E., Ph.D., Infrastructure Works Design Manager
ERGA OSE S.A., Greece

Abstract .The beneficial effect of a composite material anchorage system (CMAS) on the ductility and stiffness of reinforced concrete (RC) beams strengthened with Carbon (CFRP) or Glass (GFRP) sheets is investigated .The system consists of a tuft of high-strength glass fibers, anchored into the lateral faces of the reinforced concrete beam by penetrating the accompanied GFRP or CFRP "U"-shaped strips (U-strips) used to prevent delamination of the FRP sheets that are placed in the lower face of the beam for strengthening. Experimental analysis of RC beams strengthened with one GFRP sheet and anchored at its ends by GFRP U-strips shows an increase of the ultimate bending capacity as well as a ductility enhancement due to the CMAS. Similarly, a CFRP sheet on the bottom face of the beam anchored at its ends by CFRP U-strips with the aid of CMAS provides, not only a significant strength increase, but also, its failure mode becomes ductile without considerable variation of the beam stiffness. This behavior is experimentally verified by measuring the deflection and deformation of the FRPs and of the concrete beam at the middle point of its bottom and upper faces.

1. INTRODUCTION

The need for strengthening reinforced and prestressed concrete structures is becoming more apparent, particularly when there is an increase in load requirement, earthquake induced damage of the structural members, or design/construction defects. Various solutions apply to this problem, ranging from complete or partial replacement of the structure to structural member strengthening with a variety of techniques. Laminated advanced composite materials, such as fiber-reinforced plastics (FRP), can be successfully used for this purpose. Sheets in the form of woven or stitch flexible carbon or glass fibers, externally bonded to concrete surfaces with appropriate epoxies, were considered in this paper. They are made of unidirectional dry fibers of carbon or glass and are attached to the bottom face of the beam, parallel to the main reinforcement.

The use of FRP materials for structural repair presents several advantages and has been recently investigated all over the world [1-7]. The writers have so far

E.E. Gdoutos and Z.P. Marioli-Riga (eds), Recent Advances in Composite Materials, 345-355

concentrated their investigations on the strengthening of "virgin" RC beams, i.e., beams with no previously induced damage of any kind. It has been shown that, as a result of the FRP application, the mode of failure of a flexural member that has not been appropriately design may change from ductile to brittle. For instance, concrete shear failure may substantially reduce the nominal flexural capacity predicted with standard design computations. By changing the thickness of the FRP sheet, changing its bonded length, or adding shear reinforcement, the crack distribution along the beam can be significantly modified and the failure mechanism be changed. Failure mechanisms that have been experimentally verified are: FRP rupture, concrete crushing, shear failure, peeling at the adhesive – concrete interface (delamination), and so on. Further studies [8-9] have shown that carefully designed external anchorages of the FRP sheets can lead to increase in load capacity with substantial regain of lost beam structural ductility as well as modification of the brittle failure mode to a more ductile failure mode.

In this paper, experimental results are presented for the case of beams strengthened with GFRP (commercial name *XL-Wrap® UGK-200™*) or CFRP (commercial name *XL-Wrap® UCG-45™*) plates. Comparison with control beams is also presented .The experimental program consists of four – point bending tests as well as pure tension and compression tests applied to steel bars and concrete cubes to verify material properties. A ply was used for both types of FRPs at the bottom beam face, where the direction of the fibers was selected to be parallel to the axis of the beam, in order to act as flexural reinforcement. In all cases, FRP sheets were anchored at their ends by U-strips of the same material as the FRP. Specifically, U-strips were bonded to the bottom and lateral faces of the beam and tufts from high-strength glass fibers (commercial name *XL-Encore®*) were embedded through the U-strips inside the core of the beam, thus forming a composite material anchorage system (CMAS). All beams have the same rectangular section. Experimental results indicate that the brittle shear failure of the control beams or even that of the strengthened beams with bonded FRP sheets is transformed to a ductile one in the case of beams strengthened with one sheet of *XL-Wrap® UGK-200™* or *XL-Wrap® UCG-45™* and anchored with the *XL-Encore®* anchorage system. Specifically, the strength of RC beams with one *XL-Wrap® UGK-200™* sheet increases effectively, whereas that of strengthened beams with *XL-Wrap® UCG-45™* sheet increases drastically without considerable variation of the beam stiffness, indicating the beneficial effect of the *XL-Encore®* anchorage used in beams. Furthermore, by evaluating the beam deflections, the ductility of beams with anchored FRPs is enough pronounced to that encountered in beams with FRPs simply bonded on their bottom face.

2. EXPERIMENTAL PROGRAM

2.1 Design and Details of Test Beams

Five simply supported beams, 10x15 cm in cross section and 110 cm long, were tested in 4-point bending in order to avoid the appearance of shear forces at the mid-span of beam. One of them, the control or "virgin" beam, was designed to fail in flexure. All beams were reinforced with two Ø8 mm diameter steel bars at both tension and compression faces and at an effective depth of 1.5 cm as shown in Fig.

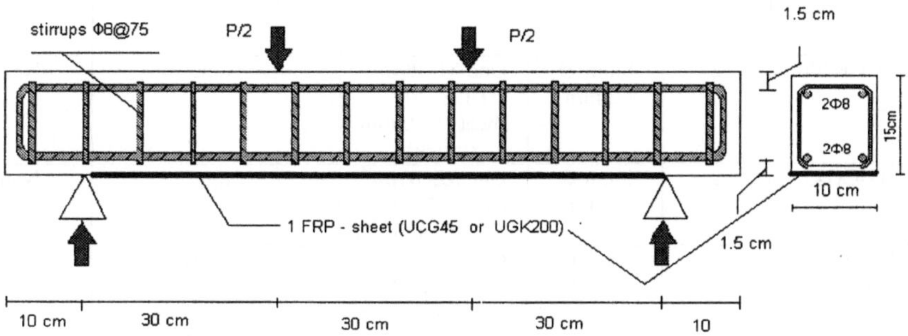

Figure 1: Specimen Dimensions (in cm) and Test Setup .

1. To avoid shear failure of the control beam as well as of the beams strengthened with FRPs due to possible insufficient anchorage of FRPs, the beams were adequately reinforced against shear with vertical stirrups Ø8 mm in diameter at 7.5 cm center to center spacing. Details of all the beams, including the configuration of the external reinforcement sheets are given in Fig. 1 and Table 1.

Beam "BV1" was the control beam. Two beams, with code names "B1/1-UCG45" and "B1/1-UGK200", were reinforced with one sheet of *XL-Wrap® UCG-45™* or *XL-Wrap® UGK-200™*, bonded at their tension face with a specially design adhesive epoxy mixture. The adhesive mixture (commercial name *XL-POXY®*) was consisted of a hardener (at a volume percentage 30%) blended with a resin (at a volume percentage 70%).

All beams were extensively instrumented to clearly establish their performance characteristics. At beam mid-span, two strain-gages were used to measure FRP strains (tension face) and strains at the extreme beam compression face. The mid-span beam deflection was measured by an LVDT of a 100 mm travel.

Tests were carried out by a TONI-TECHNIK bending machine at a loading rate of 0.2 KN/sec and the applied loads were continuously monitored through a high-accuracy load cell. All measurements were automatically recorded through a data logger.

Table 1.Structural Details and Anchorage Arrangements of Tested Beams

Beam	# of rebars	Stirrups	Tension sheet	End-anchorage of FRP	Sheet anchorage along span of beam
BV1	2	8mm @ 75mm	None	None	None
B1/1UGK200	2	8mm @ 75mm	One *XL-Wrap*® *UGK-200*™ sheet (1.30 mm thick)	None	None
B1/1UCG45	2	8mm @ 75mm	One *XL-Wrap*® *UCG-45*™ sheet (0.45 mm thick)	None	None
B1/1UGK200/ XL-ENCORE®	2	8mm @ 75mm	One *XL-Wrap*® *UGK-200*™ sheet (1.30 mm thick)	*XL-Wrap*® *UGK-200*™ "U"-strips (400 mm long, extended to compression region)	None
B1/1UCG45/ XL-ENCORE®	2	8mm@75 mm	One *XL-Wrap*® *UCG-45*™ sheet (0.45 mm thick)	*XL-Wrap*® *UCG-45*™ "U"-strips (400 mm long, extended to compression region)	None

2.2 Materia Properties

The beams were designed for mean 28-day cube strength of 20 MPa. Both main steel reinforcing bars as well as vertical stirrups exhibit average measured yield strength of 500 MPa. The *XL-Wrap*® *UGK-200*™ and *XL-Wrap*® *UCG-45*™ sheets exhibit a linear stress-strain behavior and their properties are shown in Table 2. The adhesive used for bonding the sheets was the two-component *XL-POXY*®, an epoxy resin mixture suitable for the particular FRPs.

Table 2: Properties of XL-Wrap® UCG-45™ and XL-Wrap® UGK-200™ sheets

Property	XL-Wrap® UCG-45™ laminate	XL-Wrap® UCG-45™ laminate
Ultimate tensile strength in primary fiber direction (MPa)	1,100	550
Elongation at break (%)	1.5	2.0
Tensile modulus (GPa)	75.0	27.5
Ultimate tensile strength at 90° to primary fiber (MPa)	50	50
Laminate thickness (mm)	0.45	1.30

2.3 Anchorages

Beams with code names "B1/UCG45/XL-ENCORE®" and "B1/UGK200/XL-ENCORE®" (see Table 1) were strengthened by one *XL-Wrap® UCG-45™* and one *XL-Wrap® UGK-200™* sheet, respectively, on their tension faces. In addition, in these beams the FRP sheets were anchored by means of U-strips (of the same material as the main FRP) at the laminate edges to prevent premature sheet debonding at the laminate cut-off edge (Fig. 2(a)). Furthermore, the U-strips were secured by CMAS, by drilling holes to the lateral beam faces and embedding through the strips tufts of glass fibers, *XL-Encore®*, in the core of the RC beams (Fig. 2 (b)).

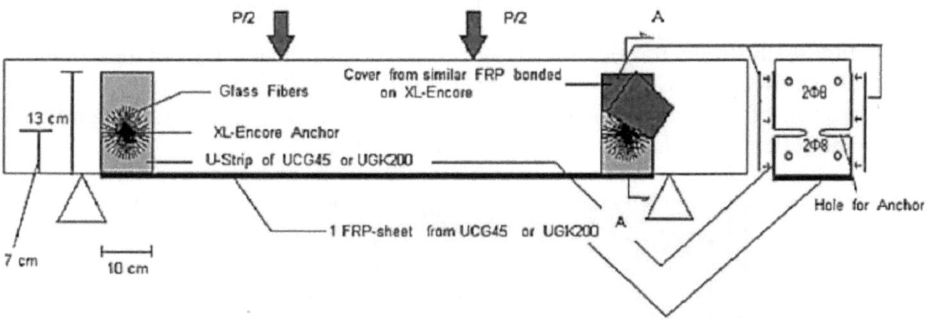

Figure 2: (a) XL-Encore® CMAS anchorage system detail used with FRPs in strengthening RC beams, and (b) section (A)-(A)

3. TEST RESULTS AND DISCUSSION

A large amount of test data was obtained during the experimental program. However, only those relevant to this paper are presented here.

3.1 Strengthened RC Beams without use of CMAS

The structural behavior of the three beams without use of CMAS described in Table 1 is represented by their load – mid-span deflection shown in Fig. 3 .The redcolor deflection curve shows that the control beam "BV1" behaved like a typical under-reinforced concrete beam.

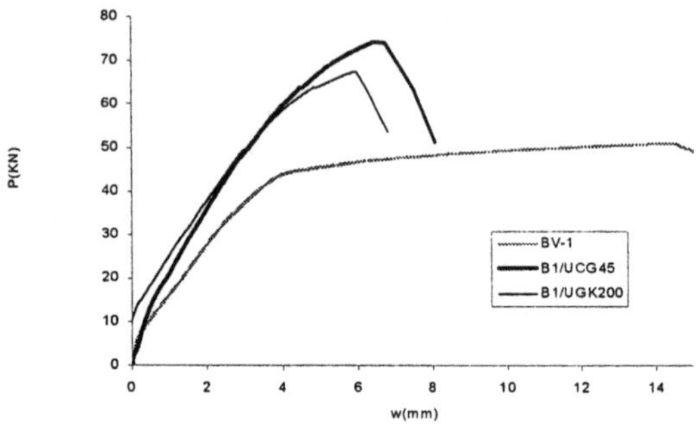

Figure 3: Load vs. mid-span deflection curves in control and strengthened beams with one sheet of XL-Wrap® UCG-45™ and XL-Wrap® UGK-200™ composite materials

Fig. 4 illustrates the flexural failure of the control beam. By bonding either a sheet of *XL-Wrap® UCG-45™* or a sheet *XL-Wrap® UGK-200™* without further consideration of the end – anchorage stresses of the FRPs, results to a significant mid-point deflection reduction at ultimate load.

Figure 4. Flexural failure of control beam " BV1"

Although the stiffness of the strengthened beams "B1/1UCG45" and "B1/1UGK200" is about the same with that of the control beam, the ultimate load of the beam strengthened by a sheet of *XL-Wrap® UCG-45™* is higher than that of the beam strengthened with a sheet of *XL-Wrap® UGK-200™*. However, the sudden drop of load at relatively small deflection levels emphasizes that beam strengthening with FRP laminates only, i.e., without the use of CMAS, can result in sudden degradation of structural behavior and that the provision of adequate anchorages at the sheet ends can significantly improve the beam's structural behavior, including its ductility levels. Figures 5 and 6 illustrate the failure mode of strengthened RC

Figure 5: Peeling of concrete layer along tension rebars in beam "B1/UGK200"

beams with no CMAS. This is exhibited by peeling of the concrete cover along the rebars, which, although it indicates good behavior of the *XL-POXY*® adhesive by successfully attaching the laminate on the tension concrete face, it also results in early degradation of the load carrying mechanism.

Figure 6: Peeling of concrete layer along tension rebars in beam "B1/UCG45"

3.2 The beneficial influence of anchorages

As mentioned earlier, beams "B1/1UGK200/XL-ENCORE®" and "B1/1UCG45/XL-ENCORE®" were designed to further improve their strength by preventing the slip between the laminate and the concrete face. For this purpose, the CMAS anchorage system, made from a tuft of high-strength glass fibers [10], was chosen and applied (see Fig. 7(a)) by penetrating the U-strips into the core of the beam at a depth of about 6 cm. The protruding tips of the anchorage fibers were bonded to the U-strips (see Fig. 7(b)) and a patch of FRP covered the anchored U-strip (see Fig. 2(a)).

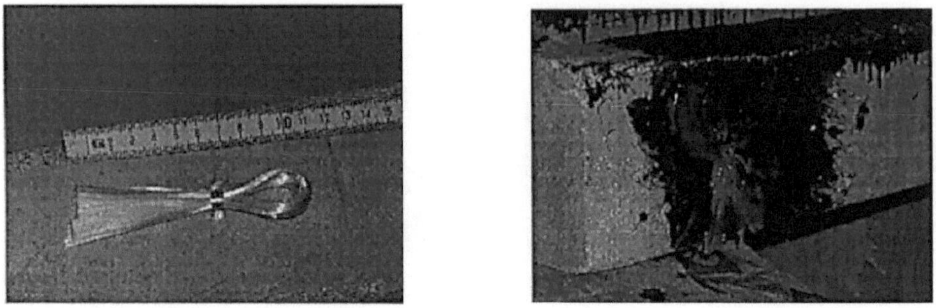

(a) (b)

Figure 7: (α) XL – Encore® anchorage system made from a tuft of high-strength glass fibers, and (b) the fiber edges bonded to the U-strip

Increased strength levels were achieved, as Fig. 8 illustrates, since the ultimate load was increased by 8.2% in the case of beam "B1/1UGK200/XL-ENCORE®" and by 17.5% for the "B1/1UCG45/XL-ENCORE®" beam, when compared to the respective loads of the strengthened beams without use of CMAS. This increase in load capacity was also reflected to the failure mode, which was changed from a brittle to a more ductile one, as shown by the load - deflection

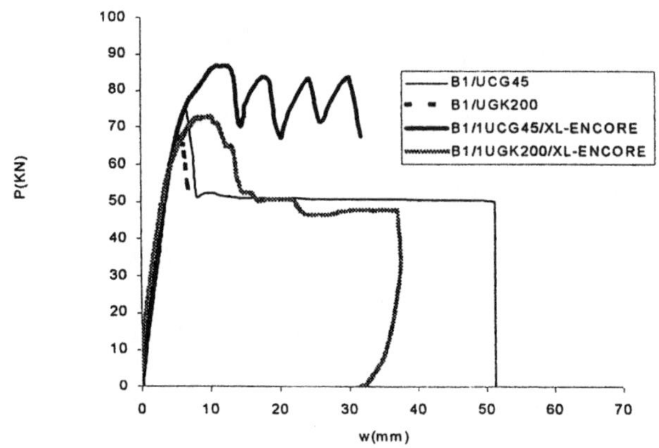

Figure 8: Load vs. mid-span deflection curves for strengthened RC beams with or without use of the CMAS XL-Encore® anchor system

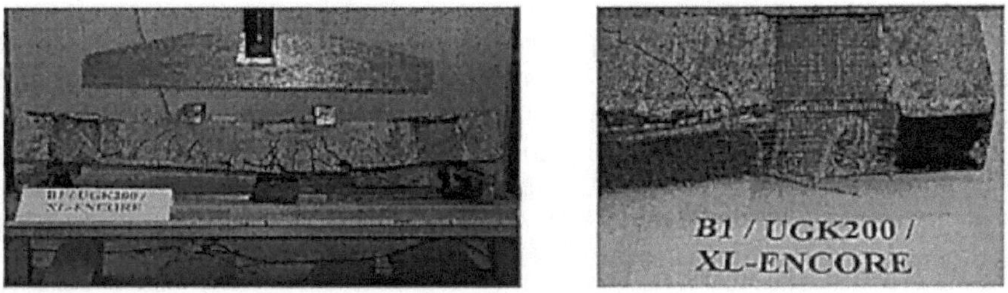

(a) (b)

Figure 9: (a) Failure of the "B1/1UGK200/XL-ENCORE®" beam resulting to failure of the XL-Encore® (b)

curves (see Fig. 8). The fluctuations of the red- and orange-colored curves (Fig.8) represent the load response of the strengthened beams with use of CMAS anchors indicating the progressive anchorage failure (Fig. 9(b) and 10(b)), which is accompanied by successive noises during the tests.

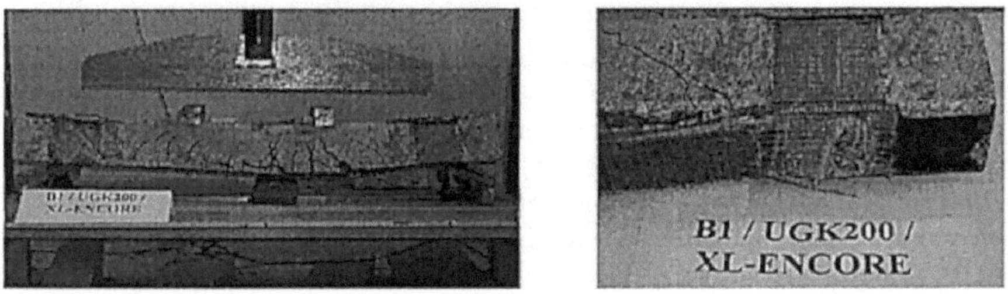

(a) (b)

Figure 10: (a) Failure of the "B1/UCG45/XL-ENCORE®" beam followed the rupture of the XL-Encore® (b)

To further quantify the role and effectiveness of the CMAS, Table 3 provides a comparison of the structural response of the control, strengthened with FRPs and strengthened with FRPs and CMAS beams.

Table 3: Comparing control and strengthened beam behavior with and without use CMAS

Property	BV1	B1/ 1UGK200	B1/1UGK200/ XL-ENCORE®	B1/ 1UCG45	B1/1UCG45/ XL-ENCORE®
Load at 1st flexural crack (KN)	7.5	10.0	13.0	13.0	14.0
Load at tension steel yielding (KN)	36.0	45.0	46.0	49.0	54.0
Ultimate load (KN)	52.0	67.5	73.0	74.5	87.5
Deflection at yielding point (mm)	2.84	2.83	2.50	3.01	3.70
Deflection at ultimate load (mm)	14.3	6.0	10.3	6.6	11.1
Concrete strain near ultimate load (μs)	2.90	----	4.42	------	2.82
FRP strain near ultimate load (μs)	-----	3.3	9.4	2.4	9.9

The results clearly show the increase of the load carrying capacity of the strengthened beams with anchored FRPs. On the other hand, an improvement in the ductility levels of the beams is recognized by the larger deflection levels achieved at ultimate loads. Finally, the complete utilization of FRPs and, thus, the effectiveness of the proposed CMAS anchorage system, *XL-Encore®*, can be shown by the measured strains at the *XL-Wrap® UGK-200™* and *XL-Wrap® UCG-45™* sheets which approach the value of 1.0%.

4. CONCLUSIONS

The major conclusions from this study are given next:

1. Bonding a CFRP (*XL-Wrap® UCG-45™*) or a GFRP (*XL-Wrap® UGK-200™*) sheet to function as external reinforcement at the tension face of a RC beam without consideration of the end-anchorage stresses of the sheet, although it improves the structural beam behavior, it results in an early and sudden drop of the load vs. deflection curve and, thus, it does not fully utilize the benefits of this type of strengthening technique. With this type of premature failure, the FRP sheet is not used to its maximum potential.
2. Although the ultimate load of the strengthened beams without use of the XL-Encore® anchorage system is significantly increased from that attained in the control beam, the deflection at this load is about the half of that developed in the control beam due to the fact that the failure mechanism changes from a flexural (of the control beam) to a shear type (of the strengthened beams).
3. By choosing to anchor the FRPs at their ends through the XL-Encore® anchorage system, it is possible to increase the ultimate load capacity of the unstrengthened beam more than 70%. In addition, the brittle failure mode of the strengthened beams with no CMAS is changing to a ductile one when these anchors are used.

Acknowledgements

The authors wish to thank "exelTEk S.A. – Advanced Structural Technology" for the financial support of this research program as well as for providing the composite materials and the CMAS system as well as its specialized personnel for the FRP applications. Also, they are grateful to Mr. D. Chatzisofianos, student at the Technological Educational Institution of Piraeus, for preparing the test set-up and conducting the tests as part of his Thesis Dissertation.

5. REFERENCES

1. Baaza, I.M., Missihoun, M. and Labossiere, P. Strengthening of reinforced concrete beams with CFRP sheets , *Proc. 1st Int. Conf on Compos. In Infrastruct. ICCI 96* Univ. of Arizona, Tucson , Ariz. , 746-759 ,1996.
2. Chajes , M. ,J. , Thomson,T. A., Januszka , T. F. , and Fin, W. :Flexural strengthening of concrete beams using externally bonded composite materials ,*Constr. And Build. Mat.* **8(3)** (1994), 191-201.
3. Crasto , A. S. , Kim , R. Y. , Fowler, C. , and Mistretta , J.P.:Rehabilitation of concrete bridge beams with externally-bonded composite plates. Part I, *Proc. 1st Int. .Conf. on Compos. In Infrastruct ICCI 96,* , 857-869, 1996.
4. Nakamura , M. , Sakai , H. , Yagi , K. and Tanaka , T. :Experimental studies on the f flexural reinforcing effect of carbon fiber sheet bonded to reinforced concrete beam , *Proc. ,1st Int Conf. on Cmpos. In Infrastruct. ICCI 96,* 760-773, 1996 .
5. Saadatmanesh , H. , and Eshani ,M.:RC beams strengthening with GFRP plates : experimental study ,*J. Struct. Engng., ASCE* **117(11)** (1991a),3417-3433.
6. Saadatmanesh , H. , and Eshani ,M.: RC beams strengthening with GFRP plates :analytical and parametric studies,*J. Struct. Engng.,ASCE* **117(11)** (1991b) ,3432-3455.
7. Varastehpour ,H. , and Hamelin , P. :Structural behavior of reinforced concrete beams strengthened by epoxy bonded FRP plates , *Proc., 2nd Int. RILEM Symp. 2FRPRCS,*Reunion Internationale des Laboratoires d' Essais et de Recherches sur les Materiaux et les Constructions, 1995.
8. Mukhopadhyaya, P., Swamy, N. and Lynsdale ,C. :Optimizing response of beams strengthened with GFRP plates, *J. of Comp. for Construcion,ASCE* **2(2)** (1998),87-95.
9. Spadea , G.,Bencardino,F., and Swamy , R.N. :Structural behavior of composite RC beams ith externally bonded CFRP , *J. of Comp. for Construcion,ASCE* **2(3)** (1998),132- 137.
10. Design Manual for Composite Material Design and Application, exelTEk S.A., 2000.

ON THE OPTIMUM DESIGN OF FIBRE REINFORCED CONCRETE - STEEL COMPOSITE COLUMNS

G.D. HATZIGEORGIOU and D.E. BESKOS
Department of Civil Engineering,
University of Patras
Patras University Campus
GR- 265 00 Patras, Greece

Abstract. This paper presents a procedure for the optimum design of fibre reinforced concrete - filled steel tube columns. It takes into account the effects of confinement of concrete and its steel fibre reinforcement. The constraints of the optimum design, prescribed by the strength and stability requirements, are adopted from the provisions of modern structural codes. A computer program for the analysis has been developed, which emphasizes the dimensioning of the composite column and identification of the optimum quantity of fibres in matrix material. The paper concludes with two representative numerical examples, which show that in-filled composite columns can be designed both economically and effectively using the proposed methodology.

1. INTRODUCTION

Steel-concrete composite columns are commonly used in structural engineering because they appropriately combine the advantages of their constituent materials [1, 2]. The increased use of composite columns has created a great research interest in analytical, numerical and experimental work [3]. There are two types of composite columns: a steel section, usually an I-section, encased fully or partially in concrete, and a steel tube section in-filled with concrete. Extensive studies on composite columns have been done for the case of encased columns [4]. On the other hand, in-filled composite columns have received less attention compared to encased columns [5].

Concrete-filled steel tube columns are used as, e.g., high-rise building columns, bridge girders and piers, as well as piles. Typical representative cross-sections of in-filled composite columns are shown in Fig. 1. The steel tube performs effectively in bending as well as in axial tension and compression. The concrete represents a high-quality core to endure compressive loading. On the other hand, the steel tube is used in construction as formwork, ich can attend the concrete and this leads to a more economical, easier and faster erection

E.E. Gdoutos and Z.P. Marioli-Riga (eds), Recent Advances in Composite Materials, 357-370

procedure than in conventional reinforced concrete structures. Furthermore, in-filled composite columns can provide excellent monotonic and seismic resistance due to confinement of the concrete and continuous bracing of the steel tube to delay local buckling, improved damping behaviour in comparison to traditional steel frames, and improved ductility [3].

Figure 1: Typical cross-sections of concrete -filled steel tube columns

This paper represents an effort towards the optimum (minimum cost) design of concrete-filled steel tube columns. It examines only circular composite columns but the proposed methodology can also be applied to other sections as well. The structure of this paper is as follows: at first, basic information on fibre reinforced concrete -filled steel tube columns is presented and the composite action of the concrete-filled steel tube columns is examined. Then, the analysis of composite columns under axial force and bending moment as well as under local and global buckling follows. Afterwards, the optimum (minimum cost) design methodology and its constraints are presented. The target of this section is to optimize the dimensions of the composite column and identify the most appropriate quantity of fibres in matrix material. The constraints are adopted from the Eurocode 4 [6] provisions and related to bearing capacity in terms of ultimate strength and structural stability. The paper concludes with the analysis of two representative problems to illustrate the proposed methodology and demonstrate that in-filled composite columns can be designed both economically and effectively.

2. FIBRE-REINFORCED CONCRETE

In this section the steel fibre reinforced concrete is examined. It is a cement-based material reinforced with short steel fibres. When steel fibres are added to a concrete mix, they are randomly distributed and act as crack stemmers. Debonding and pulling out of fibres require more energy, giving a considerable increase in resistance and toughness under static or dynamic, monotonic or cyclic loading [7]. Furthermore, fibre-reinforced concrete columns also present better fire resistance than plain concrete composite columns because of the active confinement of the concrete due to the fibres [8]. It has been shown that is possible to relate the mechanical properties of fibre-reinforced concrete to the fundamental properties of fibres, such as, diameter, length, percentage in the matrix material, shape, bond characteristics and tensile strength. The percentage of fibres in the matrix material ranges from 0 to 6%, which corresponds to

about 0 to 135 kg of fibres per cubic meter of concrete. Finally, the tensile strength σ_{fu} of steel fibres ranges from 1.0 to 1.2 GPa.
The compressive strength of fibre-reinforced concrete f_{cf} can be obtained from the relation [9]

$$f_{cf} = f_c + 2.1604\left(W_f \frac{l_f}{d_f} \right) \tag{1}$$

where f_c is the compressive strength of plain concrete in MPa, W_f the percentage of fibres, l_f the length of fibres and d_f the diameter of fibres. Under tensile loading, fibre-reinforced concrete presents at the ultimate stage a residual tensile strength f_{tu} given by [10]

$$f_{tu} = 2\eta_l \eta_0 \tau W_f \frac{l_f}{d_f} \tag{2}$$

where τ is the bond strength of fibres, η_0 a distribution factor (=0.41 assuming random distribution in 3-D space) and η_l an effectiveness factor of the form

$$\eta_l = \frac{\tau}{\sigma_{fu}} \frac{l_f}{d_f} \tag{3}$$

with σ_{fu} being the strength of fibres. For the average elastic tangent modulus in compression E_{cf} of fibre-reinforced concrete it can be assumed [7]

$$E_{cf} = \gamma W_f E_f + (1 - W_f) E_c \tag{4}$$

where E_f and E_c are the modulus of elasticity of fibres and concrete, respectively, while the correlation factor γ is given by

$$\gamma = \eta \left[1 - \frac{\tanh(n_r l_f / d_f)}{n_r l_f / d_f} \right] \quad \text{where} \quad n_r = \sqrt{\frac{2E_c}{E_f (1 + \nu_c) \ln(1/W_f)}} \tag{5}$$

with ν_c being the Poisson ratio of plain concrete and the fibre distribution factor η being given by

$$\eta = \begin{cases} 1/6 & \text{random distribution in 3-D space} \\ 1/3 & \text{random distribution in 2-D space} \\ 1/2 & \text{trimmed in 2-D space} \end{cases} \tag{6}$$

In the absence of experimental data, the modulus of elasticity of unconfined concrete can be obtained from the following empirical relation of Eurocode 2 [11]:

$$E_c = 9500\ (f_c + 8)^{1/3} \qquad (E_c \text{ and } f_c \text{ are in MPa}) \tag{7}$$

3. COMPOSITE ACTION

The composite action between steel and concrete results from mechanical interlock, friction and adhesion[12]. During a low level of compressive loading, Poisson's ratio of concrete is 0.15 - 0.20, which is lower than the value $\nu=0.3$ for steel. Therefore, the steel tube has no confining effect on the concrete core. As the loading is increased, longitudinal strain increases and Poisson's ratio of concrete increases above to 0.50 [5,13]. For this reason, the lateral expansion of unconfining concrete becomes greater than that of steel and a radial pressure gradually develops at the steel-concrete interface. Thus, the concrete core is confined and a hoop tension develops in the tube at this phase. Thus, the steel tube is stressed biaxially and the concrete core triaxially. For hollow steel circular sections filled with plain concrete, it has been shown that the confinement of the concrete core significantly increases the bearing capacity of composite members. Thus, the load corresponding to the ultimate bearing capacity is higher than the sum of loads that can be achieved by steel and concrete acting independently.

The confined uniaxial strength of concrete f_{cc} under triaxial stress state results from [13]

$$f_{cc} = f_c + m f_{rp} \tag{8}$$

where f_c is the unconfined strength, f_{rp} the maximum radial pressure on concrete and m an empirical coefficient. Extensive experimental work has shown that m can be taken equal to 4. The maximum radial pressure can be expressed by the semi-empirical relation [14]

$$f_{rp} = (\nu_1 - \nu_2)\left(\frac{t}{r-t}\right) f_y \tag{9}$$

where t is the thickness of the steel tube, r its outer radius and f_y its yield stress. Furthermore, ν_1 and ν_2 are the Poisson ratios of the steel tube with and without filled-in concrete, respectively, at the ultimate stage. Ratio, ν_1 is given by

$$\nu_1 = 0.2312 + 0.3582k - 0.1524\left(\frac{f_c}{f_y}\right) + 4.8430k\left(\frac{f_c}{f_y}\right) - 9.1690\left(\frac{f_c}{f_y}\right)^2 \tag{10}$$

where k is a factor of geometry of the form

$$k = 0.4011 + 3.9060 \cdot 10^{-2}\left(\frac{r}{t}\right) - 1.0320 \cdot 10^{-3}\left(\frac{r}{t}\right)^2 + 7.0480 \cdot 10^{-6}\left(\frac{r}{t}\right)^3 \quad (11)$$

While Poisson' ratio v_2 is taken as equal to 0.50 at the maximum strength point.

4. CROSS-SECTIONAL STRENGTH

In order to compute the bearing capacity of a cross-section, a distribution for strains must be assumed on this section. It is assumed that full interaction exists between the steel tube and the fibre-reinforced concrete infill and plane sections remain plane after loading. It is worth noticing that experimental results show that sometimes there is relative slip between concrete and steel [12]. This fact is insignificant on the computation of ultimate strength where it is assumed that tensile and compressive stresses of concrete and steel grow to fixed values. In [15] was suggested that the results of analysis of a composite column can be determined by assuming a full plastic stress distribution (i.e., rigid plastic assumption) in all the materials (concrete and structural steel). This assumption gives very good results and simplifies substantially the problem, and for these reasons it is adopted in this work. Figure 2 shows the four regions of the composite column with regard to material and stress sign: steel tube in tension (S_T), steel tube in compression (S_C), concrete in tension (C_T) and concrete in compression (C_C). It is evident that the axial force portion of each region results from the multiplication of its area with its applied stress. Furthermore, the respective flexural moment results from the multiplication of the axial force by the relative distance between the centre of column and the centre of gravity of the region.

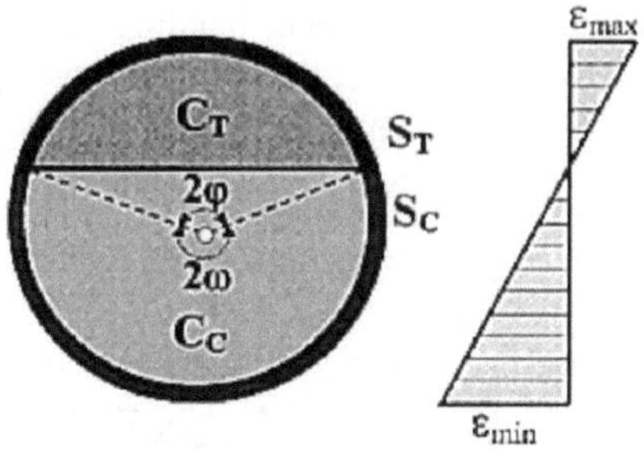

Figure 2: Cross-section analysis

Steel tube in tension: the axial force N_{ST} and the bending moment M_{ST} result from

$$N_{ST} = \left[\left(\varphi\left(r^2 - (r-t)^2\right)\right)\cdot f_y\right] \quad M_{ST} = \left[\left(\varphi\left(r^2 - (r-t)^2\right)\right)\cdot f_y\right]\cdot\left[\frac{(2r-t)\cdot\sin(\varphi)}{2\varphi}\right] \tag{12}$$

Steel tube in compression: the axial force N_{SC} and the bending moment M_{SC} are given by

$$N_{SC} = \left[\left(\omega\left(r^2 - (r-t)^2\right)\right)\cdot f_y\right] \quad M_{SC} = \left[\left(\omega\left(r^2 - (r-t)^2\right)\right)\cdot f_y\right]\cdot\left[\frac{(2r-t)\cdot\sin(\omega)}{2\omega}\right] \tag{13}$$

Concrete in tension: the axial force N_{CT} and the bending moment M_{CT} result from

$$N_{CT}=\left[\left((2\varphi-\sin(2\varphi))\frac{r^2}{2}\right)\cdot f_{tu}\right] \quad M_{CT}=\left[\left((2\varphi-\sin(2\varphi))\frac{r^2}{2}\right)\cdot f_{tu}\right]\cdot\left[\left(\frac{2}{3}\right)\frac{(r-t)\sin^3(\varphi)}{\varphi-\sin(\varphi)\cos(\varphi)}\right] \tag{14}$$

Concrete in compression: the axial force N_{CC} and the bending moment M_{CC} are given by

$$N_{CC}=\left[\left((2\omega-\sin(2\omega))\frac{r^2}{2}\right)\cdot f_c\right] \quad M_{CC}=\left[\left((2\omega-\sin(2\omega))\frac{r^2}{2}\right)\cdot f_c\right]\cdot\left[\left(\frac{2}{3}\right)\frac{(r-t)\sin^3(\omega)}{\omega-\sin(\omega)\cos(\omega)}\right] \tag{15}$$

The total axial force N_T and the total bending moment M_T result from

$$N_T=N_{ST}-N_{SC}+N_{CT}-N_{CC} \quad \& \quad M_T=M_{ST}+M_{SC}+M_{CT}+M_{CC} \tag{16}$$

The above analysis requires the knowledge of the angle φ (or ω). For a variety of values of this angle, the axial force versus bending moment interaction diagram results from Eq (16). It is worth noticing that the bearing capacity is supplied if the N-M interaction diagram encloses the set of loading combinations.

5. STABILITY ANALYSIS

It is well known that the presence of the concrete matrix reduces significantly the local buckling phenomena in the steel tube. The provisions of Eurocode 4 [6], dictate that local buckling in the steel tube can be excluded if the radius-to-thickness ratio is appropriately limited. Thus, according to [6] one has that

$$\frac{r}{t} \le 45\sqrt{\frac{235}{f_y}} \tag{17}$$

where f_y is the yield strength of the steel in MPa. Eurocode 4 [6] dictates that a composite column is protected against global buckling if the set of loading combinations is located in a 'safe region' SR within the N-M interaction diagram. Initially, Eurocode 4 [6] requires the knowledge of a reduction factor X_1, which considers the various imperfections and residual stresses, i.e.,

$$X_1 = \frac{1}{f + \sqrt{f^2 - \lambda^2}} \quad \text{where} \quad f = 0.5\left[1 + \alpha(\lambda - 0.2) + \lambda^2\right] \tag{18}$$

where α is the imperfection factor, equal to 0.21 for concrete-filled cross-sections, and λ the relative slenderness ratio given by

$$\lambda = \sqrt{N_u / N_{cr}} < 2 \tag{19}$$

where N_u is the ultimate bearing capacity and N_{cr} is the critical buckling load of the composite column given by

$$N_{cr} = \frac{\pi^2 (EI)_e}{L^2} \tag{20}$$

where L is the equivalent buckling length of the column and $(EI)_e$ its equivalent flexural stiffness given by

$$(EI)_e = E_s I_s + \beta E_c I_c \tag{21}$$

with the factor β being equal to 0.59 to account for the reduced elasticity modulus of concrete due to its non-linear behaviour.

Additionally, Eurocode 4 [6] requires knowledge of the reduction factor X_2, which is applied to consider the variation of bending moment along the composite steel-concrete column, i.e.,

$$X_2 = X_1 \frac{1-r}{4} \tag{22}$$

where r is the ratio between the values of bending moments at the member ends and satisfies the inequality $-1 \leq r \leq 1$. Figure 3 shows the M-N interaction diagram of a composite column.

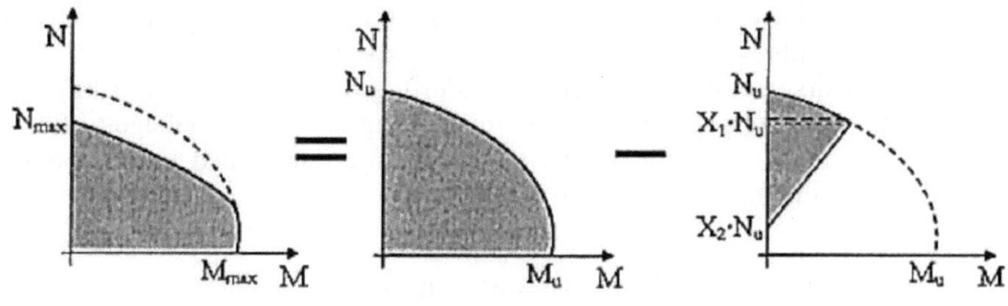

Figure 3: Effective axial force – bending moment interaction diagram

Eurocode 4 [6] also requires that $M \leq 0.9M_u$. Therefore, the abscissas of the N-M interaction diagram must be multiplied by the factor 0.9. Finally, Eurocode 4 [6] takes into account the second order phenomena multiplying the first order bending moment M by a factor α given by

$$\alpha = MAX\left[1\ ,\ \left(\frac{MAX[0.44\ ,\ 0.44r + 0.66]}{1 - N/N_{cr}}\right)\right] \tag{23}$$

6. OPTIMUM DESIGN

In the design of structures, it is crucial to select an efficient and economical design from the available alternatives. For this purpose, optimization techniques have been extensively used in recent times to optimize an objective function without violating the constraints of the problem [16, 17].

In structural optimization, the volume or weight of the material that represents the objective function is generally minimized. In this work, the most economical concrete -filled steel tube column is examined. The most important modules required in structural optimization problems are: analysis module, optimization module and interface module.

The analysis module uses analytical or numerical techniques to compute the structural response due to the applied loads. This module has already been developed in Sections 4 - 5.

The optimization module is related to the optimization method. Initially, one must define the design variables, which for this problem are the dimensions of the column (external radius and steel tube thickness) and the percentage of steel fibres. The objective function for this problem is the total material cost C given by

$$C = \underbrace{\pi(r-t)^2(1-V_f)C_C}_{concrete} + \underbrace{\pi\left[r^2-(r-t)^2\right]\gamma_S C_S}_{steel} + \underbrace{\pi(r-t)^2 V_f \gamma_F C_F}_{fibres} \tag{24}$$

where V_f is the volume percentage of fibres, and γ_S and γ_F are the material density of structural steel and fibres ($\gamma_S = \gamma_F = 7850$ kg/m^3), respectively. Furthermore, C_C, C_S and C_F are the market cost of concrete (euros/m^3), structural steel (euros/kg) and steel fibres (euros/kg), respectively. It is worth noticing that the percentage of fibres with respect to weight and with respect to volume are interrelated by the weight equality which has the form

$$V_f \gamma_F = \gamma_C W_f \Rightarrow V_f = \frac{\gamma_C}{\gamma_F} W_f \approx 0.318 W_f \tag{25}$$

where γ_C ($\approx$2500 kg/m^3) is the density of fibre reinforced concrete. For the minimum cost design of composite columns problem, the Feasible Design Method [16,18] is applied, as described below.

1. Select a set of (r_i,t_i,W_{fi}) inherent in the feasible region, where i=1 (first iteration). Evaluate total cost C_i using Eq. (24).

2. Find the most effective direction S_i of cost minimization in (r,t,W_f) space, which is determined:
   a) If a constraint is activated, S_i is parallel with it, else,
   b) from minimum ∇C

3. Find a suitable step length λ along the direction S_i and obtain a new value of total cost C_{i+1} by

$$C_{i+1}=C_i+\lambda_i S_i \tag{26}$$

   The value of λ is chosen such that C_{i+1} lies in the feasible region.

4. Test for convergence of the method. If $|C_{i+1}-C_{i+1}|$ is very nearly equal to zero then terminate the iteration by taking $C_{OPT} = C_{i+1}$, else, set new iteration number i=i+1 and go to step 2.

The interface module combines the analysis and the optimization modules. This is a very important phase in the optimization procedure, which involves processes, such as, sensitivity analysis and constraint approximation. The derivatives of the physical response of the column with respect to a design variable can be evaluated from sensitivity analysis. The sensitivity analysis is carried out to calculate the gradients of the constraints with respect to each design variable, which are necessary for constructing constraint approximation functions. The sensitivity value of a constraint function g(x) with respect to the design variable x, is defined with the aid of the forward difference approach as

$$\frac{\partial g(x)}{\partial x} \cong \frac{g(x+\Delta x)-g(x)}{\Delta x} \tag{27}$$

Where Δx is a perturbation of the design variable x.

The constraints of the problem refer to physical and design restrictions. The physical restrictions are based on common sense and for this problem are given by

$$0<r \tag{28}$$

$$0<t<r \tag{29}$$

$$0\leq W_f<1 \tag{30}$$

The design restrictions are related to bearing capacity, stability and practice. In the previous sections, where the bearing capacity and the stability of the column are examined, it has been shown that the set of loading combinations (N_i, M_i) must be located in the 'safe region' SR within the N-M interaction diagram, i.e.,

$$(N_i, M_i) \in R \tag{31}$$

Since bearing capacity and buckling resistance are functions of the set of design variables, x, the above equation can be expressed as

$$(N_i, M_i) \in R(x) \tag{32}$$

In common engineering practice, the percentage of fibres with respect to weight has an upper limit of about 6%. Therefore,

$$0 \leq W_f \leq 0.06 \tag{33}$$

It is evident that Eq. (33) represents a stronger restriction than Eq. (30). Moreover, an additional restriction arises from the combination of Eqs (17) and (29) which reads

$$\frac{r}{45}\sqrt{\frac{f_y}{235}} \leq t < r \tag{34}$$

Finally, the provisions of Eurocode 4 [6] require for the relative slenderness λ, which is a function of the set of design variables x, the following restriction

$$\lambda(x) \leq 2 \tag{35}$$

7. APPLICATIONS

The above analysis has been implemented in a computer program in FORTRAN named OPCOL. Two representative numerical examples are examined in order to illustrate the use and demonstrate the advantages of the proposed optimum design methodology of concrete-filled steel tube columns.

7.1 Optimization of a composite column

A plain concrete-filled steel tube column subjected to an axial loading N = -8.0 MN is examined. The material parameters are: concrete compressive strength f_c = 20 MPa and steel yield stress f_y = 360 MPa. The market cost (Greece - November 2002) of materials is: 180 euros/m^3 for concrete and 0.7 euros/kg for steel. The column has an effective length equal to 6.0 m. The external radius must be greater than 10.0 cm and less than 30.0 cm and the tube thickness must be greater than 1.0 mm and less than 20.0 mm. The Starting Point (SP) is (r, t) = (30 cm, 20 mm). The analysis of this problem appears in Fig. 4.

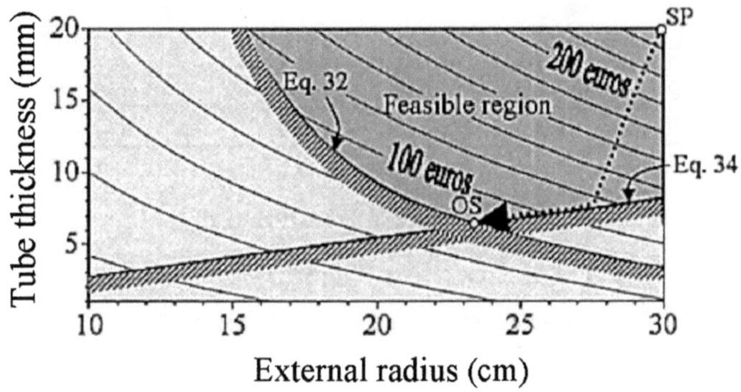

Figure 4: Optimum Solution of Example 1

The Optimum Solution (OS) is obtained when R=23.0 cm and t = 6.7 mm with the total cost being equal to C_T=80.48 euros per column meter.

7.2 Optimization of a composite bridge pier

A composite bridge pier is subjected to the following loading combinations [N, M_1, M_2]:

- **A** : (-32.0 MN, 3.6 MNm, -5.4 MNm), and
- **B** : (-30.0 MN, 4.5 MNm, 4.5 MNm).

The material parameters are: concrete uniaxial compressive strength f_c = 40 MPa, steel yield stress f_y=360 MPa, fibres bond strength τ=10 MPa, steel fibres strength σ_{fu}=1200 MPa. The fibres dimensions are: l_f= 40 mm and d_f=0.5 mm. The market cost of materials is (Greece - November 2002): 220 euros/m^3 for (plain) concrete, 0.7 euros/kg for structural steel and 0.9 euros/kg for fibres. The equivalent buckling length of pier is equal to 17.0 m. The fibres percentage varies from 0 to 6%, the external radius must be greater than 50.0 cm and less than 75.0 cm and the thickness of steel tube must be greater than 10.0 mm and less than 20.0 mm.

The Starting Point (SP) of analysis is (r,t,W_f) = (55 cm, 20 mm, 0 %). The minimum cost is C_T=497.37 euros per column meter and will materialize for R=52.0 cm, t = 14.6 mm and W_f = 3.83 %. Figure 5 shows the optimum design process and the Optimum Solution (OS) in the (r,t) space and for the optimum fibre percentage.

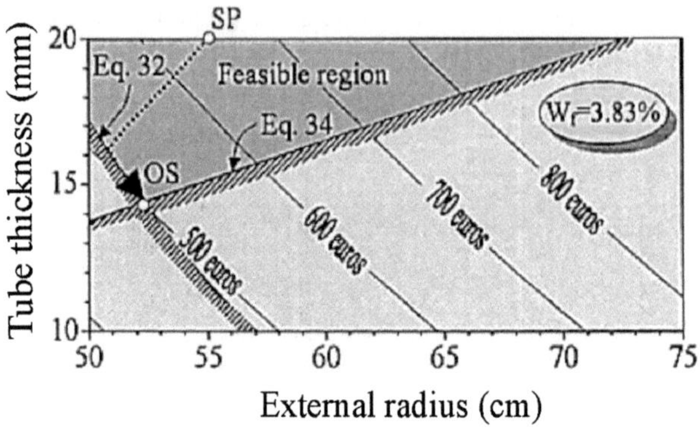

Figure 5. Optimum Solution of Example 2

Finally, the axial force - bending moment interaction diagrams for the optimum composite column and for loading combination A and B appear in Fig. 6.

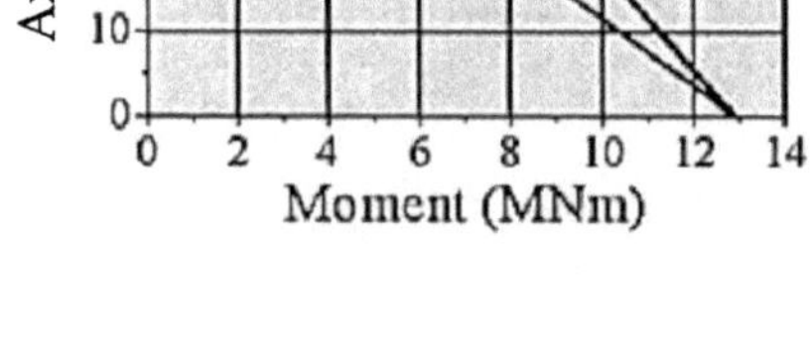

Figure6. Axial force – bending moment interaction diagrams

It is worth noticing that the most critical loading combination is not a priori evident because of different N-M allowed ('safe') region.

8. CONCLUSIONS

The optimum design of fibre reinforced concrete-filled steel tube columns was examined in this paper. The study took into account the effects of confinement of concrete and its steel fibre reinforcement. The design variables of the problem were the dimensions of the column (external radius and steel tube thickness) and the percentage of steel fibres. A neat optimization methodology was applied to find the most appropriate values of variables. The constraints of the optimum design were prescribed by the strength and stability requirements by means of up-to-date structural codes. Applications using numerical examples demonstrated that fibre reinforced concrete-filled steel tube columns can be designed both economically and effectively using the proposed methodology. However, it is difficult to construct practical rules for the optimum design of composite columns since it is a native market cost dependent procedure.

9. REFERENCES

1. Oehlers, D.J. and Bradford, M.A.: *Composite Steel and Concrete Structural Members, Fundamental Behavior,* Pergamon Press, 1995.
2. Cosenza, E. and Zandonini, R.: Composite Construction, in *Structural Engineering Handbook*, Chen W.F. (ed.), CRC Press, Boca Raton, 1999.
3. Hajjar, J.F., Composite steel and concrete structural systems for seismic engineering, *Journal of Constructional Steel Research*, **58** (2002), 703-723.
4. Munoz, P.R. and Hsu, C.T.: Behaviour of biaxially loaded concrete-encased composite columns, Journal of Structural Engineering, ASCE, **123** (1997), 1163–1171.
5. Shanmugam, N.E. and Lakshmi, B.: State of the art report on steel–concrete composite columns, *Journal of Constructional Steel Research*, **57** (2001), 1041–1080.
6. Eurocode 4: *ECS - European Committee for Standardisation, ENV 1994-1-1, Design of Composite Steel and Concrete Structures, Part 1.1 General Rules and Rules for Buildings*, Brussels, Belgium, 1994.
7. Bentur, A. and Mindess, S.: *Fibre Reinforced Cementitious Composites*, Elsevier, London, 1990.
8. Lie, T.T.: Fire resistance of circular steel columns filled with bar-reinforced concrete, *Journal of Structural Engineering*, ASCE, **120** (1994), 1489–1509.
9. Nataraja, M.C., Dhang, N. and Gupta, A.P.: Stress-strain curves for steel-fiber reinforced concrete under compression, *Cement & Concrete Composites*, **21** (1999), 383-390.
10. Kaneko, Y., Connor, J.J., Triantafillou, T.C. & Leung, C.K.: Fracture mechanics approach for failure of concrete shear key. I: Theory, *Journal of Engineering Mechanics*, **119** (1993), 681-700.
11. Eurocode 2: *ECS European Committee for Standardisation, ENV 1991, Design of Concrete Structures, Part 1: General Rules and Rules for Buildings*, Brussels, Belgium, 1991.
12. Galambos, T.V.: *Guide to Stability Design Criteria for Metal Structures*, John Wiley & Sons, New York, 1998.
13. Susantha, K.A.S., Ge, H. and Usami, T.: Uniaxial stress–strain relationship of concrete confined by various shaped steel tubes, *Engineering Structures*, **23** (2001), 1331-1347.
14. Tang, J., Hino, S., Kuroda, I. and Ohta T.: Modelling of stress–strain relationships for steel and concrete in concrete filled circular steel tubular columns, *Steel Construction Engineering*: *JSSC*, **3** (1996), 35-46.

15. Roik, K. and Bergmann, R.: Composite Columns, in *Constructional Steel Design, An International Guide,* Dowling, P.J., Harding, J.E., and Bjorhovde, R. (eds.), Elsevier Applied Science, 1992.

16. Arora, J.S.: *Introduction to Optimum Design,* Mc Graw Hill, New York, 1989.

17. Haftka, R.T. and Gurdal, Z.: *Elements of Structural Optimization*, 3rd Edition, Kluwer Academic Publishers, The Netherlands, 1999.

18. Vanderplaats, G.N.: *Numerical Optimization Techniques for Engineering Design with Applications*, McGraw-Hill Inc., New York, 1993.

CONTACT DEMAMINATION OF COMPOSITE PATCHES REPAIRING METAL STRUCTURES

G. TSAMASPHYROS
National Technical University Of Athens
Department of Theoretical and Applied Mechanics
5, Iroon Polytechniou Str,
GR- 157 73 Athens, Greece

Abstract. The increasing demand for life extension of both military and civilian aircrafts leaded to significant advances in repair technology of cracked metallic structures. Thus composite patch repair of metallic structures became a rapidly grown technology in the field of aerospace. However, because of the different nature and properties of the material that form a repair (metals composites adhesives), side effects like debonding due to high stress concentration in the vicinity of the crack or thermal residual stresses because of different thermal expansion coefficients of the adherents may occur.

In this paper the important phenomenon of delamination is examined taking into account the possibility of frictionless contact that appears between laminates. Buckling delamination is examined in particular, because in-place compressive loading is a typical and most damaging type of delamination. To that end, the concavity of initial imperfections is important in connection with contact. Two types of contact can be distinguished, local and global. This is because delamination can be thought both as a plate system, in the macro-behavior, but also as a crack near the delamination front. The global contact appears as the laminates try to interpenetrate. The local contact relates to the impending interpenetration of crack – tips (negative k_I stress intensity factors). Both local and global contact conditions were implemented in a FEM code using non-linear plate theory. An example of thin film delaminations is examined.

1. INTRODUCTION

Current economic world conditions are forcing to the operation of both military and civilian aircrafts well beyond their original design life, resulting in innovative repair techniques. One of them, the adhesively bonded composite patch repair of metallic aircraft components, is becoming a well-established technology. Bonded repairs are mechanically efficient, cost effective and can de applied rapidly to produce an inspectable damage tolerant repair. Compared to metals advanced fiber composites have the advantages of formability, tailorability of stiffness, high specific strength and immunity to corrosion and fatigue. Composite patches can be procured and secondarily bonded on cracked structures or cocured in situ [1].

The greatest concerns with mechanical repairs are the danger of crack initiation from one of the new fastener holes as well as the difficulty in detecting this crack by standard NDI procedures until it emerges from under the repair. The crack may initiate because of high stress concentrations (usually at the first row of fasteners) or

E.E. Gdoutos and Z.P. Marioli-Riga (eds), Recent Advances in Composite Materials, 371-379

because of poor quality hole drilling. The danger of crack initiation from hidden corrosion, which can develop under a poor mechanical repair also, exists. On the other land, cracks growing under bonded patches and readily detected by standard NDI procedures. Furthermore, the bonded patch, if correctly designed has a relatively small influence on the stress field, so no crack initiation occurs in adjacent regions. Another significant advantage of composite patch repairs is the ability to tailor the geometry and the material properties of the patch to specific strength and stiffness requirements in order to achieve the desired redaction of stress intensity factors [2,3]. More data on design and application of composite patch can be found in [4,5].

On the other hand there are certain drawbacks in the application of this kind of repairs coming mainly from the elevated temperature required to cure and/or bond the patch on the structure, in conjunction to the significant differences between the thermal expansion coefficients of the composites and the aluminum. A potential crack area is generated on the substrate near the patch edges, if the later is not well designed and if stepping of composite layers is not adequate, because of the increased shear stresses caused by the adhesive. Moreover, a debonding area is usually induced by fatigue around the crack, this changing the stress distribution in the area [6].

In this paper a representative cracked cylindrical fuselage section of thickness t_f is studied. The radius is two meters, representing average commercial aircrafts. Loads are caused by the differences between internal and external pressures and consider to be applied vertically to the skin from the internal side. The 1 bar load applied is an extreme case (two times up) of the load carried by a compressed fuselage of a commercial aircraft flying in high altitude. The skin is constrained by stringers at its edges. A composite patch Ω_p of thickness t_p is considered to be placed on top of a crack 2a and a debonding Ω_d ($\Omega_d \subset \Omega_p$) is induced around the crack. Since the normal curing temperature of the adhesive is 80°C and the temperature in high altitude is –20°C a uniform temperature between –20°C and -130°C must to applied on the area Ω_p of the model to simulate post-curing conditions, in order to study the effects of thermal expansion coefficients mismatch. Increase of stresses in aluminum near the patch edges is calculated to predict crack propagation or a possible growth of the delamiration.

We deal with the post-buckling range of a laminated patch, which contains a delamination. The interaction between post-buckling and local delamination growth is analyzed within a non linear finite element plate model, using an analytical formula for computing the energy release rate along the delamination front [7].

No matter how, the delamination portion may buckle under compression, causing high interlaminar stresses which tend to enlarge it. This may lead to severe reduction of the overall component function. The numerical simulation of such configuration is a complex problem that involves non-linear mechanics and fracture considerations. A complete local approach with fracture mechanics needs 3-D computations. However, we assume that a non-linear plate theory leads to a good evaluation of the stored energy in the whole laminate and therefore the delamination growth can be analyzed with energetic considerations. We can then compute the evolution of the energy release rate of pointwise crack growth along the delamination front. What we can gain with such an approach is additional computational power in order to include many more accompanying phenomena such as contact.

We assume that the aspect ratio (length to thickness) is large and that the plates consisting the delaminations undergo fairly small lateral deflections, but moderate rotations. Hence, the framework of the von Karman kinematic approximation is suitable for this study. The material can be non-linear elastic. Buckling mode shapes that may have a mode-line (line of zero displacements) indicate simultaneous occurrence of contact. We will exclude such cases from the present paper.

The important phenomenon of delamination buckling is examined in particular, because in-place compressive loading is typical and detrimental. Two types of contact can be distinguished, local and global. The latter may occur everywhere in the plate while the local contact is limited to the crack-front (negative K_I stress intensity factors). Both local and global contact conditions were considered using a finite element scheme, which employed nonlinear plate theory. The global contact problem is formulated as it appears in post-buckling of delamination. The case of simultaneous buckling and contact is also addressed in this paper. Two particularly interesting examples of thin film delaminations are presented. In the first, the contact at buckling is due to the material anisotropy. In this case the bucking load and the postbucking analysis were very well supported by experiments. In the second example, contact at buckling arises indicated that contact may significantly effect the fracture parameters along the delamination front and is therefore, important for delamination arrest.

2. GLOBAL CONTACT

A single delamination Ω_d is considered. The points of the middle-plane of the undeformed, delaminated flat plate are related to a fixed right-hand Cartesian system (Ox_1,x_2,x_3), with the out of plane coordinate $x_3=0$ (fig.1). In planeload is imposed along the delamination boundary line $\partial\Omega_d$ which suffices to be Lipschitzian.

Regarding global contact, the candidate contact surface Ω_c is included in the delaminated area Ω_d (Fig. 1). Unilateral contact without friction will be examined (Signorini-Fichera problem) under quasi–static conditions. Both delaminated plate and rigid support are constrained against rigid body motion. The problem may develop multiple connected contact locations.

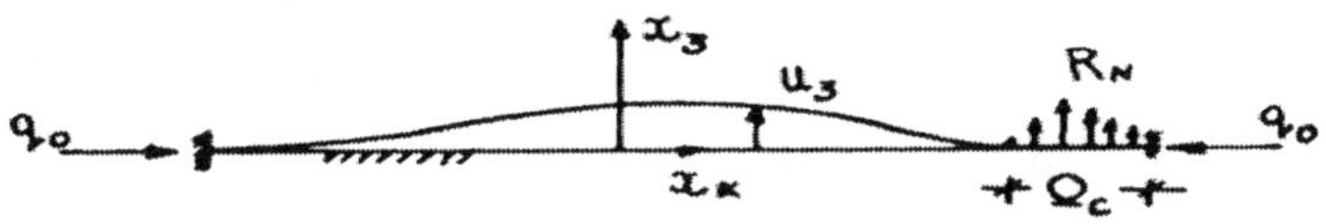

Fig. 1.: The contact zone model. Materials 1 and 2 are linear elastic.

We denote by $u_i(i=1,2,3)$ the displacements of the delaminated area and by $v_i(i=1,2,3)$ the displacements of the fuselage. Solution of the problem entails the satisfaction of equilibrium together with the following constraint condition, which concerns the kinematics of the mid-surface of the plate:

$$u_3(x_1,x_2,0) - v_3(x_1,x_2,0) \geq 0 \tag{1}$$

Due to plate continuity, there is contact of at least first order. That is, at the contact region Ω_c,

$$u_3(x_1,x_2,0)-v_3(x_1,x_2,0)=0 \quad u_{3,\beta}(x_1,x_2,0)-u_{3,\beta}(x_1,x_2,0)=0 \quad ;x_\alpha \in \Omega_c (\alpha,\beta=1,2) \tag{2}$$

(the comma indicates differentiation with respect to x_β – plane coordinates).

The assumption of a contact domain leads to contact reactions acting on the delaminated plate. The normal reactions R_N (fig. 1) associated with contact are:

$$R_N(u_3-v_3)=0, \; R_N(x_a) \geq 0 \; ;x_a \in \Omega_c \tag{3}$$

and are assumed to be maximal monotone. Note that R_N is smooth according to smoothness of u_3 and the resultant of all R_N is in the interior of Ω_c.

The plate is initially unconstrained ($\overset{0}{\Omega}_c=\emptyset$) and the contact region is permitted to advance and recede freely until equilibrium is achieved without interpenetration (spontaneous contact is excluded). A useful global contact relation (essentially a negative L^1 – norm), suitable for a FEM contact convergence criterion is:

$$\left| \int_{\Omega_d} (u_3-v_3)d\Omega \right| = a \int_{\Omega_d} |u_3-v_3| \, d\Omega, \tag{4}$$

where a=1 means no contact and $0 \leq a<1$ indicates contact.

At $\partial\Omega_d$ the clamped kinematic constraints $u_3-v_3=0$ and $u_{3,\alpha}-v_{3\alpha}=0$ hold. Within von Karman theory, the non-zero generalized strains read as

$$\overset{m}{\epsilon}_{\alpha\beta} = (u_{\alpha,\beta}+u_{\beta,\alpha}+u_{3,\alpha}\, u_{3,\beta})/2 \tag{5}$$

$$\kappa_{\alpha\beta}=-u_{3,\alpha\beta} \quad ;\alpha,\beta=1,2, \tag{6}$$

Where u_α are the in-plane displacements that must be H_1 – continuous.

The membrane forces, $N_{\alpha\beta}$, and bending moments $M_{\alpha\beta}$, are work conjugate function of the membrane strains $\overset{m}{\epsilon}_{\alpha\beta}$, and the curvatures $\kappa_{\alpha\beta}$ respectively with a convex, positive definite, strain energy density $W(\overset{m}{\epsilon}_{\alpha\beta}\ \kappa_{\gamma\delta})$. The constitutive equations, $N_{\alpha\beta}=\partial W/\partial \overset{m}{\epsilon}_{\alpha\beta}$, $M_{\alpha\beta}=\partial W/\partial \kappa_{\alpha\beta}$, satisfy the usual symmetry and ellipticity conditions.

The equilibrium equations are:

$$N_{\alpha\beta,\beta}=0 \text{ (in – plane)} \tag{7}$$

$$M_{\alpha\beta,\alpha\beta} + N_{\alpha\beta}\, u_{3,\alpha\beta}=0 \text{ (out-of-plane)} \tag{8}$$

Repeated indices imply summation. The transmition relations at the contact boundary $\partial\Omega_c$ are eqs. The normal contact tractions are:

$$R_N=-(M_{\alpha\beta,\alpha\beta} + N_{\alpha\beta}\, u_{3,\alpha\beta}), \tag{9}$$

With $R_N=0$ at $x_\alpha \epsilon\, \partial\Omega_c$

According to the plate model, the stress field is not singular at the delamination front $\partial\Omega_d$ but discontinuous. This discontinuity is transported with the delamination front as it grows. However, a crack – type singularity has still to be dealt with. To this end, a crack-tip criterion for contact has to be considered as proposed by Chai [8]. This situation is called contact in the small (or local), to distinguish it from contact in the large (or global). Attention will be directed to isotropic, linear fracture analysis of bimaterial beam configurations, in accord with overall elastic plate analysis. As for the global contact, no frictional dissipation will be accounted for the local contact.

It will be assumed that the local energy release rate can be decomposed into mode I, II, III. Mode III can be decomposed for linear elastic materials, if the mode III shear resultant does not couple with the other plate resultant forces and moments [9]. Mode decomposition, in the Irwin's spirit, may not be always possible, or may be possible in an approximate way, depending on the elastic properties of the material close to the interface. The mode decomposition, in absence of local contact, is given by Suo and Hutchinson [10]:

$$\frac{K_{II}}{K_I} = \frac{\cos\omega-(t_pP/(M\sqrt{12}))\sin\omega}{-\sin\omega-(t_pP/(M\sqrt{12}))\cos\omega} \tag{10}$$

Where P and M are the axial force and bending moment at the vicinity of the delamination front, and t_p is the plate's thickness (Fig. 2).

In general, ω depends on the Dundurs' elastic mismatch parame Expressions analogous to (14) can be stated for linear elastic, orthotropic mate with dissimilar elastic properties across the delamination plane.

The contact zone model, which is used in this work, has been examinec Comninou [11] and has been shown to give a uniqe solution. Following Comnin local closure requires K_I=0; this may not be true for certain orthotropic bimater [12]. In this formulation, the energy release rate is preserved and in case of l contact, is entirely attributed to moden II. Note that the extent of local cor cannot be found from the present consideration.

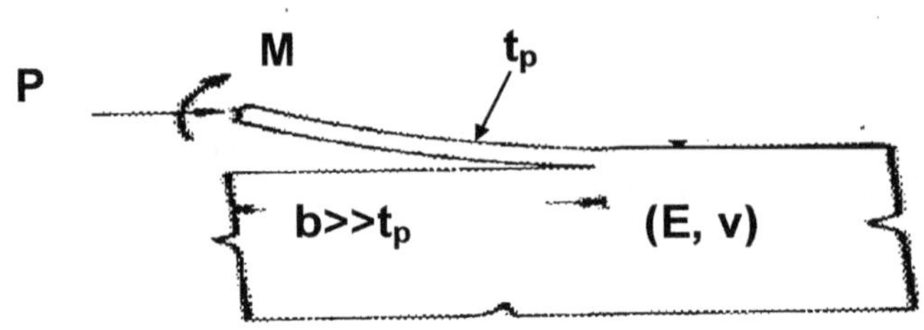

Fig. 2.: Axial Force P and moment M acting close to the delamination front in the normal direction.

The solution of the delamination problem subjected to a frictionless contact is obtained by a programming method which is suggested by the Kuhn-Tucker contact conditions (1) and (3). The problem is solved numerically as a sequence of bilateral problems by tracking only the vertical displacements. In turn, the interface normal tractions follow as a consequence of the procedure, and ensure subsequent possible separation (reverse contact).

Loading is imposed incrementally. The growth of the contact region is monitored at the end of each loading step by imposing or lifting nodal kinematic boundary restrictions corresponding to eqs (2). This stage of the numerical proceeding is called contact prediction. The global stiffness matrix is updated at the end of the load step to account for the activation of the new contact constraints.

The interface normal traction in the contact region are the corresponding reactions to the kinematic constrains. If among the nodes with prescribed contact constraints there are some with negative normal reaction (inconsistent with contact conditions (3)), then these constraints are released. This stage is called contact correction. A final optimality test is performed to ensure that the imposed final constrained nodes are the minimum possible ones. The contact algorithm at buckling is discussed in detail in the Appendix.

3.EXAMPLES

As a first try of our method we consider the case of a non cracked fuselage and we suppose that the patch is completely symmetric about the delamination plane examined. The material is linear elastic and isotropic with Poisson's ratio ν=0.3. The patch thickness is t_p and the Young's modulus is E=70Gpa. Due to symmetry, only a quarter of the delamination is discretized. The delamination front, used in the analysis was normalized to vary from 0 to 1, as shown in Fig 3. In these examples it can be shown [11] that $K_{III} \approx 0$, which was also verified by the computation. From a broad variety of examples we solved, we depicted four charecteristic cases for the purpose of illustrating the local and the global contact effect at delamination.

An elliptical delamination with b/a=0.5 and $t_p/a=1/20$ was examined. The FEM mesh is shown in Fig. 3.

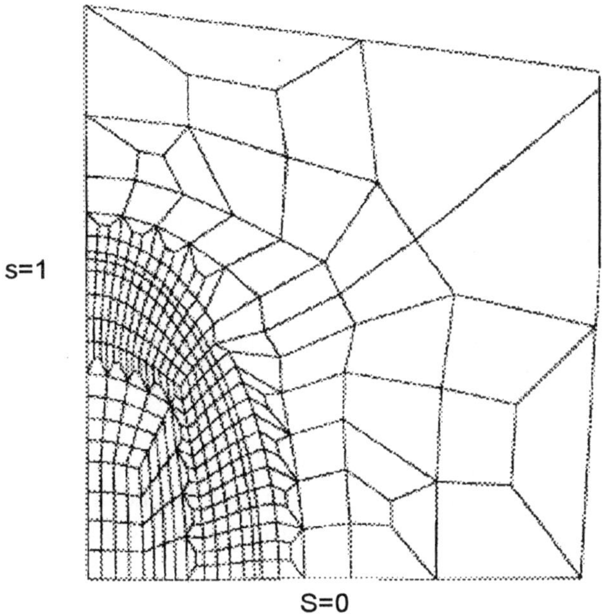

Fig. 3.: FEM mesh used for the elliptical delamination (b/α=0.5) under uniform strain.

The buckling strain was computed to be $\epsilon_y^b=0.665(h/\alpha)^2=(\alpha_p-\alpha_f)\ T_b$ where T_b is the buckling difference of temperature and α_p, α_f are the thermal expansion coefficients of the patch and the fuselage respectively. Local overlap develops immediately upon buckling. No global overlap was found, as was also shown by Chai.

The normalized energy release rate, as well as the stress intensity factors, were computed along the normalized delamination quarter front line, s ϵ[0.1] for ϵ_y^0 =1.235, 2497, 4.917. It was found that G_{min} (Fig. 4) is always at s=1, but G_{max} is not always at s=0. In the case, our G-distribution results were in departure with Chai's [12] results, especially at high loads. The plots are not shown herein for brevity.

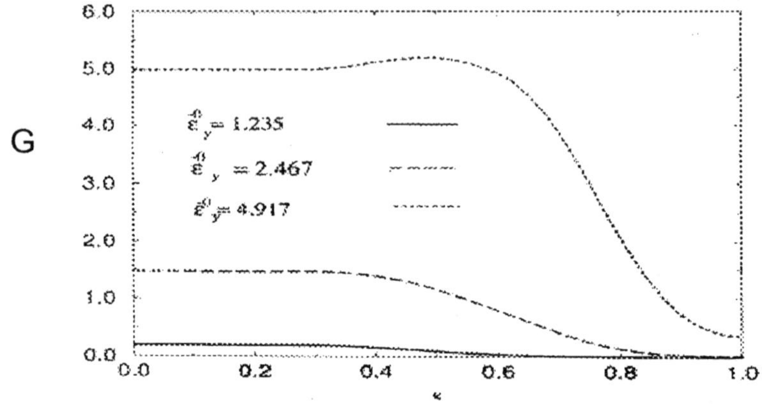

Fig. 4.: Normalized energy release rate, G, along the normalized delamination front line, s, for loads ϵ^0_y=1.235, 2497, 4.917 (elliptical delamination)

$|K_{II}|$ is always larger than K_I for all sϵ[0,1] and $\epsilon^0_y \geq 1$ (Fig. 5). What is interesting in this case is that local contact appears initially near the end of the long axis, and disappears with load. The minima of K_I, $|K_{II}|$ always occurred at s=1.

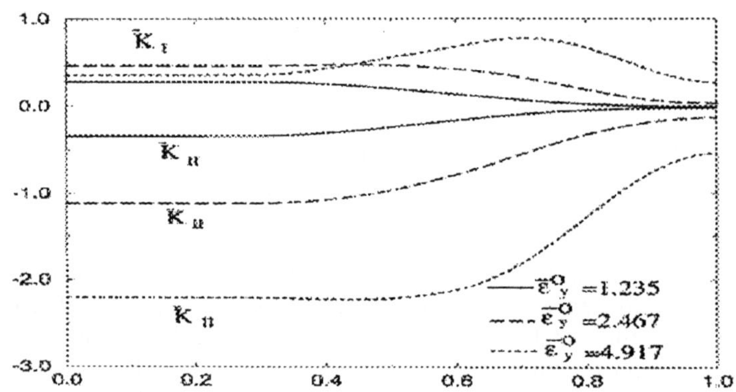

Fig.5.: Normalized stress intensity factors, $K_{I,II}$, along the normalized delamination front line, s, for ϵ^0_y=1.235, 2497, 4.917 (elliptical delamination)

4. REFERENCES

1. Baker A.A. Bonded Composite Repair of Metallic Aircraft Components – Overview of Australian Activities. AGARD, CP-550.
2. Z.P. Marioli Riga, G.J. Tsamasphyros, G.N. Kanderakis, Development of a Method for A/C Emergency Repairs by Composite Patches, S.A. Paipetis, E.E. Gdoutos, (Xanthi 1997), Vol II 143-156.
3. Duong G.N, Yu J, The Stress Intensity Factor for a Cracked Stiffened Sheet Repaired with an Adhesively Bonded Composite Patch, International Journal of Fracture **84** (1997) p. 37-60.
4. Baker A.A., Jones R., Bonded Repair of Aircraft Structures, Martinus Nijhoff Pubishers, Dordrecht. (1998).
5. Composite Repairs of Cracked Metallie Airframe Structures, U.S. Department of Transportation Federal Aviation Administration, CT-92/32 (May 1993).
6. Fredell R., Mazza J., Fatigue in New and Ageing Aircraft – The Development of an Integrated Bonded Repair System, 19 International Committee on Aeromantical Fatigue Symposium, Edinburgh, Scotland (June 1997).
7. Storakers B. and Andersson B., Nonlinear Plate Theory Applied to Delamination in Composites, J. Mech. Phys. Solids, (1998), **36**, p. 689-718.
8. Chai H. Three – Dimensional Fracture Analysis of Thin-Film Debonding, Int. J. Appl. Mech., (1990) **57**,p. 989-994.
9. Nilsson K.-F. and Storakers B., On Interface Crack Growth in Composite Plates, J. Appl. Mech., (1992) **59**, p. 530-538.
10. Suo Z. and Hutchinson J. W., Interface Crack Between Two Elastic Layers, Int. J. of Fracture, (1998), **43**, p. 1-22.
11. Comninou M., An Overview of Interface Cracks, Eng. Fracture Mechanics, 1990, **37**, p. 197-208.
12. Chai H, Buckling and Post- Buckling Behavior of Elliptical Plates : Part II – Results, J. Appl. Mech., 1990, **57**, 989-994.

Author Index

MIX
Papier aus verantwortungsvollen Quellen
Paper from responsible sources
FSC® C105338

If you have any concerns about our products,
you can contact us on
ProductSafety@springernature.com

In case Publisher is established outside the EU,
the EU authorized representative is:
Springer Nature Customer Service Center GmbH
Europaplatz 3, 69115 Heidelberg, Germany

Printed by Libri Plureos GmbH
in Hamburg, Germany